This work has arisen from lecture courses given by the authors on important topics within functional analysis. The authors, who are all leading researchers, give introductions to their subjects at a level ideal for beginning graduate students, as well as others interested in the subject. The collection has been carefully edited to form a coherent and accessible introduction to current research topics.

The first part of the book, by Professor Dales, introduces the general theory of Banach algebras, which serves as a background to the remaining material. Dr Willis then studies a centrally important Banach algebra, the group algebra of a locally compact group. The remaining chapters are devoted to Banach algebras of operators on Banach spaces: Professor Eschmeier gives all the background for the exciting topic of invariant subspaces of operators, and discusses some key open problems; Dr Laursen and Professor Aiena discuss local spectral theory for operators, leading into Fredholm theory.

LONDON MATHEMATICAL SOCIETY STUDENT TEXTS

Managing editor: Professor W. Bruce, Department of Mathematics
University of Liverpool, United Kingdom

3 *Local fields*, J. W. S. CASSELS
4 *An introduction to twister theory, second edition*, S. A. HUGGET & K. P. TOD
5 *Introduction to general relativity*, L. P. HUGHSTON & K. P. TOD
7 *The theory of evolution and dynamical systems*, J. HOFBAUER & K. SIGMUND
8 *Summing and nuclear norms in Banach space theory*, G. J. O. JAMESON
9 *Automorphisms of surfaces after Nielson and Thurston*, A. CASSON & S. BLEILER
11 *Spacetime and singularities*, G. NABER
12 *Undergraduate algebraic geometry*, M. REID
13 *An Introduction to Hankel operators*, J. R. PARTINGTON
15 *Presentations of groups, second edition*, D. L. JOHNSON
17 *Aspects of quantum field theory in curved spacetime*, S. A. FULLING
18 *Braids and coverings: Selected topics*, V. LUNDSGAARD HANSEN
19 *Steps in commutative algebra*, R. Y. SHARP
20 *Communication theory*, C. M. GOLDIE & R. G. E. PINCH
21 *Representations of finite groups of Lie type*, P. DIGNE & J. MICHEL
22 *Designs, graphs, codes, and their links*, P. J. CAMERON & J. H. VAN LINT
23 *Complex algebraic curves*, F. KIRWAN
24 *Lectures on elliptic curves*, J. W. S. CASSELS
25 *Hyperbolic geometry*, B. IVERSEN
26 *An introduction to the theory of L-functions and Eisenstein series*, H, HIDA
27 *Hilbert space: Compact operators and the trace theorem*, J. R. RETHERFORD
28 *Potential theory in the complex lane*, T. RANSFORD
29 *Undergraduate cummutative algebra*, M. REID
31 *The Laplacian on a Rlemannian manifold*, S. ROSENBERG
32 *Lectures on Lie groups and Lie algebras*, R. CARTER, G. SEGAL & J. MACDONALD
33 *A primer of algebraic D-modules*, S. C. COUTINHO
34 *Complex algebraic surfaces*, A. BEAUVILLE
35 *Young tableaux*, W. FULTON
37 *A mathematical introduction to wavelets*, P. WOJTASZCZYK
38 *Harmonic maps, loop groups and integrable systems*, M. GUEST
39 *Set theory for the working mathematician*, K. CIESIELSKI
40 *Ergodic theory and dynamical systems*, M. POLLICOTT & M. YURI
41 *The algorithmic resoution of diophantine equations*, N. P. SMART
42 *Equlibrium states in ergodic theory*, G. KELLER
43 *Fourier analysis on finite groups and applications*, A. TERRAS
44 *Classical invariant theory*, P. J. OLVER
45 *Purmutation groups*, P. J. CAMERON
46 *Rlemann surfaces: A primer*, A. BEARDON
47 *Introductory lectures on rings and modules*, J. BEACHY
48 *Set theory*, A. HAJNAL & P. HAMBURGER
49 *An introduction to K-theory for C^*-algebras*, M. RØRDAM, F. LARSEN & N. J LAUSTSEN
50 *A brief guide to algebraic number theory*, H. P. F. SWINNERTON-DYER
51 *Steps in commutative algebra*, R. Y. SHARP
52 *Finite Markov chains and algorithmic applications*, O. HAGGSTROM
53 *The prime number theorem*, G. J. O. JAMESON
54 *Topics in graph automorphisms and reconstruction*, J. LAURI & R. SCAPELLATO
55 *Elementary number theory, group theory and Ramanujan graphs*, P. SARNAK, G. DAVIDOFF & A. VALETTE
56 *Logic, induction and sets*, T. FORSTER

INTRODUCTION TO BANACH ALGEBRAS, OPERATORS, AND HARMONIC ANALYSIS

H. GARTH DALES
University of Leeds, UK

PIETRO AIENA
Università degli Studi, Palermo, Italy

JÖRG ESCHMEIER
Universität des Saarlandes, Saarbrücken

KJELD LAURSEN
University of Copenhagen, Denmark

GEORGE A. WILLIS
University of Newcastle, New South Wales, Australia

CAMBRIDGE
UNIVERSITY PRESS

CAMBRIDGE
UNIVERSITY PRESS

University Printing House, Cambridge CB2 8BS, United Kingdom

One Liberty Plaza, 20th Floor, New York, NY 10006, USA

477 Williamstown Road, Port Melbourne, VIC 3207, Australia

314-321, 3rd Floor, Plot 3, Splendor Forum, Jasola District Centre, New Delhi - 110025, India

79 Anson Road, #06-04/06, Singapore 079906

Cambridge University Press is part of the University of Cambridge.

It furthers the University's mission by disseminating knowledge in the pursuit of education, learning and research at the highest international levels of excellence.

www.cambridge.org
Information on this title: www.cambridge.org/9780521535847

First published 2003

A catalogue record for this publication is available from the British Library

Library of Congress Cataloging in Publication data
Introduction to Banach algebras, operators, and harmonic analysis /
H. Garth Dales . . . [et al.].
p. cm. – (London Mathematical Society student texts ; 57)
Includes bibliographical references and index.
ISBN 0 521 82893 7 – ISBN 0 521 53584 0 (paperback)
1. Banach algebras. 2. Operator theory. 3. Harmonic analysis.
I. Dales, H. G. (Harold G.) 1944– II. Series.
QA326.I59 2003
512′.55 – dc21 2003043947

ISBN 978-0-521-53584-7 Paperback

Contents

Preface

This volume is based on a collection of lectures intended for graduate students and others with a basic knowledge of functional analysis. It surveys several areas of current research interest, and is designed to be suitable preparatory reading for those embarking on graduate work. The volume consists of five parts, which are based on separate sets of lectures, each by different authors. Each part provides an overview of the subject that will also be useful to mathematicians working in related areas. The chapters were originally presented as lectures at instructional conferences for graduate students, and we have maintained the styles of these lectures.

The sets of lectures are an introduction to their subjects, intended to convey the flavour of certain topics, and to give some basic definitions and motivating examples: they are certainly not comprehensive accounts. References are given to sources in the literature where more details can be found.

The chapters in Part I are by H. G. Dales. These are an introduction to the general theory of Banach algebras, and a description of the most important examples: $\mathcal{B}(E)$, the algebra of all bounded linear operators on a Banach space E; $L^1(G)$, the group algebra of a locally compact group G, taken with the convolution product; commutative Banach algebras, including Banach algebras of functions on compact sets in $\mathbb{C}$ and radical Banach algebras. Chapters 3–6 cover Gelfand theory for commutative Banach algebras, the analytic functional calculus, and, in a chapter on 'automatic continuity', the lovely results that show the intimate connection between the algebraic and topological structures of a Banach algebra. Chapters 6 and 7 are an introduction to the cohomology theory of Banach algebras, at present a very active area of research; we concentrate on the basic structure, that of derivations into modules.

The chapters in Part II, by G. A. Willis, develop the theory of one of the examples discussed by Dales: these are the group algebras $L^1(G)$. Chapters 8 and 9 give a description of locally compact groups G and their structure theory, and then describe the algebras $L^1(G)$, and the related measure algebra $M(G)$, as a Banach algebra. The Gelfand theory for general commutative Banach algebras, as described by Dales, becomes Fourier transform theory in the special case of the algebras $L^1(G)$. In Chapter 10, Willis discusses compact groups, abelian groups, and free groups, and then, in Chapter 11, moves to a very important class, that of amenable groups: many characterizations of amenability arise in diverse areas of mathematics. Willis then expands a notion from Part I by discussing the automatic continuity of linear maps from group algebras.

Parts III–V of this book develop the theory of another example mentioned by Dales: this is the algebra $\mathcal{B}(E)$ for a Banach space E. However, they also concentrate on the properties of single operators of various types within $\mathcal{B}(E)$.

A seminal question in functional analysis is the 'invariant subspace problem'. Let E be a Banach space, and let $T \in \mathcal{B}(E)$. A closed subspace F of E is invariant for T if $Ty \in F$ $(y \in F)$; F is trivial if $F = \{0\}$ or $F = E$. Does such an operator T always have a non-trivial invariant subspace? A positive answer to this question in the case where E is finite-dimensional (of dimension at least 2) is the first step in the structure theory of matrices. The question for Banach spaces has been the spur for a huge amount of research in operator theory since the question was first raised in the 1930s. The question is still open in the case where E is a Hilbert space – this is one of the great problems of our subject – but counter-examples are known when E is an arbitrary Banach space. Nevertheless, there are many positive results for operators $T \in \mathcal{B}(E)$ which belong to a special class.

The chapters in Part III, by J. Eschmeier, discuss in particular one very important technique for establishing positive results: it descends from original work of Scott Brown in 1978. One class of operators considered is that of subdecomposable operators. Part III concludes with remarks about the extensions, mainly due to the author, of the positive results to n-tuples of commuting operators.

As explained by Dales, every element a of a Banach algebra has a spectrum, called $\sigma(a)$; this is a non-empty, compact subset of the complex plane $\mathbb{C}$. In particular, each operator $T \in \mathcal{B}(E)$ has such a spectrum, $\sigma(T)$. In the case where E is finite-dimensional, $\sigma(T)$ is just the set of eigenvalues of T. The notion of the spectrum for a general operator T is at the heart of the remaining chapters, by K. B. Laursen and P. Aiena.

Laursen discusses the spectral theory of operators in several different classes; these include in particular the decomposable operators, which were also introduced by Eschmeier. We understand the nature of an operator T by looking at the decomposition of $\sigma(T)$ into subsets with special properties and also at special closed subspaces of E on which T acts 'in a nice way'. In particular, Laursen discusses super-decomposable and generalized scalar operators. In Chapter 25, Laursen relates his description to notions introduced by Dales by discussing when multiplication operators on commutative Banach algebras have the various properties that he has introduced. A valuable appendix to Part IV sketches the background theory, involving distributions, to the 'functional model' of Albrecht and Eschmeier that is the natural setting for many of the duality results that have been obtained.

The final chapters, those of P. Aiena, are closely related to those of Laursen. The basic examples of the decomposable operators of Eschmeier's and Laursen's chapters are compact operators on a Banach space and normal operators on a Hilbert space. It is natural to study the decomposable operators which have similar properties to those of these important specific examples: we are led to the class of 'Fredholm operators' and related classes, a main topic of Aiena's lectures.

In these chapters, we see again, from a different perspective, some of the key ideas – decomposition of the spectrum, invariant subspace, single-valued extension property, actions of analytic functions, divisible subspaces – that have featured in earlier chapters. The final chapter by Aiena summarizes recent work of the author and others.

The lectures on which this book is based were given at two conferences. The first was held in Mussomeli, Sicily, from 22 to 29 September 1999. We are very grateful to Dr Gianluigi Oliveri, who organized this conference, and to the Associazone Culturale Archimede of Sicily who sponsored it. The lectures given at this conference were those of Dales, Eschmeier, Laursen, and Aiena. The second conference was held at the Sadar Patel University, Vallabh Vidyanagar, Gujarat, India, from 8 to 15 January 2002. We are very grateful to Professor Subhashbhai Bhatt and Dr Haresh Dedania for organizing this conference, and to the Indian Board for Higher Mathematics and to the London Mathematical Society, who supported the conference financially. The lectures given in Gujarat were those of Dales, Willis, and Laursen.

As we said, the original lectures were intended for graduate students and others with a basic knowledge of functional analysis and with a background in complex analysis and algebra typical of a first degree in mathematics. In both cases the students were enthusiastic and helpful; their suggestions led to many improvements in the exposition, and we are grateful to them for this.

In fact, the actual lectures as given did not include all that is written down here: modest additions have been made subsequently. There is more in a 'lecture' than can easily be absorbed in one hour. However, we have maintained the fairly informal style of the lecture theatre. At various points, the reader is invited to check statements that are made: these are all routine, and follow in a few lines from facts given in the lectures. There are also exercises at the end of each lecture; the answers to all the exercises are contained in the references that are specified. We hope that readers will work through the exercises as a step towards the gaining of familiarity with the subject.

There are various cross-references between the sets of lectures; indeed topics of one set of lectures often reappear, perhaps in a different guise, in other lectures. All the book depends somewhat on the first six chapters, and Part V follows from Part IV. However, otherwise the various parts of the book can be read independently. The references to each set of lectures are contained at the end of the relevant part of this book, and not at the end of the whole book. However, there are two indices for the whole book at the end (pp. 319–326): these are the symbol index and the index of terms.

Some attempt has been made to make the notation consistent between the various sets of lectures, but we have not always achieved this; we give a resumé of some standard notations at the end of this preface.

We very much enjoyed giving the original lectures and discussing the theory and associated examples in the classes that were given in the same week as the lectures. We hope that you enjoy reading them and, especially, working through the examples.

In rather more detail, we expect the reader to be familiar with the following topics:

- the definition of a Banach space and a locally convex space, weak topologies on dual spaces;
- standard theorems of functional analysis such as the Hahn–Banach theorem, closed graph theorem, open mapping theorem, and uniform boundedness theorem;
- the theory of bounded linear operators on a Banach space, duals of such operators, compact operators;
- the elementary theory of Hilbert spaces;
- undergraduate complex analysis, including Liouville's theorem;
- undergraduate algebra, including the theory of ideals, modules, and homomorphisms.

Throughout we adopt the following notation:
$\mathbb{N} = \{1, 2, 3, \ldots\}$;
$\mathbb{Z} = \{0, \pm 1, \pm 2, \ldots\}$;
$\mathbb{Z}^+ = \{0, 1, 2, \ldots\}$;
$\mathbb{Q}$ is the field of rational numbers;

$\mathbb{R}$ is the field of real numbers;
$\mathbb{C}$ is the field of complex numbers;
$\mathbb{I} = [0, 1]$;
$\mathbb{T} = \{z \in \mathbb{C} : |z| = 1\}$;
$\mathbb{D}\,(z; r) = \{w \in \mathbb{C} : |w - z| < r\}$;
$\mathbb{D} = \mathbb{D}(0; 1)$;
Z is the coordinate functional on $\mathbb{C}$, or on a subset of $\mathbb{C}$;
E' is the dual space of a topological linear space E;
$E_{[1]}$ is the closed unit ball of a Banach space E;
$[x, y]$ is the inner product of $x, y \in H$, where H is a Hilbert space;
$\langle x, \lambda \rangle$ is the action of $\lambda \in E'$ on $x \in E$, where E is a Banach space.

H. G. D., Leeds

Part I

Banach algebras

H. GARTH DALES
University of Leeds, UK

1

Definitions and examples

1.1 Definitions

A Banach algebra is first of all an algebra. We start with an algebra A and put a topology on A to make the algebraic operations continuous – in fact, the topology is given by a norm.

Definition 1.1.1 *Let E be a linear space. A* norm *on E is a map $\|\cdot\| : E \to \mathbb{R}$ such that:*

(i) $\|x\| \geq 0 \ (x \in E)$; $\|x\| = 0$ *if and only if* $x = 0$;
(ii) $\|\alpha x\| = |\alpha| \, \|x\| \quad (\alpha \in \mathbb{C}, \, x \in E)$;
(iii) $\|x + y\| \leq \|x\| + \|y\| \quad (x, y \in E)$.

Then $(E, \|\cdot\|)$ is a normed space. *It is a* Banach space *if every Cauchy sequence converges, i.e., if $\|\cdot\|$ is complete.*

Definition 1.1.2 *Let A be an algebra. An* algebra norm *on A is a map $\|\cdot\| : A \to \mathbb{R}$ such that $(A, \|\cdot\|)$ is a normed space, and, further:*

(iv) $\|ab\| \leq \|a\| \, \|b\| \quad (a, b \in A)$.

The normed algebra $(A, \|\cdot\|)$ is a Banach algebra *if $\|\cdot\|$ is a complete norm.*

In Chapters 1–7, we shall usually suppose that a Banach algebra A is *unital*: this means that A has an identity e_A and that $\|e_A\| = 1$. Let A be a Banach algebra with identity. Then, by moving to an equivalent norm, we may suppose that A is unital. It is easy to check that, for each normed algebra A, the map $(a, b) \mapsto ab, \quad A \times A \to A$, is continuous.

H. G. Dales, P. Aiena, J. Eschmeier, K. B. Laursen, and G. A. Willis, *Introduction to Banach Algebras, Operators, and Harmonic Analysis*. Published by Cambridge University Press.

1.2 Examples

Let us give some elementary examples.

(i) Let S be any non-empty set. Then $\mathbb{C}^S$ is the set of functions from S into $\mathbb{C}$. Define *pointwise* algebraic operations by

$$(\alpha f + \beta g)(s) = \alpha f(s) + \beta g(s),$$
$$(fg)(s) = f(s)g(s),$$
$$1(s) = 1,$$

for each $s \in S$, each $f, g \in \mathbb{C}^S$, and each $\alpha, \beta \in \mathbb{C}$. Then $\mathbb{C}^S$ is a commutative, unital algebra. We write $\ell^\infty(S)$ for the subset of bounded functions on S, and define the *uniform norm* $|\cdot|_S$ on S by

$$|f|_S = \sup\{|f(s)| : s \in S\} \quad (f \in \ell^\infty(S)).$$

Check that $(\ell^\infty(S), |\cdot|_S)$ is a unital Banach algebra.

(ii) Let X be a topological space (e.g., think of $X = \mathbb{R}$). We write $C(X)$ for the algebra of all continuous functions on X, and $C^b(X)$ for the algebra of bounded, continuous functions on X. *Check* that $(C^b(X), |\cdot|_X)$ is a unital Banach algebra.

Now take Ω to be a compact space (e.g., $\Omega = \mathbb{I} = [0, 1]$). Then we have $C^b(\Omega) = C(\Omega)$, and so $(C(\Omega), |\cdot|_\Omega)$ is a unital Banach algebra. This is a very important example.

(iii) Let $\mathbb{D} = \{z \in \mathbb{C} : |z| < 1\}$, the open unit disc. The *disc algebra* is

$$A(\overline{\mathbb{D}}) = \{f \in C(\overline{\mathbb{D}}) : f \text{ is analytic on } \mathbb{D}\}.$$

Check that $A(\overline{\mathbb{D}})$ is a unital Banach algebra. (You just have to show that $A(\overline{\mathbb{D}})$ is closed in $C(\overline{\mathbb{D}})$: why is this?)

Each $f \in A(\overline{\mathbb{D}})$ has a Taylor expansion about the origin:

$$f = \sum_{n=0}^{\infty} \alpha_n Z^n = \sum_{n=0}^{\infty} \frac{f^{(n)}(0)Z^n}{n!}.$$

Here Z is the *coordinate functional*, so that $Z : z \mapsto z$ on $\mathbb{C}$. Some functions in $A(\overline{\mathbb{D}})$ have the further property that

$$\sum_{n=0}^{\infty} |\alpha_n| < \infty.$$

(Are there any functions f in $A(\overline{\mathbb{D}})$ without this property?) The subset of functions with this extra property is called $A^+(\overline{\mathbb{D}})$. *Check* that $A^+(\overline{\mathbb{D}})$ is a unital Banach algebra for the norm $\|\cdot\|_1$, where

$$\|f\|_1 = \sum_{n=0}^{\infty} |\alpha_n| \quad \left(f = \sum_{n=0}^{\infty} \alpha_n Z^n\right).$$

(iv) Let X be a compact set in the space $\mathbb{C}^n$. Then $P(X)$ is the family of functions that are the uniform limits on X of the restrictions to X of the polynomials (in n-variables). *Check* that $(P(X), |\cdot|_X)$ is a unital Banach algebra. In fact, $A(\overline{\mathbb{D}}) = P(\overline{\mathbb{D}})$. We shall also be interested in $P(\mathbb{T})$, where $\mathbb{T} = \{z \in \mathbb{C} : |z| = 1\}$ is the unit circle.

(v) Let X be a compact set in the complex plane $\mathbb{C}$ (or in $\mathbb{C}^n$). Then $A(X)$ is the closed subalgebra of $(C(X), |\cdot|_X)$ consisting of the functions which are analytic on the interior of X, $\operatorname{int} X$. Clearly $A(X) = C(X)$ if and only if $\operatorname{int} X = \emptyset$. Also $R(X)$ is the family of functions on X which are the uniform limits on X of the restrictions to X of the rational functions: these are functions of the form p/q, where p and q are polynomials and $0 \notin q(X)$. Clearly we have

$$P(X) \subset R(X) \subset A(X) \subset C(X).$$

The question of the equality of various of these algebras encapsulates much of the classical theory of approximation.

(vi) Let $n \in \mathbb{N}$. Then $C^{(n)}(\mathbb{I})$ consists of the functions f on $\mathbb{I}$ such that f has n derivatives on $\mathbb{I}$ and $f^{(n)} \in C(\mathbb{I})$. *Check* that $C^{(n)}(\mathbb{I})$ is a Banach algebra for the pointwise operations and the norm

$$\|f\|_n = \sum_{k=0}^{n} \frac{1}{k!} |f^{(k)}|_{\mathbb{I}} \quad (f \in C^{(n)}(\mathbb{I})).$$

(vii) Let E and F be linear spaces. Then $\mathcal{L}(E, F)$ is the collection of all linear maps from E to F; it is itself a linear space for the standard operations.

Now let E and F be Banach spaces. Then $\mathcal{B}(E, F)$ is the family of all bounded (i.e., continuous) linear operators from E to F; it is a subspace of $\mathcal{L}(E, F)$ and $\mathcal{B}(E, F)$ is itself a Banach space for the operator norm given by

$$\|T\| = \sup\{\|Tx\| : x \in E, \ \|x\| \le 1\}.$$

We write $\mathcal{L}(E)$ and $\mathcal{B}(E)$ for $\mathcal{L}(E, E)$ and $\mathcal{B}(E, E)$, respectively. The product of two operators S and T in $\mathcal{L}(E)$ is given by composition:

$$(ST)(x) = (S \circ T)(x) = S(Tx) \quad (x \in E).$$

Then trivially $\|ST\| \leq \|S\| \, \|T\| \quad (S, T \in \mathcal{B}(E))$, and $(\mathcal{B}(E), \|\cdot\|)$ is a unital Banach algebra; the identity of $\mathcal{B}(E)$ is the identity operator I_E. This is our first non-commutative example.

For example, let E be the finite-dimensional space $\mathbb{C}^n$ (say with the Euclidean norm $\|\cdot\|_2$). Then $\mathcal{L}(E) = \mathcal{B}(E)$ is just the algebra $\mathbb{M}_n = \mathbb{M}_n(\mathbb{C})$ of all $n \times n$ matrices over $\mathbb{C}$ (with the usual identifications).

(viii) The algebra $\mathbb{C}[[X]]$ of *formal power series in one variable* consists of sequences

$$(\alpha_n : n = 0, 1, 2, \dots),$$

where $\alpha_n \in \mathbb{C}$, with coordinatewise linear operations and the product

$$(\alpha_r)(\beta_s) = (\gamma_n),$$

where $\gamma_n = \sum_{r+s=n} \alpha_r \beta_s$. It helps to think of elements of $\mathbb{C}[[X]]$ as formal series of the form

$$\sum_{n=0}^{\infty} \alpha_n X^n,$$

with the product suggested by the symbolism. This algebra contains as a subalgebra the algebra $\mathbb{C}[X]$ of polynomials in one variable – these polynomials correspond to the sequences (α_n) that are eventually zero.

A *weight on* $\mathbb{Z}^+$ is a function $\omega : \mathbb{Z}^+ \to \mathbb{R}^+ \setminus \{0\}$ such that $\omega(0) = 1$ and

$$\omega(m+n) \leq \omega(m)\omega(n) \quad (m, n \in \mathbb{Z}^+).$$

Check that $\omega_n = e^{-n}$ and $\omega_n = e^{-n^2}$ define weights on $\mathbb{Z}^+$. For such a weight ω, define

$$\ell^1(\omega) = \left\{ (\alpha_n) \in \mathbb{C}[[X]] : \|\alpha\|_\omega = \sum_{n=0}^{\infty} |\alpha_n| \, \omega_n < \infty \right\}.$$

Check that $\ell^1(\omega)$ is a subalgebra of $\mathbb{C}[[X]]$, and that $(\ell^1(\omega), \|\cdot\|_\omega)$ is a commutative, unital Banach algebra.

(viii) Let G be a group, and let

$$\ell^1(G) = \left\{ f \in \mathbb{C}^G : \|f\|_1 = \sum_{s \in G} |f(s)| < \infty \right\}.$$

Then $(\ell^1(G), \|\cdot\|_1)$ is a Banach space. We can think of an element of $\ell^1(G)$ as

$$\sum_{s\in G} \alpha_s \delta_s ,$$

where $\sum |\alpha_s| < \infty$; here $\delta_s(s) = 1$ and $\delta_s(t) = 0 \ (t \neq s)$.

We define a product on $\ell^1(G)$ that is not the pointwise product; it is denoted by $\star$ and is sometimes called *convolution multiplication*. In this multiplication,

$$\delta_s \star \delta_t = \delta_{st} \quad (s, t \in G),$$

where st is the product in G. (Actually this formula defines the product.) Thus

$$(f \star g)(t) = \sum\{f(r)g(s) : rs = t\} \quad (t \in G). \tag{1.2.1}$$

Check that $\ell^1(G)$ is a unital Banach algebra for this product and the norm $\|\cdot\|_1$. It is commutative if and only if G is an abelian group. Special case: take $G = \mathbb{Z}$, a group with respect to addition.

(ix) (Strictly, this example needs the theory of the Lebesgue integral on $\mathbb{R}$.) The Banach space $L^1(\mathbb{R})$ has the norm $\|\cdot\|_1$ given by

$$\|f\|_1 = \int_{-\infty}^{\infty} |f(t)|\,\mathrm{d}t .$$

For functions $f, g \in L^1(\mathbb{R})$, define their convolution product $f \star g$ by

$$(f \star g)(t) = \int_{-\infty}^{\infty} f(t-s)g(s)\mathrm{d}s \quad (t \in \mathbb{R}).$$

Integration theory shows that $f \star g$ is defined almost everywhere (a.e.) and that $f \star g$ gives an element of $L^1(\mathbb{R})$; further, $\|f \star g\|_1 \leq \|f\|_1 \|g\|_1$, and so we obtain a commutative Banach algebra (which does not have an identity).

This example is central to the theory of Fourier transforms.

(x) Let U be a non-empty, open set in $\mathbb{C}$ (or in $\mathbb{C}^n$). Then $H(U)$ denotes the set of analytic (or holomorphic) functions on U. Clearly $H(U)$ is an algebra for the pointwise operations. However the algebra $H(U)$ is *not* a Banach algebra. For each compact subset K of U, define

$$p_K(f) = |f|_K \quad (f \in H(U)).$$

Then each p_K is an algebra seminorm on $H(U)$. The space $H(U)$ is a Fréchet space with respect to the family of these seminorms; in this topology, $f_n \to f$ if and only if (f_n) converges to f uniformly on compact subsets of U. The algebra is a *Fréchet algebra* because $p_K(fg) \leq p_K(f)p_K(g)$ in each case.

A related algebra is $H^\infty(U)$, the algebra of bounded analytic functions on U. *Check* that this algebra is a Banach algebra with respect to the uniform norm $|\cdot|_U$.

1.3 Philosophy of why we study Banach algebras

There are several reasons why we study Banach algebras. They:

- cover many examples;
- have an abstract approach that leads to clear, quick proofs and new insights;
- blend algebra and analysis;
- have beautiful results on intrinsic structure.

1.4 Basic properties

We begin our study of general Banach algebras by considering invertible elements in such algebras.

Definition 1.4.1 *Let A be a unital algebra. An element $a \in A$ is* invertible *if there exists an element $b \in A$ with $ab = ba = e_A$. The element b is unique; it is called the* inverse *of a, and written a^{-1}. The set of invertible elements of A is denoted by* $\operatorname{Inv} A$.

Check that $a, b \in \operatorname{Inv} A \Rightarrow ab \in \operatorname{Inv} A$ and $(ab)^{-1} = b^{-1}a^{-1}$.

Now let $(A, \|\cdot\|)$ be a unital Banach algebra. *Check* that, for each $a \in A$, we have

$$\lim_{n\to\infty} \|a^n\|^{1/n} = \inf\{\|a^n\|^{1/n} : n \in \mathbb{N}\} \leq \|a\|\,.$$

Theorem 1.4.2 *Let $(A, \|\cdot\|)$ be a unital Banach algebra.*

(i) *Suppose that $a \in A$ and* $\lim \|a^n\|^{1/n} < 1$. *Then $e_A - a \in \operatorname{Inv} A$.*
(ii) $\operatorname{Inv} A \supset \{b \in A : \|e_A - b\| < 1\}$.
(iii) $\operatorname{Inv} A$ *is an open subset of A.*
(iv) *The map $a \mapsto a^{-1}$, $\operatorname{Inv} A \to \operatorname{Inv} A$, is continuous.*

Proof (i) The series $e_A + \sum_{n=1}^{\infty} a^n$ converges to $(e_A - a)^{-1}$.

(ii) This is immediate from (i).

(iii) Take $a \in \operatorname{Inv} A$, and then take $b \in A$ with $\|b\| < \|a^{-1}\|^{-1}$. Note that $a - b = a(e_A - a^{-1}b)$ and $\|a^{-1}b\| < 1$. By (i), $e_A - a^{-1}b \in \operatorname{Inv} A$. Hence $a - b \in \operatorname{Inv} A$.

(iv) Exercise: use the inequality that

$$\|b^{-1} - a^{-1}\| \leq 2\|a^{-1}\|^2\|b - a\|$$

whenever $a, b \in \operatorname{Inv} A$ with $\|b - a\| \leq 1/2\|a^{-1}\|$. □

1.5 Exercises

1. Check the details of the examples.
2. Prove Theorem 1.4.2(iv).
3. Identify $\operatorname{Inv} A$ for as many as possible of the examples A given in §1.2. (Easy for $A = C(\Omega)$, $A = A(\overline{\mathbb{D}})$, $A = H(U)$, $A = \mathcal{B}(E)$; harder for the algebra $A = A^+(\overline{\mathbb{D}})$; not possible in general for $\ell^1(G)$.) Show that $\operatorname{Inv} \mathbb{C}[[X]] = \{(\alpha_n) : \alpha_0 \neq 0\}$.
4. For $f \in L^1(\mathbb{T})$ (in particular for $f \in C(\mathbb{T})$), the *Fourier coefficients* are

$$\widehat{f}(k) = \frac{1}{2\pi} \int_{-\pi}^{\pi} f(e^{i\theta}) e^{-ik\theta} \, d\theta \quad (k \in \mathbb{Z}) .$$

Let $s_n(\theta) = \sum_{-n}^{+n} \widehat{f}(k) e^{ik\theta}$ and set

$$\sigma_n(f) = \frac{1}{n+1}(s_0 + \cdots + s_n) .$$

Then *Féjer's theorem* says that: for each $f \in C(\mathbb{T})$, $\sigma_n(f) \to f$ uniformly on $\mathbb{T}$.

Deduce that the following are equivalent for $f \in C(\mathbb{T})$:

(a) $f \in P(\mathbb{T})$;

(b) $f = F \mid \mathbb{T}$ for some $F \in A(\overline{\mathbb{D}})$;

(c) $\widehat{f}(-k) = 0 \ (k \in \mathbb{N})$.

We can now identify $A(\overline{\mathbb{D}})$ with $P(\mathbb{T})$ (why?), and regard $A(\overline{\mathbb{D}})$ as a closed subalgebra of $C(\mathbb{T})$ – if we should wish to do this!

5. Let

$$W(\mathbb{T}) = \{f \in C(\mathbb{T}) : \|f\|_1 = \sum_{k=-\infty}^{\infty} |\widehat{f}(k)| < \infty\} .$$

Check that $(W(\mathbb{T}), \|\cdot\|_1)$ is a commutative, unital Banach algebra (for the pointwise operations). *Check* that the map

$$\sum_{n=-\infty}^{\infty} \alpha_n \delta_n \mapsto \sum_{n=-\infty}^{\infty} \alpha_n Z^n, \quad \ell^1(\mathbb{Z}) \to W(\mathbb{T}) ,$$

is an isometric isomorphism. (W stands for N. Wiener, who was the first to study these algebras.)

6. Let $L^1(\mathbb{I})$ be the Banach space of (equivalence classes of) integrable functions on $\mathbb{I}$, with the norm

$$\|f\|_1 = \int_0^1 |f(t)|\,\mathrm{d}t \quad (f \in L^1(\mathbb{I}))\,.$$

For $f, g \in L^1(\mathbb{I})$, define $f \star g$ by

$$(f \star g)(t) = \int_0^t f(t-s)g(s)\,\mathrm{d}s \quad (t \in \mathbb{I})\,.$$

Show that $L^1(\mathbb{I})$ is a Banach algebra for this product. It is called the *Volterra algebra*, and is denoted by $\mathcal{V}$.

Set $u(t) = 1 \ (t \in \mathbb{I})$, so that

$$(u \star f)(t) = \int_0^t f(s)\,\mathrm{d}s\,.$$

Calculate $u^{\star n}$ and $\|u^{\star n}\|_1$, where $u^{\star n}$ denotes the nth power of u in the algebra $\mathcal{V}$. The map $V : f \mapsto u \star f$ on $L^1(\mathbb{I})$ is the *Volterra operator*, discussed in later chapters.

1.6 Additional notes

1. By an *algebra* A, we always mean a linear space over $\mathbb{C}$ together with a multiplication such that $a(bc) = (ab)c$, $a(b+c) = ab + ac$, $(a+b)c = ac + bc$, and $\alpha(ab) = (\alpha a)b = a(\alpha b)$ for $a, b, c \in A$ and $\alpha \in \mathbb{C}$. The algebra has an *identity* e_A if $e_A a = a e_A = a \ (a \in A)$. Suppose that A does not have an identity. Then $A^\# = \mathbb{C} \odot A$ is a unital algebra for the product

$$(\alpha, a)(\beta, b) = (\alpha\beta, \alpha b + \beta a + ab) \quad (\alpha, \beta \in \mathbb{C},\ a, b \in A)\,;$$

if A is a Banach algebra, then so is $A^\#$ for the norm $\|(\alpha, a)\| = |\alpha| + \|a\|$.

2. For $f \in \mathbb{C}^S$, define $\overline{f}(s) = \overline{f(s)}$, the complex conjugate of $f(s)$. Then the map $f \mapsto \overline{f}$ is an involution on $\mathbb{C}^S$ and on $C(\Omega)$. *Check* that $|f|_\Omega^2 = |f\overline{f}|_\Omega$ in the latter case. The algebra $C(\Omega)$ with this involution is the canonical example of a commutative, unital C^*-algebra; see §3.5.

3. Let Ω be a locally compact space (e.g., $\mathbb{R}$). For a continuous function f on Ω, supp f, the *support* of f, is the closure of the set $\{x \in \Omega : f(x) \neq 0\}$. We write $C_{00}(\Omega)$ for the algebra of functions of compact support, and $C_0(\Omega)$ for the algebra of functions f that *vanish at infinity*, i.e., $\{x \in \Omega : |f(x)| \geq \varepsilon\}$ is compact for each $\varepsilon > 0$. *Check* that $(C_0(\Omega), |\cdot|_\Omega)$ is a Banach algebra. Is $(C_{00}(\Omega), |\cdot|_\Omega)$ also a Banach algebra? Is it dense in $(C_0(\Omega), |\cdot|_\Omega)$?

4. A closed, unital subalgebra A of an algebra $(C(\Omega), |\cdot|_\Omega)$ such that, for each $x, y \in \Omega$ with $x \neq y$ there exists $f \in A$ with $f(x) \neq f(y)$, is called a *uniform algebra*.

5. In the text, we defined $\ell^1(G)$ for a group G. *Check* that the construction (with the product being defined in (1.2.1)) also works for a semigroup S instead of G – save that $\ell^1(S)$ is unital only if S has an identity.

6. There is a common generalization of $L^1(\mathbb{R})$ and $\ell^1(G)$. Each locally compact group G has a *left Haar measure* m, and $L^1(G)$, consisting of measurable functions f on G with

$$\|f\|_1 = \int_G |f(t)|\, \mathrm{d}m(t) < \infty ,$$

becomes a Banach algebra for the product

$$(f \star g)(t) = \int_G f(s)g(s^{-1}t)\, \mathrm{d}m(s) .$$

This is the *group algebra* of G. Note that G need not be abelian. See Part II

7. There is no norm $\|\cdot\|$ on $H(U)$ such that $(H(U), \|\cdot\|)$ is a Banach algebra: see Dales (2000, 5.2.33(ii)).

8. Most of the above is in Rudin (1973, 10.1–10.7) and Rudin (1996, 18.1–18.4). For uniform algebras, including the disc algebra $A(\overline{\mathbb{D}})$, see Gamelin (1969). The disc algebra is utilized in Part III, Theorem 14.12. All the examples are given in substantial detail in Dales (2000). See, for example, Dales (2000, §2.1). Uniform algebras and group algebras are discussed in §4.3 and §3.3 of Dales (2000), respectively. The group algebras $L^1(G)$ are a main topic of Part II of this book; for the related measure algebra $M(G)$, see Proposition 9.1.2. For the theory of topological algebras, including Fréchet algebras, see Dales (2000, §2.2).

2

Ideals and the spectrum

We now establish some basic results about Banach algebras. A key idea is that of the *spectrum* of an element. Throughout A is a unital algebra with identity e_A.

It is pleasing to see, first, that the basic ideas of our subject can be proved so quickly, and that the proofs are an attractive blend of basic results from complex analysis and functional analysis.

2.1 The spectrum

Let us first look at the spectrum of an element in a Banach algebra. The concept generalizes that of the eigenvalues of a matrix.

Definition 2.1.1 *Let A be a unital algebra, and let $a \in A$. The* resolvent set *of a is*

$$\rho_A(a) = \{z \in \mathbb{C} : ze_A - a \in \operatorname{Inv} A\}\,;$$

the spectrum *of a is $\sigma_A(a)$, the complement of $\rho_A(a)$ in $\mathbb{C}$, so that*

$$\sigma_A(a) = \mathbb{C} \setminus \rho_A(a)\,;$$

the resolvent function *of a is the function*

$$R_a : z \mapsto (ze_A - a)^{-1}, \quad \rho_A(a) \to \operatorname{Inv} A\,.$$

Usually we write $\rho(a)$ for $\rho_A(a)$, etc.

We shall use the following, easily checked identity: for each $z, w \in \rho(a)$, we have

$$R_a(w) - R_a(z) = (z - w)R_a(z)R_a(w)\,. \qquad (2.1.1)$$

Now suppose that A is a Banach algebra, and let $a \in A$. It follows from Theorem 1.4.2(ii) that

$$\sigma(a) \subset \{z \in \mathbb{C} : |z| \leq \|a\|\}.$$

Definition 2.1.2 *Let A be a unital algebra, and let $a \in A$. The* spectral radius *of a is*

$$\nu_A(a) = \nu(a) = \sup\{|z| : z \in \sigma(a)\}.$$

The element a is quasi-nilpotent *if $\nu(a) = 0$ (i.e., $\sigma(a) = \{0\}$ or $\sigma(a) = \emptyset$); the set of quasi-nilpotents is denoted by $\mathfrak{Q}(A)$.*

Check that, if a is nilpotent (i.e., $a^n = 0$ for some $n \in \mathbb{N}$), then $a \in \mathfrak{Q}(A)$;

for $T \in \mathcal{B}(\mathbb{C}^n) = \mathcal{L}(\mathbb{C}^n) \cong \mathbb{M}_n$, the spectrum of the matrix T is the (finite) set of eigenvalues of T;

for $f \in C(\Omega)$, $\sigma(f)$ is equal to $f(\Omega)$, the range of f, and $\nu(f) = |f|_\Omega$, so that the only quasi-nilpotent in $C(\Omega)$ is 0.

The following is the key basic theorem of our subject.

Proposition 2.1.3 *Let A be a unital Banach algebra, and let $a \in A$.*

(i) *The resolvent set $\rho(a)$ is open in $\mathbb{C}$.*
(ii) *For each $\lambda \in A'$, the function $\lambda \circ R_a$ is analytic on $\rho(a)$.*
(iii) *The spectrum $\sigma(a)$ is compact and non-empty.*
(iv) *For each $n \in \mathbb{N}$ and $r > \nu(a)$, we have*

$$a^n = \frac{1}{2\pi \mathrm{i}} \int_{|\zeta|=r} \zeta^n (\zeta e_A - a)^{-1} \, \mathrm{d}\zeta .$$

(v) $\nu(a) = \lim_{n\to\infty} \|a^n\|^{1/n}$.

Proof (i) The map

$$\theta : z \mapsto z e_A - a, \quad \mathbb{C} \to A,$$

is continuous, and $\mathrm{Inv}A$ is open. So the set $\rho(a) = \theta^{-1}(\mathrm{Inv}A)$ is open in $\mathbb{C}$.

(ii) Fix $z \in \rho(a)$, and let $w \in \rho(a) \setminus \{z\}$. Set $f = \lambda \circ R_a$. Then

$$\frac{f(w) - f(z)}{w - z} = \lambda\left(\frac{R_a(w) - R_a(z)}{w - z}\right) = \lambda(-R_a(w)R_a(z))$$

$$\to -\lambda(R_a(z)^2) \quad \text{as } w \to z,$$

using 1.4.2(iv). Thus f is analytic on $\rho(a)$.

(iii) By (i), the spectrum $\sigma(a)$ is closed. We know that $\sigma(a)$ is bounded. So $\sigma(a)$ is compact.

Assume towards a contradiction that $\sigma(a) = \emptyset$. Take $\lambda \in A'$. By (ii), the function $\lambda \circ R_a$ is entire. But

$$R_a(z) = z^{-1}(e_A - z^{-1}a)^{-1} \to 0 \quad \text{as } |z| \to \infty .$$

By Liouville's theorem, $\lambda \circ R_a = 0$. Hence $\lambda(R_a(0)) = 0$. This is true for each $\lambda \in A'$, and so $R_a(0) = 0$ by the Hahn–Banach theorem. But this is a contradiction. Thus $\sigma(a)$ is non-empty.

(iv) For $r > \|a\|$, the series $\sum_{k=0}^{\infty} a^k/z^{k+1}$ is uniformly convergent to $(ze_A - a)^{-1}$ on $\{z \in \mathbb{C} : |z| = r\}$, and so the equation holds for this value of r. Since R_a is analytic on $\rho(a)$, the equation holds for all $r > \nu(a)$ by Cauchy's theorem.

(v) Let $z \in \sigma(a)$ and $n \in \mathbb{N}$. We check easily that $z^n \in \sigma(a^n)$. Hence we have $|z|^n \le \|a^n\|$, and so $\nu(a) \le \inf \|a^n\|^{1/n}$.

Take $r > \nu(a)$, and set $M_r = \sup\{\|R_a(z)\| : |z| = r\}$. Then, by (iv), we have

$$\|a^n\| \le r^{n+1} M_r \quad (n \in \mathbb{N}) .$$

This shows that $\limsup \|a^n\|^{1/n} \le r$. The result follows. □

Part (iii) of the above result is the *fundamental theorem of Banach algebras*; part (v) is the *spectral radius formula*.

Consequences: $a \in \mathfrak{Q}(A)$ if and only if $\|a^n\|^{1/n} \to 0$;

$$a^n \to 0 \text{ as } n \to \infty \quad \text{if and only if} \quad \nu(a) < 1 .$$

Example Let $A = \ell^1(\omega) \subset \mathbb{C}[[X]]$ for a weight ω. Then $\|X^n\| = \omega_n$ and

$$\nu(X) = \lim_{n \to \infty} \omega_n^{1/n} .$$

So X is quasi-nilpotent if and only if $\omega_n^{1/n} \to 0$ as $n \to \infty$; this is the case when $\omega_n = \exp(-n^2)$, for example.

Recall that a unital algebra A is a *division algebra* if $\operatorname{Inv} A = A \setminus \{0\}$.

Theorem 2.1.4 (Gelfand–Mazur) *Let A be a unital normed algebra which is a division algebra. Then $A = \mathbb{C}e_A$.*

Proof Define $\theta : z \mapsto ze_A, \mathbb{C} \to A$. Then θ is a monomorphism. Take $a \in A$. By Proposition 2.1.3(iii), $\sigma(a) \neq \emptyset$ (see the additional notes for the case where

A is not a Banach algebra), and so there exists $z \in \mathbb{C}$ with $ze_A - a \notin \operatorname{Inv} A$. By hypothesis, $\operatorname{Inv} A = A \setminus \{0\}$. Hence $\theta(z) = a$, and θ is a surjection. □

2.2 Ideals and the radical

Let A be an algebra. For subsets S and T of A, we write

$$S \cdot T = \{ab : a \in S,\ b \in T\} \quad \text{and}$$

$$ST = \left\{ \sum_{j=1}^{n} \alpha_j a_j b_j : \alpha_j \in \mathbb{C}, a_j \in S,\ b_j \in T \right\},$$

so that $ST = \operatorname{lin} S \cdot T$, where 'lin' denotes the linear span. We write $S^{[2]}$ for $S \cdot S$ and S^2 for $\operatorname{lin} S^{[2]}$. A linear subspace I of A is a *left ideal* if $AI \subset A$, and an *ideal* if $AI \cup IA \subset I$. A left ideal M is *maximal* if $M \neq A$ and if there are no left ideals I with $M \subsetneq I \subsetneq A$. Every left ideal is contained in a maximal left ideal (in the case where A is unital).

For an ideal I in A, A/I is the quotient algebra: of course,

$$(a + I)(b + I) = ab + I \quad (a, b \in A).$$

Check that, if A is a normed [Banach] algebra, and I is a closed ideal in A, then A/I is a normed [Banach] algebra for the quotient norm.

Example Let Ω be a compact space, with a closed subset F. Define:

$$I(F) = \{f \in C(\Omega) : f \mid F = 0\};$$

$$J(F) = \{f \in C(\Omega) : f = 0 \text{ on a neighbourhood of } F\}.$$

Check that $I(F)$ and $J(F)$ are ideals in $C(\Omega)$, that $I(F)$ is closed, and that $J(F)$ is dense in $I(F)$. When is $I(F)$ a maximal ideal?

Proposition 2.2.1 *Let A be a unital Banach algebra.*

(i) *Let I be a* [*left*] *ideal in A. Then $\overline{I}$ is also a* [*left*] *ideal in A.*
(ii) *Let M be a maximal* [*left*] *ideal in A. Then M is closed.*

Proof (i) This is immediately checked.

(ii) Set $I = \overline{M}$. Assume that $I = A$. It follows from Proposition 1.4.2(iii) that we have $M \cap \operatorname{Inv} A \neq \emptyset$, a contradiction of the fact that $M \neq A$. Hence $I = M$ because M is maximal. □

The *radical* of an algebra A is defined to be the intersection of the maximal left ideals of $A^\#$; it is denoted by rad A. A (necessarily non-unital) algebra A is *radical* if rad $A = A$. The algebra A is *semisimple* if rad $A = \{0\}$. It is easily checked that, in the unital case,

$$\operatorname{rad} A = \{a \in A : e_A - ba \in \operatorname{Inv} A \quad (b \in A)\}\,.$$

In fact rad A is an ideal in A, and $A/\operatorname{rad} A$ is a semisimple algebra.

Proposition 2.2.2 *Let A be a unital Banach algebra. Then* rad A *is a closed ideal and* $A/\operatorname{rad} A$ *is a semisimple Banach algebra.* □

Some people think that semisimple Banach algebras are 'good', and rad A is the 'bad' bit; we should like to remove the bad bit by writing

$$A = (A/\operatorname{rad} A) \oplus \operatorname{rad} A$$

in some sense (but this is not always possible).

Proposition 2.2.3 *Let A be a unital Banach algebra.*

(i) rad $A \subset \mathfrak{Q}(A)$.
(ii) *Suppose that I is a left ideal of A with* $I \subset \mathfrak{Q}(A)$. *Then* $I \subset \operatorname{rad} A$.
(iii) *In the case where A is commutative,* $\operatorname{rad} A = \mathfrak{Q}(A)$.

Proof (i) Let $a \in \operatorname{rad} A$. Then $e_A - a/z \in \operatorname{Inv} A$ for all $z \neq 0$, and so we have $\sigma(a) = \{0\}$.

(ii) Let $a \in I$. For each $b \in A$, $ba \in I$ because I is a left ideal. So $ba \in \mathfrak{Q}(A)$ by (i) and $e_A - ba \in \operatorname{Inv} A$. Hence $a \in \operatorname{rad} A$.

(iii) Take $a \in \mathfrak{Q}(A)$ and $b \in A$. We have $(ba)^n = b^n a^n \quad (n \in \mathbb{N})$ because A is commutative, and so

$$\|(ba)^n\|^{1/n} = \|b^n a^n\|^{1/n} \le \|b\| \|a^n\|^{1/n} \to 0 \quad \text{as } n \to \infty\,.$$

Hence $\nu(ba) = 0$ and $ba \in \mathfrak{Q}(A)$. Thus $\mathfrak{Q}(A)$ is an ideal, so that $\mathfrak{Q}(A) = \operatorname{rad} A$ by (i) and (ii). □

For example, the commutative Banach algebra $C(\Omega)$ is semisimple.

Notation: for a Banach space E, and for $x_0 \in E$ and $\lambda_0 \in E'$, we define $x_0 \otimes \lambda_0 \in \mathcal{B}(E)$ by

$$x_0 \otimes \lambda_0 : x \mapsto \langle x, \lambda_0 \rangle x_0, \quad E \to E\,.$$

These maps show that, given $x_0, y_0 \in E \setminus \{0\}$, there exists $S \in \mathcal{B}(E)$ such that $Sx_0 = y_0$. An operator $T \in \mathcal{B}(E)$ is *finite-rank* if $T(E)$ is finite-dimensional. Each finite-rank operator is a linear combination of rank-one operators of the form $x_0 \otimes \lambda_0$ (where $x_0 \in E$ and $\lambda_0 \in E'$). The collection of finite-rank operators on E is denoted by $\mathcal{F}(E)$; it is clearly an ideal in $\mathcal{B}(E)$.

Proposition 2.2.4 *Let E be a non-zero Banach space. Then $\mathcal{B}(E)$ is semi-simple.*

Proof Take $T \neq 0$ in $\mathcal{B}(E)$. There exist $x_0, y_0 \in E \setminus \{0\}$ with $Tx_0 = y_0$. Choose $S \in \mathcal{B}(E)$ with $Sy_0 = x_0$. Then clearly $(I_E - ST)x_0 = 0$, and so $I_E - ST \notin \operatorname{Inv} \mathcal{B}(E)$. Hence $T \notin \operatorname{rad} \mathcal{B}(E)$, and so $\operatorname{rad} \mathcal{B}(E) = \{0\}$. □

However, the algebra $\mathcal{B}(E)$ contains many non-zero nilpotent operators, and hence $\operatorname{rad} \mathcal{B}(E) \neq \mathfrak{Q}(\mathcal{B}(E))$.

Let E be a Banach space. An operator $T \in \mathcal{B}(E)$ is *compact* if $\overline{T(E_{[1]})}$ is compact in E, where $E_{[1]}$ denotes the closed unit ball of E. The collection $\mathcal{K}(E)$ of compact operators is a closed ideal in $\mathcal{B}(E)$. Clearly $\mathcal{F}(E) \subset \mathcal{K}(E)$; the closure $\mathcal{A}(E) = \overline{\mathcal{F}(E)}$ of $\mathcal{F}(E)$ in $\mathcal{B}(E)$ is the ideal of *approximable* operators. For many Banach spaces, we have $\mathcal{A}(E) = \mathcal{K}(E)$ (technically, this is true at least when E has the approximation property), but Banach spaces E such that $\mathcal{A}(E) \subsetneq \mathcal{K}(E)$ are known.

In many ways, the ideal $\mathcal{K}(E)$ is the natural 'infinite-dimensional' generalization of the algebras $\mathbb{M}_n$ of all $n \times n$ matrices over $\mathbb{C}$. (Of course, $\mathcal{K}(E) = \mathcal{B}(E) = \mathbb{M}_n$ when the Banach space E has finite dimension n.) It was an early achievement of functional analysis to fully analyse the spectrum of a compact operator.

Theorem 2.2.5 *Let E be an infinite-dimensional Banach space, and suppose that $T \in \mathcal{K}(E)$. Then:*

(i) *$\sigma(T)$ is a countable compact set of the form $\{z_n : n \in \mathbb{N}\} \cup \{0\}$, where $\{z_n : n \in \mathbb{N}\}$ is either finite or a sequence which converges to 0;*
(ii) *each $z \in \sigma(T) \setminus \{0\}$ is an eigenvalue of T, and the corresponding eigenspace $\{x \in E : Tx = zx\}$ is finite-dimensional;*
(iii) *for each $z \in \mathbb{C} \setminus \{0\}$, the subspace $(zI_E - T)(E)$ is closed in E and*

$$\dim \ker(zI_E - T) = \dim(E/(zI_E - T)(E)),$$

so that, in particular, $zI_E - T$ is surjective if and only if it is injective. □

2.3 Exercises

Throughout, A is a unital Banach algebra.

1. Let B be a closed subalgebra of A with $e_A \in B$. Let $a \in A$. Show that $\sigma_B(a) \supset \sigma_A(a)$ and $\partial\sigma_B(a) \subset \partial\sigma_A(a)$. (Here ∂ denotes the topological frontier with respect to $\mathbb{C}$.) What does this mean for the way $\sigma_B(a)$ can be obtained from $\sigma_A(a)$? Calculate $\sigma_A(Z)$ and $\sigma_B(Z)$ when $A = C(\mathbb{T})$ and $B = A(\overline{\mathbb{D}})$.
2. Let $a \in A$. Show that there is a maximal (with respect to inclusion) commutative subalgebra B of A containing a. Show that B is unital, that B is closed, and that $\sigma_B(a) = \sigma_A(a)$.
3. Let $a \in A$, and let U be an open neighbourhood of $\sigma(a)$ in $\mathbb{C}$. Show that there exists $\delta > 0$ such that $\sigma(b) \subset U$ whenever $b \in A$ with $\|a - b\| < \delta$. (This says that the map σ from A is *upper-semicontinuous* in a certain sense.)
4. Let

$$\mathbb{C}((X)) = \left\{ \sum_{n=n_0}^{\infty} \alpha_n X^n : \alpha_n \in \mathbb{C},\ n_0 \in \mathbb{Z} \right\}.$$

Show that $\mathbb{C}((X))$ is a field (use the identification of $\operatorname{Inv} \mathbb{C}[[X]]$); it is the quotient field of $\mathbb{C}[[X]]$, called the *Laurent field* (why?). Show that it is not a normed algebra for any norm.

It can be shown that $\mathbb{C}[[X]]$ is not a Banach algebra for any norm. (This is quite hard.) Is it obvious whether or not it is a normed algebra for some norm?

5. Let $a, b \in A$. Show that $\sigma(ab)$ and $\sigma(ba)$ differ by at most the point 0. (Hint: if c inverts $e_A - ab$, consider $e_A + bca$.) Calculate $\sigma(L)$, $\sigma(R)$, $\sigma(LR)$, and $\sigma(RL)$ when R and L are, respectively, the right and left shifts on the usual Banach space $\ell^1(\mathbb{N})$.
6. Let M be the operator $(\alpha_n) \mapsto (\alpha_n/(n+1))$ on ℓ^2, and let R be the right shift operator. Set $T = MR$. Prove that T is compact and that $\sigma(T) = \{0\}$. Does T have any eigenvalues? Calculate $\|T^n\|$ for $n \in \mathbb{N}$, and check that $\lim_{n\to\infty} \|T^n\|^{1/n} = 0$. (For these operators, and many others, see Part IV.)
7. Take $A = \ell^1(\omega)$ for the weight $\omega = (\omega_n)$, where $\omega_n = e^{-n^2}$ $(n \in \mathbb{Z}^+)$. Show that

$$\operatorname{rad} A = \{(\alpha_n) : \alpha_0 = 0\} \quad \text{and} \quad \operatorname{Inv} A = \{(\alpha_n) : \alpha_0 \neq 0\}.$$

In particular, the commutative, unital Banach algebra A is not semisimple.

8. Let $\mathcal{V}$ be the Volterra algebra, and let $u \in \mathcal{V}$ be as in Exercise 1.5.6. Calculate the spectral radius $\nu(u)$. Deduce that $\mathcal{V}$ is a radical Banach algebra.

2.4 Additional notes

1. The definition of spectrum, etc., makes sense for an arbitrary unital algebra. For example, calculate $\sigma(X)$ for the algebra $A = \mathbb{C}[[X]]$. Let $H(\mathbb{C})$ be the algebra of entire functions on $\mathbb{C}$. What are the possibilities for $\sigma(f)$ when $f \in H(\mathbb{C})$?
2. Let A be a normed algebra. Then A has a completion, say B (Bonsall and Duncan 1973, 1.12; Dales 2000, §2.1). Clearly $\sigma_B(a) \subset \sigma_A(a)$ for $a \in A$. So $\sigma_A(a) \neq \emptyset$ for each $a \in A$ even for a normed algebra A.
3. Suppose that A is not unital. Then $\sigma_A(a)$ is defined to be the spectrum of $(0, a)$ in $\mathbb{C} \odot A = A^{\#}$. The spectrum can also be defined directly by using quasi-inverses (Bonsall and Duncan 2000, §3; Dales 2000, 1.3.4): this is actually quite important.
4. When A is non-unital, you must deal with maximal *modular* ideals, rather than maximal ideals. A left ideal I in A is *modular* if there exists $u \in A$ such that $a - au \in I$ for each $a \in A$.
5. Our radical is strictly the *Jacobson* radical. There are many other radicals of an algebra, but the Jacobson radical is certainly the most important one for Banach algebra theory. For the algebraic theory of $\operatorname{rad} A$, see Dales (2000, §1.5).
6. For §2.1, see Rudin (1973, Chapter 10) and Dales (2000, §2.3). For §2.2, see Bonsall and Duncan (1973, Chapter 3) and Dales (2000, §2.3). For the properties of $\mathcal{B}(E)$ and $\mathcal{K}(E)$ as Banach algebras, see Dales (2000, §2.5); for individual operators, see the later chapters. Compact operators are extensively discussed in Part III.

3

Gelfand theory

The wonderful 'Gelfand theory' of commutative Banach algebras was laid down by Gelfand in (1941), more than 60 years ago. We still essentially follow the route of the master.

3.1 Characters

Definition 3.1.1 *Let A be an algebra. A* character *on A is an epimorphism $\varphi : A \to \mathbb{C}$.*

Thus a character φ is a non-zero linear functional on A such that

$$\varphi(ab) = \varphi(a)\varphi(b) \quad (a, b \in A).$$

(Characters are sometimes called *multiplicative linear functionals.*) We write Φ_A for the set of all characters on A; Φ_A is called the *character space* of A. If A has an identity e_A, then clearly $\varphi(e_A) = 1 \ (\varphi \in \Phi_A)$.

Example Let $A = C(\Omega)$ and $\varepsilon_{x_0}(f) = f(x_0) \ (f \in A)$ for some $x_0 \in \Omega$. Then $\varepsilon_{x_0} \in \Phi_A$.

Let A be an algebra, and let $\varphi \in \Phi_A$. We write

$$M_\varphi = \ker \varphi = \{a \in A : \varphi(a) = 0\}.$$

Proposition 3.1.2 *Let A be a unital algebra. Then the map $\varphi \mapsto M_\varphi$ is a bijection from Φ_A onto the set of maximal ideals of codimension 1 in A.*

Proof Clearly each M_φ is a maximal ideal of codimension 1.

Suppose that $M_\varphi = M_\psi$. Then $\varphi = \psi$ because $\varphi(e_A) = \psi(e_A)$.

Let M be a maximal ideal of codimension 1. Then the quotient map

$$\varphi : A \to A/M = \mathbb{C}$$

is a character with $M_\varphi = M$. □

Now let A be a unital Banach algebra. The next (easy) result is the key to much of Banach algebra theory.

Theorem 3.1.3 *Let $\varphi \in \Phi_A$. Then φ is continuous and $\|\varphi\| = 1$.*

Proof First proof. The maximal ideal M_φ is closed by Proposition 2.2.1(ii), and so φ is continuous.

Second proof. Assume towards a contradiction that there exists $a \in A$ such that $\|a\| < 1$ and $|\varphi(a)| = 1$. By replacing a by $a/\varphi(a)$, we can suppose that $\varphi(a) = 1$. By Theorem 1.4.2(i), there exists $b \in A$ with $b(e_A - a) = e_A$. We have

$$\varphi(b) - \varphi(b)\varphi(a) = \varphi(e_A)$$

because φ is a homomorphism. But $\varphi(a) = \varphi(e_A) = 1$, and so we have a contradiction. Thus φ is continuous and $\|\varphi\| \leq 1$. Since $\varphi(e_A) = 1$, we have $\|\varphi\| = 1$. □

We thus see that Φ_A is a subset of the space

$$\{\lambda \in A' : \lambda(e_A) = 1 = \|\lambda\|\},$$

which is called the *state space* of A.

The weak-$*$ topology on A' is denoted by $\sigma = \sigma(A', A)$, so that $\lambda_\nu \to \lambda$ in (A', σ) if and only if $\lambda_\nu(a) \to \lambda(a)$ for each $a \in A$. Recall that the unit ball of A', written $A'_{[1]}$, is weak-$*$ compact. The space Φ_A is taken to have the relative weak-$*$ topology from A': we shall call it the *Gelfand topology*.

Take $a \in A$. We define $\widehat{a}$ on A' by

$$\widehat{a}(\lambda) = \lambda(a) \quad (\lambda \in A').$$

The definition of the weak-$*$ topology shows immediately that $\widehat{a}$ is continuous on (A', σ).

Theorem 3.1.4 *The space Φ_A is compact (in the weak-$*$ topology).*

Proof For each $a, b \in A$, define

$$K_{a,b} = \{\lambda \in A' : \lambda(ab) = \lambda(a)\lambda(b)\} ;$$

also define $K_{e_A} = \{\lambda \in A' : \lambda(e_A) = 1\}$. Then each $K_{a,b}$ and the set K_{e_A} are all compact in A' [why?], and clearly

$$\Phi_A = \bigcap\{K_{a,b} : a, b \in A\} \cap K_{e_A} .$$

Thus Φ_A is compact. □

We now regard $\widehat{a}$ as being defined on Φ_A; for each $a \in A$, we have $\widehat{a} \in C(\Phi_A)$ and $|\widehat{a}|_{\Phi_A} \leq \|a\|$.

For a general Banach algebra, the character space Φ_A may be empty; it does not have a significant role. But we shall now see that it is highly significant for commutative Banach algebras.

3.2 Gelfand theory

Throughout this subsection A is a *commutative*, unital Banach algebra.

Proposition 3.2.1 *Every maximal ideal of A has the form M_φ for some $\varphi \in \Phi_A$.*

Proof By Proposition 3.1.2, we must show each maximal ideal M of A has codimension 1. We know that M is closed, and so A/M is a Banach algebra. But, in a commutative algebra, A/M is a field (this is a commutative division algebra). By the Gelfand–Mazur theorem, 2.1.4, $A/M = \mathbb{C}$, and so indeed we have $\dim(A/M) = 1$. □

Thus Φ_A corresponds to the set of maximal ideals of A; for this reason Φ_A is sometimes called the *maximal ideal space* of A.

Recall that $C(\Phi_A)$ is a Banach algebra for the uniform norm on Φ_A.

Theorem 3.2.2 (Gelfand representation theorem) *Let A be a commutative, unital Banach algebra. Then:*

(i) *the map $\mathcal{G} : a \mapsto \widehat{a}$, $A \to C(\Phi_A)$, is a norm-decreasing homomorphism;*
(ii) $\sigma(a) = \widehat{a}(\Phi_A) = \{\widehat{a}(\varphi) : \varphi \in \Phi_A\}$;
(iii) $\nu_A(a) = |\widehat{a}|_{\Phi_A}$;
(iv) $a \in \operatorname{Inv} A$ *if and only if* $\widehat{a} \in \operatorname{Inv} C(\Phi_A)$;
(v) $\operatorname{rad} A = \mathfrak{Q}(A) = \ker \mathcal{G}$.

Proof We show that (ii) holds. Indeed,

$$\begin{aligned} z \in \sigma(a) &\Leftrightarrow ze_A - a \notin \operatorname{Inv} A \\ &\Leftrightarrow ze_A - a \text{ belongs to some maximal ideal} \\ &\Leftrightarrow ze_A - a \in M_\varphi \text{ for some } \varphi \in \Phi_A \\ &\Leftrightarrow z = \varphi(a) \text{ for some } \varphi \in \Phi_A \\ &\Leftrightarrow z \in \widehat{a}(\Phi_A)\,. \end{aligned}$$

The rest is immediately checked. □

Corollary 3.2.3 *The following are equivalent:*

(a) *A is semisimple;*
(b) *$\mathcal{G}$ is a monomorphism;*
(c) *ν is a norm on A.* □

Thus a semisimple, commutative, unital Banach algebra is identified as a subalgebra of $C(\Omega)$ for the compact space $\Omega = \Phi_A$. Such algebras are called *Banach function algebras*. Recall that there are commutative, unital Banach algebras which are not semisimple.

3.3 Examples

Let A be a unital Banach function algebra contained in $C(\Omega)$ for some compact Ω. We suppose that, if $x \neq y$ in Ω, then there exists $f \in A$ with $f(x) \neq f(y)$. For $x \in \Omega$, define

$$\varepsilon_x(f) = f(x) \quad (f \in A)\,.$$

The functionals ε_x are the *evaluation maps* on A.

Lemma 3.3.1 *The map $\eta : x \mapsto \varepsilon_x$, $\Omega \to \Phi_A$, is an embedding.*

Proof Certainly $\varepsilon_x \in \Phi_A$. If $\varepsilon_x = \varepsilon_y$, then $x = y$. Thus the map η is an injection.

Let $x_\nu \to x_0$ in Ω. Then $f(x_\nu) \to f(x_0)$ for each $f \in A$, and so $\varepsilon_{x_\nu} \to \varepsilon_{x_0}$ in Φ_A (for the weak-$*$ topology). Thus the map η is continuous, and so it is a homeomorphism by an elementary result from topology. □

Now regard Ω as a closed subset of Φ_A. We say that A is *natural* if $\Omega = \Phi_A$, so that the only characters on A are the evaluation maps.

Notation: for $f \in C(\Omega)$, we define

$$\mathbf{Z}(f) = \{x \in \Omega : f(x) = 0\},$$

the *zero set* of f.

Proposition 3.3.2 *Let A be a unital Banach function algebra on Ω. Then A is natural on Ω if and only if, given $f_1, \ldots, f_n \in A$ with $\bigcap_{j=1}^n \mathbf{Z}(f_j) = \emptyset$, there exist $g_1, \ldots, g_n \in A$ with*

$$f_1 g_1 + \cdots + f_n g_n = 1. \tag{3.3.1}$$

Proof Take $f_1, \ldots, f_n \in A$, and set

$$I = \left\{ \sum_{j=1}^n f_j g_j : g_1, \ldots, g_n \in A \right\},$$

an ideal in A.

Suppose that A is natural and $\bigcap_{j=1}^n \mathbf{Z}(f_j) = \emptyset$. Then, for each $\varphi \in \Phi_A$, there exists f_j with $f_j(\varphi) \neq 0$. Hence $I \not\subset M_\varphi$ for any φ. So $I = A$ and there exist $g_1, \ldots, g_n$ satisfying (3.3.1).

Conversely, suppose that A is not natural, so that there exists $\varphi \in \Phi_A \setminus \Omega$. For each $x \in \Omega$, take $f_x \in M_\varphi$ with $f_x(x) = 1$; we have $f_x(y) \neq 0$ for y in a neighbourhood of x. By an elementary compactness argument, there exist $f_1, \ldots, f_n \in M_\varphi$ with $\bigcap_{j=1}^n \mathbf{Z}(f_j) = \emptyset$, and so there cannot exist $g_1, \ldots, g_n \in A$ to satisfy (3.3.1). □

Corollary 3.3.3 *For each compact space Ω, the Banach function algebra $C(\Omega)$ is natural.*

Proof Take $f_1, \ldots, f_n \in C(\Omega)$ with $\bigcap_{j=1}^n \mathbf{Z}(f_j) = \emptyset$, and set

$$g = \sum_{j=1}^n f_j \overline{f}_j = \sum_{j=1}^n |f_j|^2,$$

so that $g(x) > 0$ $(x \in \Omega)$ and $g \in \operatorname{Inv} C(\Omega)$. Set $g_j = \overline{f}_j g^{-1}$ $(j = 1, \ldots, n)$. Then (3.3.1) is satisfied. □

3.4 Commutative group algebras

We show here that, for commutative group algebras, the Gelfand transform coincides with the well-known Fourier transform.

Let G be a locally compact abelian (or LCA) group. (Think of the cases where $G = \mathbb{Z}$ or $G = \mathbb{T}$ or $G = \mathbb{R}$.) A *character on G* is a group morphism from G onto $\mathbb{T}$. The set Γ of all continuous characters is itself an abelian group with respect to pointwise multiplication

$$\langle s, \gamma_1 + \gamma_2 \rangle = \langle s, \gamma_1 \rangle \langle s, \gamma_2 \rangle \quad (s \in G, \gamma_1, \gamma_2 \in \Gamma),$$

where we write $\langle s, \gamma \rangle$ for $\gamma(s)$. Let Γ have the topology of uniform convergence on compact subsets of G: then Γ is also an LCA group, called the *dual group* of G. The *Pontryagin duality theorem* asserts that the dual of Γ is G, in a natural way.

For example: the dual of $\mathbb{Z}$ is $\mathbb{T}$, in the sense that every continuous character on $\mathbb{Z}$ has the form $n \mapsto z^n$ for some $z \in \mathbb{T}$; the dual of $\mathbb{T}$ is $\mathbb{Z}$; and the dual of $\mathbb{R}$ is $\mathbb{R}$, but the dual group is best thought of as a 'different copy of $\mathbb{R}$'.

Let G be an LCA group. For $f \in L^1(G)$, the *Fourier transform* of f is $\widehat{f}$, defined on Γ by the formula

$$\widehat{f}(\gamma) = \int_G f(s)\langle -s, \gamma \rangle \, \mathrm{d}m(s) \quad (\gamma \in \Gamma).$$

For example, in the case where $G = \mathbb{T}$, we have already defined $\widehat{f}$ on $\mathbb{Z}$ in Exercise 1.5.4. Now suppose that $G = \mathbb{R}$. Then

$$\widehat{f}(y) = \int_{-\infty}^{\infty} f(t)\,\mathrm{e}^{-\mathrm{i}yt}\,\mathrm{d}t \quad (y \in \mathbb{R}).$$

In the general case, set

$$A(\Gamma) = \{\widehat{f} : f \in L^1(G)\},$$

the range of the Fourier transform. The *Riemann–Lebesgue Lemma* shows that

$$A(\Gamma) \subset C_0(\Gamma).$$

In fact, $A(\Gamma)$ is a self-adjoint, translation-invariant, dense subalgebra of $C_0(\Gamma)$: it is a Banach function algebra on the locally compact space Γ.

Theorem 3.4.1 *Let G be an LCA group. Then the Fourier transform*

$$\mathcal{F} : f \mapsto \widehat{f}, \quad L^1(G) \to A(\Gamma) \subset C_0(\Gamma),$$

is a continuous homomorphism with $\|\mathcal{F}\| = 1$. □

We can now identify the character space of $L^1(G)$ with the dual group Γ (so that 'characters on G' correspond to 'characters on $L^1(G)$', and this is why the two meanings of the word 'character' become somewhat blurred).

Theorem 3.4.2 *Let G be an LCA group. Then each character on $L^1(G)$ has the form $f \mapsto \widehat{f}(\gamma)$ for some $\gamma \in \Gamma$. Thus $A(\Gamma)$ is a natural Banach function algebra on Γ, the algebra $L^1(G)$ is semisimple, and the Gelfand transform coincides with the Fourier transform in this case.* □

3.5 C^*-algebras

Definition 3.5.1 *An* involution *on an algebra A is a map $* : A \to A$ such that:*

(i) $(\alpha a + \beta b)^* = \overline{\alpha} a^* + \overline{\beta} b^*$ $(a, b \in A,\ \alpha, \beta \in \mathbb{C})$;
(ii) $(a^*)^* = a$ $(a \in A)$;
(iii) $(ab)^* = b^* a^*$ $(a, b \in A)$.

Definition 3.5.2 *A* C^*-algebra *is a Banach algebra A with an involution such that*

$$\|a^* a\| = \|a\|^2 \quad (a \in A).$$

Examples (i) The algebras $C(\Omega)$ are commutative, unital C^*-algebras for the involution $f \mapsto \overline{f}$.

(ii) Let H be a Hilbert space with inner product denoted by $[\cdot, \cdot]$. For $T \in \mathcal{B}(H)$, take T^* to be the *adjoint* of T, defined by the formula:

$$[Tx, y] = [x, T^* y] \quad (x, y \in H).$$

Then $\mathcal{B}(H)$ is a unital C^*-algebra for the involution $T \mapsto T^*$.

(iii) Let A be a closed subalgebra of $\mathcal{B}(H)$ such that $T^* \in A$ whenever $T \in A$ (so that A is *$*$-closed*). Then A is a C^*-algebra.

Let A be a commutative, unital, semisimple Banach algebra. We know that the Gelfand transform embeds A in $C(\Phi_A)$. When is this map a surjection? The answer is given by the (commutative) *Gelfand–Naimark theorem.*

Theorem 3.5.3 (Gelfand–Naimark) *Let A be a commutative, unital C^*-algebra. Then the Gelfand transform*

$$a \mapsto \widehat{a}, \quad A \to C(\Phi_A),$$

is an isometric $*$-isomorphism. □

3.6 Exercises

1. Let A be a natural Banach function algebra on Ω, and let I be an ideal in A. The *hull* of I is

$$\mathfrak{h}(I) = \bigcap \{\mathbf{Z}(f) : f \in I\}.$$

Show that $\mathfrak{h}(I)$ is a non-empty, closed subset of Ω. For a closed subset F of Ω, define

$$\mathfrak{k}(F) = \{f \in A : f \mid F = 0\},$$

the *kernel* of F. Show that $I \subset \mathfrak{k}(\mathfrak{h}(I))$.

Prove that the map $F \mapsto \mathfrak{k}(\mathfrak{h}(F))$ is a closure operation on the family of Gelfand-closed subsets of Φ_A. The topology it defines is the *hull-kernel topology*. By considering the disc algebra, show that this topology need not be Hausdorff. Prove that the hull-kernel topology coincides with the Gelfand topology if and only if A is *regular*, in the sense that, for each closed subset F of Φ_A and each $\varphi \in \Phi_A \setminus F$, there exists $f \in A$ with $f(F) = \{0\}$ and $f(\varphi) = 1$. Prove that the algebras $C(\Omega)$ and $L^1(G)$ are regular, but the disc algebra. $A(\overline{\mathbb{D}})$ is not regular. The hull-kernel topology is important in Part IV of this book.

2. Show that every closed ideal in $C(\Omega)$ has the form $\mathfrak{k}(F)$ for some closed subset F of Ω.
3. Let Ω_1 and Ω_2 be compact spaces. Show that $C(\Omega_1)$ and $C(\Omega_2)$ are isomorphic (as algebras) if and only if Ω_1 and Ω_2 are homeomorphic (as topological spaces).
4. Show that $C^{(n)}(\mathbb{I})$ is natural for each $n \in \mathbb{N}$.
5. Let A be a commutative, unital Banach algebra, and let $a \in A$. Then

$$\mathbb{C}[a] = \left\{ \sum_{j=0}^{n} \alpha_j a^j : \alpha_1, \ldots, \alpha_n \in \mathbb{C},\ n \in \mathbb{N} \right\}$$

in the smallest unital subalgebra of A containing a. We say that a is a *polynomial generator* of A if $\mathbb{C}[a]$ is dense in A. Consider the map

$$\varphi \mapsto \varphi(a), \quad \Phi_A \to \sigma(a).$$

This map is always a continuous surjection. Show that, in the case where a is a polynomial generator of A, the map is a homeomorphism, and conclude that we can then identify Φ_A with $\sigma(a)$.

6. Now suppose that A is a (possibly non-commutative) unital C^*-algebra, and that $a \subset A$. The algebra

$$\mathbb{C}[a, a^*] = \left\{ \sum_{i,j=1}^{n} \alpha_{ij} a^i (a^*)^j : \alpha_{ij} \in \mathbb{C},\ n \in \mathbb{N} \right\}$$

is the smallest unital subalgebra of A containing a and a^*. The *C^*-subalgebra generated by a* is defined to be $C^*(a) = \overline{\mathbb{C}[a, a^*]}$.

Let $B = C^*(a)$. In the case where a is *normal* (i.e., $aa^* = a^*a$), B is a commutative, unital C^*-algebra. Show that Φ_B is then homeomorphic to $\sigma(a)$. Thus the Gelfand–Naimark theorem gives an isometric unital $*$-isomorphism $C^*(a) \to C(\sigma(a))$. Its inverse is an isometric, unital $*$-isomorphism

$$\Theta_a : C(\sigma(a)) \to C^*(a) \subset A$$

with $\Theta_a(Z) = a$. We set

$$f(a) = \Theta_a(f) \quad (f \in C(\sigma(a))),$$

so that $f(a)$ is 'a continuous function of a'. The map Θ_a is a *continuous functional calculus* for a.

7. Let X be a completely regular topology space. Show that $(C^b(X), |\cdot|_X)$ is a C^*-algebra for the involution $f \mapsto \overline{f}$. Its character space is called βX, the *Stone–Čech compactification* of X. Show that the embedding

$$x \mapsto \varepsilon_x, \quad X \to \beta X,$$

is continuous and has dense range. Every bounded, continuous function on X has a continuous extension to βX.

3.7 Additional notes

1. In the case where A is a non-unital Banach algebra, we can only say that the character space Φ_A is locally compact. The Gelfand representation is then a homomorphism into $C_0(\Phi_A)$. We have seen that this happens in the case where A is $L^1(\mathbb{R})$.

2. The proof of the commutative Gelfand–Naimark theorem 3.5.3 is now an easy exercise. The (harder) non-commutative Gelfand–Naimark theorem shows that every C^*-algebra is $*$-isomorphic to a closed, $*$-closed subalgebra of $\mathcal{B}(H)$ for some Hilbert space H.

3. We proved very easily that every character on a Banach algebra is automatically continuous. It is a remarkable fact that it has been an open question for more than 50 years whether or not every character on a Fréchet algebra is continuous. This is called *Michael's problem.* For the best partial results, see Dixon and Esterle (1986). Beware of fallacious 'proofs' that appear in the literature.

4. Let A be a Banach algebra. A *bounded approxmiate identity* in A is a net (e_α) in A such that $\sup_\alpha \|e_\alpha\| < \infty$ and

$$\lim_\alpha(\|a - e_\alpha a\| + \|a - ae_\alpha\|) = 0 \quad (a \in A).$$

An important theorem of Cohen says that a Banach algebra A with a bounded approximate identity *factors*, in the sense that $A = A^{[2]}$.

It is important to note that many Banach algebras which do not have an identity do have a bounded approxmiate identity, and hence this factorization property. For example, every group algebra $L^1(G)$ and every C^*-algebra has a bounded approximate identity. For more on bounded approximate identities, see Part II, Chapter 9.

5. Let A be a natural Banach function algebra. Then A is *weakly regular* if, for each proper closed subspace F of Φ_A, there exists $f \in A$ with $f \neq 0$ such that $F \subset \mathbf{Z}(f)$, and A has the *unique uniform norm property* (UUNP) if A admits exactly one uniform norm, namely the spectral radius $\nu_A(\,\cdot\,)$. For the relationships between these properties and regularity, see Bhatt and Dedania (2002).

6. Gelfand theory is covered in Rudin (1973, Chapter 11), Bonsall and Duncan (1973, §17) and Dales (2000, §2.3), for example. There have been enormous studies of the algebras $A(\Gamma)$. For a sample, see Dales (2000, §4.5); various substantial texts on this topic are listed in the bibliography. See §9.3 of Part II for more on the Fourier transform of $L^1(G)$.

The general (non-commutative) Gelfand–Naimark theorem (involving the GNS construction) is in all the books listed on C^*-algebras. For example, look at Kadison and Ringrose (1983, §4.5). See also Dales (2000, 3.2.29). A book discussing the Stone–Čech compactification is Gillman and Jerison (1960); see also Dales (2000, §4.2). Cohen's factorization theorem for a Banach algebra with a bounded approximate identity, and several important generalizations, are given in Dales (2000, §2.9) and many other sources listed in the references.

4

The functional calculus

In Exercise 3.6.7 we described a 'continuous functional calculus' Θ_a for a normal element a of a C^*-algebra A. We now replace A by an arbitrary Banach algebra and obtain a weaker, but very important, 'analytic functional calculus'. But first we need some more examples of character spaces.

4.1 More character spaces

Definition 4.1.1 *Let X be a compact set in $\mathbb{C}^n$. Then* the polynomial convex hull *of X is*

$$\widehat{X} = \{z \in \mathbb{C}^n : |p(z)| \leq |p|_X \quad \text{for all polynomials} \quad p\}.$$

Proposition 4.1.2 *Let X be a compact set in $\mathbb{C}^n$. Then $\Phi_{P(X)} = \widehat{X}$.*

Proof Let φ be a character on $\mathbb{C}[X_1, \ldots, X_n]$ (the algebra of polynomials in n variables). Then $\varphi(p) = p(z^0)$ for some $z^0 = (\varphi(X_1), \ldots, \varphi(X_n)) \in \mathbb{C}^n$. This character extends to a (continuous) character on $P(X)$ if and only if $z^0 \in \widehat{X}$. The result follows. □

For a general compact set X in $\mathbb{C}^n$, it can be difficult to identify $\widehat{X}$. However, it is quite easy to show that, for $X \subset \mathbb{C}$, the polynomially convex hull $\widehat{X}$ is the union of X and the bounded components of $\mathbb{C} \setminus X$. Indeed, let V be a bounded component of $\mathbb{C} \setminus X$, and let $z \in V$. Then, for each polynomial $p \in \mathbb{C}[X]$, we have $|p(z)| \leq |p|_{\partial V}$ by the maximum modulus principle. But $\partial V \subset X$, and so $|p(z)| \leq |p|_X$, whence $z \in \widehat{X}$. The converse is also fairly easy.

For example $\Phi_{P(\mathbb{T})} = \overline{\mathbb{D}}$, so $P(\mathbb{T})$ really sits on $\overline{\mathbb{D}}$, not on $\mathbb{T}$. Indeed $P(\mathbb{T})$ is isometrically isomorphic to $A(\overline{\mathbb{D}})$.

Proposition 4.1.3 *The Banach function algebra $A^+(\overline{\mathbb{D}})$ is natural.*

Proof Let $A = A^+(\overline{\mathbb{D}})$. Then A is polynomially generated by Z. It is easy to check that $\sigma(Z) = \overline{\mathbb{D}}$. So $A^+(\overline{\mathbb{D}})$ is natural by Exercise 3.6.5. □

This shows that, if $f \in A^+(\overline{\mathbb{D}})$ and $f(z) \neq 0$ $(z \in \overline{\mathbb{D}})$, then $1/f$ also belongs to $A^+(\overline{\mathbb{D}})$; this is not obvious just from the definition of $A^+(\overline{\mathbb{D}})$.

Recall that $\ell^1(\mathbb{Z})$ (with convolution product) and $W(\mathbb{T})$ (with pointwise product) are isomorphic; they are the 'same' algebra, so the following is a special case of 3.4.2.

Proposition 4.1.4 *The Banach function algebra $W(\mathbb{T})$ is natural.*

Proof Let $\varphi \in \Phi_{W(\mathbb{T})}$, and set $\zeta = \varphi(Z)$. Then $|\zeta| \leq \|\varphi\| \, \|Z\| = 1$. We have $\overline{Z} = Z^{-1}$ and so $\zeta \neq 0$ and $\varphi(\overline{Z}) = \zeta^{-1}$. Also $|\zeta^{-1}| \leq \|\varphi\| \, \|\overline{Z}\| = 1$. Thus $\zeta \in \mathbb{T}$.

It follows that, for each trignometric polynomial p of the form $\sum_{k=-n}^{n} c_k Z^k$ (so that $p(e^{i\theta}) = \sum_{k=-n}^{n} c_k e^{ik\theta}$), we have $\varphi(p) = p(\zeta)$. But these polynomials are dense in $W(\mathbb{T})$, and so $\varphi = \varepsilon_\zeta$ and $W(\mathbb{T})$ is natural. □

Corollary 4.1.5 (Wiener) *Let $f \in C(\mathbb{T})$ have absolutely convergent Fourier series, and suppose that $1/f \in C(\mathbb{T})$. Then $1/f$ has absolutely convergent Fourier series.* □

Wiener's original proof of this was rather long; the Banach algebra proof is much nicer.

4.2 Analytic functional calculus

The idea of this section is to show how to define a function $f(a)$ of an element a of a Banach algebra and a certain analytic function f.

Let A be a unital algebra (not necessarily commutative), and let $a \in A$. Then we can certainly define $p(a)$ for a polynomial $p \in \mathbb{C}[X]$: indeed, in the case where $p = \sum_{j=0}^{n} \alpha_j X^j$, we set

$$p(a) = \sum_{j=0}^{n} \alpha_j a^j$$

(where $a^0 = e_A$). We see that the map

$$\Theta_a : p \mapsto p(a), \quad \mathbb{C}[X] \to A,$$

is a unital homomorphism with $\Theta_a(X) = a$.

Now suppose that r is a rational function, say r has the form p/q, where both $p, q \in \mathbb{C}[X]$ and $q \neq 0$ (and r is defined on $\{z \in \mathbb{C} : q(z) \neq 0\}$). Provided that $q(a) \in \operatorname{Inv} A$, we can sensibly define

$$r(a) = p(a)q(a)^{-1}$$

in A. *Check* that $r(a)$ is well-defined, and that the map $a \mapsto r(a)$ is a homomorphism.

The next step requires A to be a unital Banach algebra. Let f be an analytic function on the disc $U = \mathbb{D}(0; r)$, centre 0, radius $r > \nu(a)$, say

$$f(z) = \sum_{j=0}^{\infty} \alpha_j z^j \quad (|z| < r).$$

Then the series $\sum_{j=0}^{\infty} \alpha_j a^j$ converges in A because $\sum_{j=0}^{\infty} |\alpha_j| \|a^j\| < \infty$ – and we define $f(a)$ to be its sum. Clearly $f(a)$ is well-defined and the map

$$\Theta_a : f \mapsto f(a), \quad H(U) \to A,$$

is a unital homomorphism with $\Theta_a(Z) = a$.

The common generalization of these ideas is the *single-variable analytic functional calculus.*

Thus, fix a unital Banach algebra A, an element $a \in A$, and an open neighbourhood U of $\sigma(a)$ in $\mathbb{C}$.

Fact There is an open set V in $\mathbb{C}$ with $\sigma(a) \subset V \subset \overline{V} \subset U$ such that V has only finitely many components, such that the closures of these components are pairwise disjoint, and such that ∂V consists of a finite number of closed, rectifiable curves (V is called a *Cauchy domain*). Set $\Gamma = \partial V$, the frontier of V. Then the A-valued integral

$$\Theta_a^U(f) = \frac{1}{2\pi \mathrm{i}} \int_\Gamma \frac{f(\zeta)\,\mathrm{d}\zeta}{\zeta e_A - a} = \frac{1}{2\pi \mathrm{i}} \int_\Gamma f(\zeta)(\zeta e_A - a)^{-1}\,\mathrm{d}\zeta$$

exists for each $f \in H(U)$ and is independent of the choice of V.

Definition 4.2.1 *A functional calculus map for a (on U) is a unital homomorphism*

$$\theta : H(U) \to A$$

such that $\theta(Z) = a$. The map is continuous *if $\theta(f_n) \to \theta(f)$ whenever $f_n \to f$ uniformly on compact subsets of U.*

Theorem 4.2.2 *The map $f \mapsto f(a) = \Theta_a^U(f)$ is a continuous functional calculus map for a. Further:*

(i) *if $f = p/q \in H(U)$, then $q(a) \in \mathrm{Inv}A$ and $f(a) = p(a)q(a)^{-1}$;*
(ii) *if $f = \sum_{j=0}^{\infty} \alpha_j Z^j$ has radius of convergence r, where $r > \nu(a)$, then*

$$f(a) = \sum_{j=0}^{\infty} \alpha_j a^j\,;$$

(iii) *if $\varphi \in \Phi_A$, then $\varphi(f(a)) = f(\varphi(a))$; in the case where A is commutative,*

$$\widehat{f(a)} = f(\widehat{a}) \in C(\Phi_A)\,;$$

(iv) *if B is a maximal commutative subalgebra of A containing a, then also $f(a) \in B$.*

Proof Most of this is immediate.

Clearly Θ_a is linear; we check that it is a homomorphism.

Let $f, g \in H(U)$, and choose a contour Γ_1 surrounding $\sigma(a)$ in U, as above, to specify $\Theta_a^U(f)$. Let V be the open set bounded by Γ_1, and choose a contour Γ_2 surrounding $\sigma(a)$ in V to specify $\Theta_a^U(g)$. We have

$$\Theta_a^U(f)\Theta_a^U(g) = \left(\frac{1}{2\pi\mathrm{i}}\right)^2 \int_{\Gamma_2}\int_{\Gamma_1} f(\zeta)g(\eta)(\zeta e_A - a)^{-1}(\eta e_A - a)^{-1}\,\mathrm{d}\zeta\,\mathrm{d}\eta\,. \tag{4.2.1}$$

The integrand contains the term $(\zeta e_A - a)^{-1}(\eta e_A - a)^{-1} = R_a(\zeta)R_a(\eta)$, and, by (2.1.1), this is $(R_a(\zeta) - R_a(\eta))/(\eta - \zeta)$ whenever $\eta \neq \zeta$. Thus the right-hand side of (4.2.1) can be written as the sum of two double integrals. By Fubini's theorem and the fact that $\int_{\Gamma_2} f(\zeta)(\eta - \zeta)^{-1}\mathrm{d}\zeta = 0$ for each $\eta \in \Gamma_1$, one of these two integrals is 0. By Cauchy's integral formula, we have $\int_{\Gamma_1} g(\eta)(\eta - \zeta)^{-1}d\eta = 2\pi\mathrm{i}g(\zeta)$ for $\zeta \in \Gamma_2$, and so

$$\begin{aligned}\Theta_a^U(f)\Theta_a^U(g) &= \left(\frac{1}{2\pi\mathrm{i}}\right)^2 \int_{\Gamma_2} f(\zeta)(\zeta e_A - a)^{-1}\left\{\int_{\Gamma_1} g(\eta)(\eta - \zeta)^{-1}\mathrm{d}\eta\right\}\mathrm{d}\zeta \\ &= \frac{1}{2\pi\mathrm{i}}\int_{\Gamma_2} f(\zeta)g(\zeta)(\zeta e_A - a)^{-1}\mathrm{d}\zeta = \Theta_a^U(fg)\,.\end{aligned}$$

Thus $\Theta_a^U : H(U) \to A$ is a homomorphism.

Take U to be a disc around 0 with $\sigma(a) \subset U$. Then

$$\Theta_a(Z) = \frac{1}{2\pi \mathrm{i}} \int_\Gamma \frac{\zeta \,\mathrm{d}\zeta}{\zeta e_A - a} = \frac{1}{2\pi \mathrm{i}} \int_\Gamma \left(e_A + \frac{a}{\zeta} + a^2\zeta^2 + \cdots \right) \mathrm{d}\zeta = a$$

because $(1/2\pi\mathrm{i})\int_\Gamma \zeta^{-1}\mathrm{d}\zeta = 1$ if $n = 1$ and $= 0$ if $n \neq 1$. It follows that $\Theta_a^U(f) = f(a)$ whenever $f \in R(U)$, the algebra of rational functions with poles off U.

The remainder is left as an easy exercise. □

Corollary 4.2.3 (Spectral mapping theorem) *Let $a \in A$ and $f \in H(U)$ for $U \supset \sigma(a)$. Then $\sigma(f(a)) = f(\sigma(a))$.* □

Corollary 4.2.4 (Wiener–Levy) *Let $f \in C(\mathbb{T})$ have absolutely convergent Fourier series, and let F be analytic on a neighbourhood of $f(\mathbb{T}) = \sigma(f)$. Then $F \circ f$ has an absolutely convergent Fourier series.* □

Proof Apply the theorem to $W(\mathbb{T})$, and note that we have $f(\mathbb{T}) = \sigma(f)$ because $W(\mathbb{T})$ is natural. □

The functional calculus can be used to define many specific, important elements in A. For example, we have obvious definitions of

$$\exp a, \quad \sin a, \quad \cos a$$

(for each $a \in A$). Now suppose that $U = \mathbb{C} \setminus \mathbb{R}^-$, and define log and $Z^{1/2}$ on U as analytic functions (with $\log 1 = 0$ and $1^{1/2} = 1$). Then we can define $\log a$ and $a^{1/2}$ whenever $\sigma(a) \subset U$. The usual rules apply – for example, it follows immediately that $\exp(\log a) = a$ and $(a^{1/2})^2 = a$.

4.3 The idempotent theorem

Let A be an algebra. An *idempotent* in A is an element p such that $p^2 = p$. Two idempotents p and q are *orthogonal*, written $p \perp q$, if $pq = qp = 0$.

The following result follows easily from the functional calculus.

Theorem 4.3.1 *Let E be a Banach space, and let $T \in \mathcal{B}(E)$. Suppose that*

$$\sigma(T) = \sigma_1 \cup \sigma_2,$$

where σ_1 and σ_2 are disjoint, non-empty, closed subsets of $\sigma(T)$. Then there exist idempotents $P, Q \in \mathcal{B}(E)$ with $TP = PT$ and $P + Q = I_E$ such that

$$\sigma(T \mid P(E)) = \sigma_1 \quad \text{and} \quad \sigma(T \mid Q(E)) = \sigma_2.$$

Proof There exists a function f analytic on a neighbourhood of $\sigma(T)$ such that $f \mid \sigma_1 = 1$ and $f \mid \sigma_2 = 0$. Define $P = \Theta_T(f)$ and $Q = I_E - P$. Then P and Q are idempotents in $\mathcal{B}(E)$, $TP = PT$, and $P + Q = I_E$. Further, $P(E)$ and $Q(E)$ are closed linear subspaces of E.

For $\zeta \in \mathbb{C} \setminus \sigma_1$, there exists a function g analytic on a neighbourhood of $\sigma(T)$ with $(\zeta 1 - Z)g = f$. Set $S = \Theta_T(g)$. Then

$$(\zeta I_E - T)S = S(\zeta I_E - T) = P\,,$$

and so $\zeta \notin \sigma(T \mid P(E))$. Hence we have $\sigma(T \mid P(E)) \subset \sigma_1$. Similarly we see that $\sigma(T \mid Q(E)) \subset \sigma_2$.

Now take $\zeta \in \mathbb{C} \setminus (\sigma(T \mid P(E)) \cup \sigma(T \mid Q(E)))$. Then

$$(\zeta I_{P(E)} - T \mid P(E))^{-1} P + (\zeta I_{Q(E)} - T \mid Q(E))^{-1} Q$$

is the inverse of $\zeta I_E - T$ in $\mathcal{B}(E)$, and so $\zeta \in \rho(T)$. Thus we have shown that $\sigma(T \mid P(E)) = \sigma_1$ and $\sigma(T \mid Q(E)) = \sigma_2$. □

Let $T \in \mathcal{B}(E)$, and let z be an isolated point of $\sigma(T)$. Then the corresponding idempotent with spectrum $\{z\}$ is denoted by P_z. The resolvent function R_T is an analytic function near z; if R_T has a pole of order k at z, then k is exactly the *index* of z – this is the minimum $n \in \mathbb{Z}^+$ such that

$$\ker(zI_E - T)^n = \ker(zI_E - T)^{n+1};$$

k is also the minimum $n \in \mathbb{Z}^+$ such that $(zI_E - T)^n P_z = 0$. The functional calculus in the case where $\sigma(T)$ is totally disconnected is called the *Riesz functional calculus*. See the discussion of Riesz operators in Part IV, Chapter 22.

In the case where $T \in \mathcal{K}(E)$, each $z \in \sigma(T) \setminus \{0\}$ is a pole of R_T (and has finite index); the projection P_z has non-zero, finite-dimensional range, and this is $P_z(E) = \ker(zI - T)^k$, where k is the order of the pole.

There is a very well-developed theory of compact operators on a Hilbert space. For example, let H be a Hilbert space, and take $T \in \mathcal{K}(H)$ with $T = T^*$ (so that T is *self-adjoint*), say $\sigma(T) = \{z_n : n \in \mathbb{Z}^+\}$. Then $z_0 = 0$ and (z_n) is a null sequence in $\mathbb{R}$, and there is an orthonormal basis (e_n) in H such that

$$Tx = \sum_{n=1}^{\infty} z_n [x, e_n] e_n \quad (x \in H)\,,$$

where the series for T converges in the operator norm.

4.4 Exercises

1. Let ω be a weight function on $\mathbb{Z}^+$, and set $A = \ell^1(\omega)$ and $\rho = \lim_{n\to\infty} \omega_n^{1/n}$. Show that, in the case where $\rho > 0$, the character space Φ_A is homeomorphic to the closed disc $\overline{\mathbb{D}(0;\rho)}$, and that the Gelfand transform is the map

$$a = \sum_{n=0}^{\infty} \alpha_n X^n \longmapsto \widehat{a} = \sum_{n=0}^{\infty} \alpha_n Z^n \quad \text{on } \overline{\mathbb{D}(0;\rho)},$$

Note that $\widehat{a}$ is continuous on $\overline{\mathbb{D}(0;\rho)}$, and analytic on the open disc $\mathbb{D}(0;\rho)$. Deduce that A is semisimple whenever $\rho > 0$. What happens if $\rho = 0$?

2. Let A be a unital Banach algebra.

(i) Define $\exp a$ for $a \in A$. Show that, if $a, b \in A$ with $ab = ba$, then

$$\exp(a+b) = (\exp a)(\exp b).$$

Is this necessarily true without the condition that $ab = ba$?

(ii) Let $a \in A$ be such that $\sigma(a) \subset U = \mathbb{C} \setminus \mathbb{R}^-$. Note that there exists a function $\log \in H(U)$ with $\log 1 = 0$ and $\exp(\log z) = z \ (z \in U)$. Show that $a = \exp(\log a)$.

(iii) Denote by $\mathrm{Inv}_0 A$ the component of $\mathrm{Inv} A$ containing e_A. Prove that

$$\mathrm{Inv}_0 A = \{(\exp a_1) \cdots (\exp a_n) : a_1, \ldots, a_n \in A\}.$$

4.5 Additional notes

1. We have defined the functional calculus for a fixed open neighbourhood U of $\sigma(a)$. In fact, the full theory deals with $\mathcal{O}_{\sigma(a)}$, the algebra of germs of analytic functions on $\sigma(a)$ formed by varying the neighbourhoods U, where $\mathcal{O}_{\sigma(a)}$ is given the locally convex inductive-limit topology from the locally convex spaces $H(U)$, and we finally obtain a map $\Theta_a : \mathcal{O}_{\sigma(a)} \to A$.

2. In the case where A is a unital C^*-algebra and $a \in A$ is normal, we have defined (in Exercise 3.6.6) a unital homomorphism

$$f \mapsto f(a), \quad C(\sigma(a)) \to A.$$

This homomorphism extends the analytic functional calculus for a, in the sense that the two definitions of $f(a)$ agree when f is the restriction to $\sigma(a)$ of a function analytic on a neighbourhood of $\sigma(a)$. For a related functional calculus, with domain $H^\infty(\mathbb{D})$, see Part III, Corollary 14.1.14; similar several-variable versions of this functional calculus are mentioned in Part III, Chapter 20.

3. Let A be a commutative, unital Banach algebra, and let $a = (a_1, \ldots, a_n)$ belong to $A^{(n)}$, the n-fold Cartesian product of A with itself. Then the *joint spectrum* $\sigma(a)$ of a is

$$\sigma(a) = \{(\varphi(a_1), \ldots, \varphi(a_n)) : \varphi \in \Phi_A\}.$$

There is a several-variable analytic functional calculus

$$\Theta_a : f \mapsto f(a), \quad \mathcal{O}_{\sigma(a)} \to A,$$

a continuous, unital homomorphism with $\Theta_a(Z_j) = a_j$ for $j = 1, \ldots, n$.

Again the spectral mapping theorem holds: $\sigma(f(a)) = f(\sigma(a)) \subset \mathbb{C}^n$ for $a \in A^{(n)}$ and $f \in \mathcal{O}_{\sigma(a)}$. This result depends on deep results in the theory of analytic functions of several complex variables.

For another joint spectrum, see Part III, Chapter 20.

4. Theorem 4.3.1 features in Parts III and IV. The several-variable analytic functional calculus can be used to prove an extension of Theorem 4.3.1 called *Šilov's idempotent theorem.* (It is not known how to prove this without using the powerful several-variable calculus.) The theorem is as follows.

Theorem *Let A be a commutative Banach algebra, and suppose that K is a compact and open subset of Φ_A. Then there is a unique idempotent p in A such that $\widehat{p}$ is the characteristic function of K.* □

5. For polynomial convexity, see Gamelin (1969) and Stout (1971). The functional calculus is in all the references, including Dales (2000, §2.4). For the spectral theory of compact operators, see Meise and Vogt (1997), for example; the Riesz functional calculus is discussed in Part IV of this book. There are two different approaches to the several-variable functional calculus for commutative Banach algebras: for these, see Bourbaki (1960), Gamelin (1969, §III. 4), Stout (1971, Chapter 1, §8), and Palmer (1994, §3.5). A powerful extension is given in Zame (1979). For various related functional calculus maps, see Part III. More results on joint spectra and the several-variable functional calculus can be found in Eschmeier and Putinar (1996).

5

Automatic continuity of homomorphisms

There is a deep and surprising connection between the algebraic and topological properties of a Banach algebra: sometimes the algebraic properties determine the topological properties. Let A be a unital Banach algebra. Then we have already seen that a character on A (defined algebraically as an epimorphism from A onto $\mathbb{C}$) is automatically continuous. In fact many other homomorphisms between Banach algebras A and B are automatically continuous. We explore some of these ideas in the present chapter.

5.1 Automatic continuity

The first result is a very old theorem of Gelfand.

Proposition 5.1.1 *Let A be a Banach algebra, and let B be a commutative, semisimple Banach algebra. Then each homomorphism $\theta : A \to B$ is automatically continuous.*

Proof Let $a_n \to 0$ in A and $\theta(a_n) \to b$ in B. By the closed graph theorem, it suffices to prove that $b = 0$.

Take $\varphi \in \Phi_B$. Then $\varphi \circ \theta \in \Phi_A \cup \{0\}$, and so both φ and $\varphi \circ \theta$ are continuous. We have

$$(\varphi \circ \theta)(a_n) \to (\varphi \circ \theta)(0) = 0$$

and

$$(\varphi \circ \theta)(a_n) = \varphi(\theta(a_n)) \to \varphi(b)$$

as $n \to \infty$, and so $\varphi(b) = 0$. Thus $b \in \bigcap\{\ker \varphi : \varphi \in \Phi_B\}$.

Since B is commutative, $\bigcap\{\ker\varphi : \varphi \in \Phi_B\} = \operatorname{rad} B$ (by 3.2.2(v)); since B is semisimple, $\operatorname{rad} B = 0$. Hence $b = 0$ and θ is continuous. □

Definition 5.1.2 *Let $(A, \|\cdot\|)$ be a Banach algebra. Then A has a* unique complete norm *if each norm with respect to which A is a Banach algebra is equivalent to the given norm $\|\cdot\|$.*

Corollary 5.1.3 *Let $(A, \|\cdot\|)$ be a commutative, semisimple Banach algebra. Then A has a unique complete norm.*

Proof Let $|||\cdot|||$ be another complete algebra norm on A. Then the identity map $(A, \|\cdot\|) \to (A, |||\cdot|||)$ is continuous by Proposition 5.1.1. □

Even if we deviate slightly from semisimplicity by considering commutative Banach algebras with one-dimensional radicals, the above result may fail – see Exercise 5.3.1.

It was a major open question for many years whether every (perhaps non-commutative) semisimple Banach algebra has a unique complete norm. This was eventually proved in 1967 by B. E. Johnson. We shall present a proof due to B. Aupetit (1982), as simplified by T. J. Ransford (1989).

One standard tool in *automatic continuity theory* is the separating space of a linear map; it measures 'how far a linear map is from being continuous'.

Definition 5.1.4 *Let E and F be Banach spaces, and let $T : E \to F$ be a linear map. Then the* separating space of T is

$$\mathfrak{S}(T) = \{y \in F : \text{there exists } x_n \to 0 \text{ in } E \text{ with } Tx_n \to y\}.$$

It is easily checked that $\mathfrak{S}(T)$ is a closed linear subspace of F, and it follows from the closed graph theorem that T is continuous if and only if $\mathfrak{S}(T) = \{0\}$.

Let A and B be Banach algebras, and let $\theta : A \to B$ be a homomorphism with $\overline{\theta(A)} = B$. Then it is also easy to check that $\mathfrak{S}(\theta)$ is an ideal in B. Proposition 5.1.1 really shows that $\mathfrak{S}(\theta) \subset \operatorname{rad} B$ in the case where B is commutative; we shall establish this even for non-commutative B.

Lemma 5.1.5 *Let $(A, \|\cdot\|)$ be a unital Banach algebra, let $a \in A$, and let $\varepsilon > 0$. Then there is a norm $|||\cdot|||$ on A such that $|||\cdot|||$ is equivalent to $\|\cdot\|$, $|||e_A||| = 1$, and $|||a||| \le \nu(a) + \varepsilon$.*

Proof Set $b = a/(\nu(a)+\varepsilon)$. Then $S = \{b^n : n \in \mathbb{Z}^+\}$ is a bounded semigroup in $(A, \cdot)$. For $c \in A$, set

$$p(c) = \sup\{\|sc\| : s \in S\}, \quad |||c||| = \sup\{p(cd) : d \in A,\ p(d) \leq 1\}.$$

Check that this works. □

Let A be an algebra. Then $A[X]$ denotes the algebra of all polynomials with coefficients in A.

Lemma 5.1.6 *Let A be a Banach algebra, let $p \in A[X]$, and take $R > 1$. Then*

$$(\nu_A(p(1)))^2 \leq \sup_{|z|=R} \nu_A(p(z)) \cdot \sup_{|z|=1/R} \nu_A(p(z)).$$

Proof Let $q \in A[X]$, and take $\lambda \in A'$ with $\|\lambda\| = 1$ and $\lambda(q(1)) = \|q(1)\|$. Set $F = \lambda \circ q$. By the maximum modulus theorem applied to the function $z \mapsto F(z)F(1/z)$ on the annulus $\{z \in \mathbb{C} : 1/R \leq |z| \leq R\}$, we obtain

$$|F(1)|^2 \leq \sup_{|z|=R} |F(z)| \cdot \sup_{|z|=1/R} |F(z)|.$$

It follows that

$$\|q(1)\|^2 \leq \sup_{|z|=R} \|q(z)\| \cdot \sup_{|z|=1/R} \|q(z)\|. \tag{5.1.1}$$

Apply (5.1.1) with $q = p^{2^n}$, where $n \in \mathbb{N}$. By the spectral radius formula, 2.1.3 (v), we have

$$\|p^{2^n}(z)\|^{1/2^n} \to \nu_A(p(z)) \quad \text{as} \quad n \to \infty$$

for each $z \in \mathbb{C}$. The sequence $(\|p^{2^n}(z)\|^{1/2^n})$ is monotone decreasing and the function $z \mapsto \|p^{2^n}(z)\|^{1/2^n}$ is continuous, and so it follows from Dini's theorem that, for each $r \in \mathbb{R}^+$, we have

$$\sup_{|z|=r} \|p^{2^n}(z)\|^{1/2^n} \to \sup_{|z|=r} \nu_A(p(z)) \quad \text{as} \quad n \to \infty.$$

Thus the result follows from (5.1.1). □

Theorem 5.1.7 (Aupetit) *Let A and B be Banach algebras, and suppose that $T : A \to B$ is a linear map such that $\nu_B(Ta) \leq \nu_A(a)$ $(a \in A)$.*

(i) *Suppose that* $b \in \mathfrak{S}(T)$. *Then*

$$(\nu_B(Ta))^2 \le \nu_A(a)\nu_B(Ta-b) \quad (a \in A).$$

(ii) $T(A) \cap \mathfrak{S}(T) \subset \mathfrak{Q}(B)$.

Proof (*Ransford*) (i) Choose (a_n) in A with $a_n \to 0$ and $Ta_n \to b$ as $n \to \infty$, and take $a \in A$.

For each $\varepsilon > 0$, we may, by Lemma 5.1.5, choose norms on A and B which are equivalent to the given norms and which are such that

$$\|a\| \le \nu_A(a) + \varepsilon, \quad \|Ta - b\| \le \nu_B(Ta-b) + \varepsilon \quad (a \in A,\, b \in B).$$

We apply Lemma 5.1.6 in the case where $p = (Ta - Ta_n) + (Ta_n)X$ belongs to $B[X]$, where $n \in \mathbb{N}$. We have $p(1) = Ta$, and so, for each $R > 1$,

$$(\nu_B(Ta))^2 \le \sup_{|z|=R} \nu_B(p(z)) \cdot \sup_{|z|=1/R} \nu_B(p(z)). \tag{5.1.2}$$

Now $\nu_B(p(z)) \le \|Ta - Ta_n\| + |z|\,\|Ta_n\|$. Also $p(z) = T(a - a_n + za_n)$, and so, by hypothesis,

$$\nu_B(p(z)) \le \nu_A(a - a_n + za_n) \le \|a - a_n\| + |z|\,\|a_n\|\ .$$

Thus, from (5.1.2), we have

$$(\nu_B(Ta))^2 \le (\|a - a_n\| + R\,\|a_n\|)(\|Ta - Ta_n\| + \|Ta_n\|/R).$$

This holds for each $n \in \mathbb{N}$, and so, letting $n \to \infty$, we see that

$$(\nu_B(Ta))^2 \le \|a\|\,(\|Ta - b\| + \|b\|/R).$$

But this holds for each $R > 1$, and so, letting $R \to \infty$, we obtain

$$(\nu_B(Ta))^2 \le \|a\|\,\|Ta - b\| \le (\nu_A(a) + \varepsilon)(\nu_B(Ta-b) + \varepsilon).$$

Finally this holds for each $\varepsilon > 0$, and so the result follows.

(ii) This is immediate from (i). □

Theorem 5.1.8 (Johnson) *Let A and B be Banach algebras, and suppose that $\theta : A \to B$ is an epimorphism. Then* $\mathfrak{S}(\theta) \subset \operatorname{rad} B$. *If B is semisimple, then θ is automatically continuous.*

Proof It is easy to see that Theorem 5.1.7 applies with θ for T. Since $\theta(A) = B$, Theorem 5.1.7(ii) shows that $\mathfrak{S}(\theta) \subset \mathfrak{Q}(B)$. Since $\mathfrak{S}(\theta)$ is an ideal in B, it follows from Proposition 2.2.3(ii) that $\mathfrak{S}(\theta) \subset \operatorname{rad} B$. □

Corollary 5.1.9 *A semisimple Banach algebra has a unique complete norm.* □

The sharp-eyed will notice that we have not quite generalized Proposition 5.1.1 to the case where B is not necessarily commutative. To obtain an analogous result, we certainly need to suppose that $\overline{\theta(A)} = B$, but we actually assumed that $\theta(A) = B$. Thus we have the following question, which has been open for at least 30 years.

Question 5.1.10 *Let A and B be Banach algebras, and suppose that B is semisimple. Let $\theta : A \to B$ be a homomorphism with $\overline{\theta(A)} = B$. Is θ automatically continuous?*

If you can solve this, please let me know quickly.

5.2 Homomorphisms from Banach algebras

Let $\theta : A \to B$ be a homomorphism between Banach algebras. We have given some conditions on the range B that ensure that θ is automatically continuous. What about conditions on the domain algebra A? The key tool in this case is the *continuity ideal* $\mathcal{I}(\theta)$ of the homomorphism θ.

Definition 5.2.1 *Let $\theta : A \to B$ be a homomorphism. Then*

$$\mathcal{I}(\theta) = \{a \in a : \theta(a)b = b\theta(a) = 0 \quad (b \in \mathfrak{S}(\theta))\}.$$

Thus $\mathcal{I}(\theta)$ is the (two-sided) annihilator of the separating space $\mathfrak{S}(\theta)$.

It is an easy exercise to check that $\mathcal{I}(\theta)$ is an ideal in A. (However, there is no reason for $\mathcal{I}(\theta)$ to be closed in A unless θ really is continuous – which is what we are trying to show.) It is also easy to check that $\mathcal{I}(\theta)$ is just the set of elements $a \in A$ such that both of the maps $x \mapsto \theta(ax)$ and $x \mapsto \theta(xa)$ from A into B are continuous. In the case where A is unital our aim is to show that $e_A \in \mathcal{I}(\theta)$, for then $\mathcal{I}(\theta) = A$ and θ is continuous on the whole of A.

The following *main boundedness theorem* (MBT) is due to Bade and Curtis (1960).

Theorem 5.2.2 *Let A and B be Banach algebras, and let $\theta : A \to B$ be a homomorphism. Suppose that (a_n) and (b_n) are sequences in A such that*

$a_m b_n = 0$ *whenever* $m \neq n$. *Then there is a constant* $C > 0$ *such that*

$$\|\theta(a_n b_n)\| \leq C \|a_n\| \|b_n\| \quad (n \in \mathbb{N}). \tag{5.2.1}$$

Proof We may suppose that $\|a_n\| = \|b_n\| = 1 \ (n \in \mathbb{N})$.

Assume towards a contradiction that there is no C such that (5.2.1) holds. Then there is an injective map $(i, j) \mapsto n(i, j)$, $\mathbb{N} \times \mathbb{N} \to \mathbb{N}$, such that

$$\|\theta(u_{i,j} v_{i,j}\| \geq 4^{i+j} \quad (i, j \in \mathbb{N}), \tag{5.2.2}$$

where $u_{i,j} = a_{n(i,j)}$ and $v_{i,j} = b_{n(i,j)}$. Set

$$v_i = \sum_{\ell=1}^{\infty} v_{i,\ell}/2^{\ell} \quad (i \in \mathbb{N});$$

for each i, the series for v_i converges in A, and $\|v_i\| \leq 1$. For each $i \in \mathbb{N}$, choose $j(i) \in \mathbb{N}$ such that $\|\theta(v_i)\| \leq 2^{j(i)}$, and set

$$a = \sum_{k=1}^{\infty} u_{k,j(k)}/2^k$$

(again the series converges in A); we have

$$av_i = \sum_{k=1}^{\infty} \sum_{\ell=1}^{\infty} u_{k,j(k)} v_{i,\ell}/2^{k+\ell},$$

and so $av_i = u_{i,j(i)} v_{i,j(i)}/2^{i+j(i)}$ because $a_m b_n = 0$ whenever $m \neq n$. By (5.2.2), we have $\|\theta(av_i)\| \geq 2^{i+j(i)}$. However,

$$\|\theta(av_i)\| \leq \|\theta(a)\| \|\theta(v_i)\| \leq 2^{j(i)} \|\theta(a)\|.$$

We conclude that $\|\theta(a)\| \geq 2^i$ for each $i \in \mathbb{N}$, clearly a contradiction. □

Definition 5.2.3 *Let* A *be a unital algebra. A* continued bisection of the identity *for* A *is a pair* $\{(p_n), (q_n)\}$ *of sequences of idempotents such that* $e_A = p_1 + q_1$ *and such that, for each* $n \in \mathbb{N}$, *we have* $p_n \perp q_n$, $p_n = p_{n+1} + q_{n+1}$, *and* $Ap_nA = Aq_nA$.

It is easy to check in this case that $q_m \perp q_n \quad (m \neq n)$.

Theorem 5.2.4 *Let* A *be a unital Banach algebra with a continued bisection of the identity. Then every homomorphism from* A *into a Banach algebra is automatically continuous.*

Proof Let $\{(p_n), (q_n)\}$ be as in Definition 5.2.3.

Let $\theta : A \to B$ be a homomorphism into a Banach algebra B. Assume towards a contradiction that $q_n \notin \mathcal{I}(\theta)$ for infinitely many $n \in \mathbb{N}$. Then we may suppose that the map

$$x \mapsto \theta(q_n x), \quad A \to B,$$

is discontinuous for infinitely many $n \in \mathbb{N}$, and so there exists $(x_n) \subset A$ such that $\|x_n\| = 1$ and $\|\theta(q_n x_n)\| \geq n \|q_n\|^2$ for infinitely many $n \in \mathbb{N}$. Apply the MBT with $a_n = q_n$ and $b_n = q_n x_n$ (so that $a_m b_n = 0$ when $m \neq n$): there exists $C > 0$ such that

$$\|\theta(q_n x_n)\| \leq C \|q_n\|^2 \quad (n \in \mathbb{N}).$$

Thus $C \geq n$ for infinitely many $n \in \mathbb{N}$, a contradiction.

We have shown that there exists $k \in \mathbb{N}$ with $q_k \in \mathcal{I}(\theta)$. But now, successively, we see that $p_k, p_{k-1}, q_{k-1}, \ldots, p_1, q_1, e_A$ belong to $\mathcal{I}(\theta)$, and so $\mathcal{I}(\theta) = A$ and θ is continuous. □

Corollary 5.2.5 *Let E be a Banach space such that $E \cong E \oplus E$. Then all homomorphisms from $\mathcal{B}(E)$ into a Banach algebra are continuous.*

Proof We have

$$E \simeq E \oplus E \simeq (E \oplus E) \oplus E \simeq \cdots ;$$

at the nth stage, E is linearly homeomorphic to the direct sum of n copies of itself. Let P_n and Q_n be the projections of E onto the first and second of these n components, respectively. Clearly $I_E = P_1 + Q_1$ and,

$$P_n \perp Q_n, \quad P_n = P_{n+1} + Q_{n+1} \quad (n \in \mathbb{N}).$$

Let U_n be the operator on E which exchanges the first two of the n components at the nth stage of the decomposition. Then $P_n = U_n Q_n U_n$ and $Q_n = U_n P_n U_n$ for each $n \in \mathbb{N}$. Thus $\{(P_n), (Q_n)\}$ is a continued bisection of the identity. □

5.3 Exercises

1. Write out the details of the following example.

Let A be the sequence space $(\ell^2, \|\cdot\|_2)$ with coordinatewise product. *Check* that A is a commutative, semisimple Banach algebra. Set

$$\mathfrak{A} = A \oplus \mathbb{C}$$

as a linear space, with multiplication given by $(a, z)(b, w) = (ab, 0)$. Then $\mathfrak{A}$ is an algebra and $\mathrm{rad}\,\mathfrak{A} = \{0\} \oplus \mathbb{C}$, which has dimension 1. *Check* that $\mathfrak{A}$ is a Banach algebra for the norm

$$\|(a, z)\| = \|a\|_2 + |z| \quad ((a, z) \in \mathfrak{A}).$$

Let λ be a linear functional on A such that $\lambda \mid \ell^1$ is the functional

$$(\alpha_n) \mapsto \sum_{n=1}^{\infty} \alpha_n,$$

and define

$$|||(a, z)||| = \max\{\|a\|_2, |\lambda(a) - z|\} \quad ((a, z) \in \mathfrak{A}).$$

Check that $(\mathfrak{A}, |||\cdot|||)$ is a Banach algebra (even though λ is necessarily discontinuous on A), but that $|||\cdot|||$ is not equivalent to $\|\cdot\|$ on $\mathfrak{A}$.

2. We have talked about unique complete norms. Since $\mathcal{B}(E)$ is a semisimple Banach algebra, it has a unique complete norm. In fact, a beautiful result of Eidelheit (1940) shows an even stronger result. Write out the details of the following theorem.

Let E be a (non-zero) Banach space, and let $\mathfrak{A} = \mathcal{B}(E)$, a Banach algebra for the operator norm, $\|\cdot\|$. Let $|||\cdot|||$ be any algebra norm on $\mathfrak{A}$. Then there is a constant C such that $\|T\| \leq C\,|||T|||$ $(T \in \mathfrak{A})$.

To obtain a contradiction, assume that there is no such C. Take a sequence (S_n) in A such that $|||S_n||| = 1$ $(n \in \mathbb{N})$ and such that $\|S_n\| \to \infty$. Then there exists $x_0 \in F$ with the sequence $(\|S_n x_0\|)$ unbounded, and then there exists $\lambda \in E'$ such that the sequence $(|\lambda(S_n x_0)|)$ is unbounded. Now define $z_n = \lambda(S_n x_0)$ $(n \in \mathbb{N})$.

Define $Tx = \lambda(x)x_0$ $(x \in E)$, so that $T \in \mathcal{B}(E)$. *Check* that we have $TS_nT = z_nT$ $(n \in \mathbb{N})$, and obtain a contradiction.

3. Let Ω be a compact space, and let $\|\cdot\|$ be an algebra norm on $C(\Omega)$. Show that $|f|_\Omega \leq \|f\|$ $(f \in C(\Omega))$.

4. Let E and F be Banach spaces, let $T : E \to F$ be a linear map, and then let $Q : F \to F/\mathfrak{S}(T)$ be the quotient map. Show that QT is continuous.

Suppose that G is another Banach space, and that $S \in \mathcal{B}(F, G)$. Show that

$$\overline{S\mathfrak{S}(T)} = \mathfrak{S}(ST),$$

so that ST is continuous if and only if $S\mathfrak{S}(T) = \{0\}$.

5. Show that to resolve Question 5.1.10 is equivalent to solving the following question.

Let A and B be Banach algebras, let $\theta : A \to B$ be a homomorphism, and let $b \in \mathfrak{S}(\theta)$. Is it necessarily true that $\sigma(b) = \{0\}$?

(It can be shown quite easily that $\sigma(b)$ is always a connected set containing 0, but nothing further seems to be known.)

5.4 Additional notes

1. Let A be a commutative Banach algebra. We know that, in the case where A is semisimple, A has a unique complete norm. However, this does not characterize semisimple Banach algebras. For example, automatic continuity theory shows that the Banach algebras $\ell^1(\omega)$ all have a unique complete norm, but they are not necessarily semisimple. Many other examples are given in Dales (2000).

2. Despite an immense amount of work on the 'uniqueness-of-norm' problem, basic questions remain open.

For example, let $(A, \|\cdot\|)$ be a commutative Banach algebra which is an integral domain. Does A necessarily have a unique complete norm?

A Banach algebra A is *topologically simple* if the only closed ideals in A are the trivial ones $\{0\}$ and A. It is not known whether or not there is a commutative, topologically simple Banach algebra other than $\mathbb{C}$ (this could be the hardest question in Banach algebra theory). It was shown by Cusack (1977) that, if there is a commutative Banach algebra which is an integral domain and which does not have a unique complete norm, then there is a commutative, topologically simple, radical Banach algebra.

3. Let H be a Hilbert space. We have shown that all homomorphisms from $\mathcal{B}(H)$ are continuous. What about other C^*-algebras?

First note that it is a standard triviality that all $*$-homomorphisms between C^*-algebras are automatically continuous, (see Dales (2000, 3.2.4) and all books on C^*-algebras). But we are interested in homomorphisms which are not $*$-homomorphisms.

The following result is proved in Sinclair (1976, 12.4) by a different argument from the one we used in Theorem 5.2.4. Let A be a unital C^*-algebra such that A has no proper closed ideals of finite codimension. Then every homomorphism from A into a Banach algebra is continuous.

4. The above result does not cover the commutative C^*-algebras of the form $C(\Omega)$ for a compact, infinite space Ω. It was a question of Kaplansky (1949) whether or not every homomorphism from $C(\Omega)$ is automatically continuous; this is equivalent to the question whether every algebra norm $\|\cdot\|$ on $C(\Omega)$ is equivalent to the uniform norm $|\cdot|_\Omega$. (Kaplansky proved that necessarily $|f|_\Omega \leq \|f\| \quad (f \in C(\Omega))$; see Exercise 5.3.3.)

A major advance was due to Bade and Curtis (1960). Let $\theta : C(\Omega) \to B$ be a discontinuous homomorphism into a Banach algebra B. Then there is a finite subset $F = \{x_1, \ldots, x_n\}$ of Ω (the *singularity set*) such that the restriction of θ to the ideal $J(F)$ is continuous. Thus θ is continuous on a 'big subalgebra' of $C(\Omega)$. Further developments, due to Esterle (1978*a*) and Sinclair (1975) showed that there is a maximal ideal M of $C(\Omega)$, a prime ideal $P \subsetneq M$, and an embedding of the algebra M/P into a radical Banach algebra. However it was left open whether or not such embeddings exist.

Finally it was shown independently by Dales (1979) and Esterle (1978*b*) that such embeddings do exist – at least if the continuum hypothesis (CH) be assumed – for every infinite compact space Ω. It is a striking fact that this result cannot be proved in the theory ZFC – see Dales and Woodin (1987) for an exposition, and Dales and Woodin (1996) for further developments.

5. There is a non-commutative analogue of the Bade–Curtis theorem: see Sinclair (1974). It is a conjecture of Albrecht and Dales (1983) that the following are equivalent for a C^*-algebra A: (a) there is a discontinuous homomorphism from A; (b) there exists $k \in \mathbb{N}$ and infinitely many maximal ideals M_n in A such that $A/M_n \cong \mathbb{M}_k$.

For a short proof that (b) implies (a) (with CH), see Dales and Runde (1997). For proofs that (a) implies (b) for various classes of C^*-algebras see Albrecht and Dales (1983) and Ermert (1996).

6. Is every epimorphism from a C^*-algebra onto a Banach algebra automatically continuous? There is an elegant proof of this for commutative C^*-algebras in Esterle (1980) (see also Dales 2000, 5.4.27), but the general case is open.

7. Let A be a Banach algebra with finite-dimensional radical. We would like to know necessary and sufficient conditions for A to have a unique complete norm. For an attack on this and a plausible conjecture, see Dales and Loy (1997).

8. Johnson's original proof of the uniqueness-of-norm theorem is in Johnson (1967). There are several proofs in Dales (2000) – uniqueness-of-norm theorems and counter–examples are in §5.1.

All the results which are mentioned in the Additional notes are discussed in detail in Dales (2000, Chapter 5). In particular, the structure of all (discontinuous) homomorphisms from $C(\Omega)$ is given in full detail.

6

Modules and derivations

After homomorphisms between algebras, the next most elementary maps from an algebra A are derivations into an A-bimodule E. We discuss the algebraic theory, and then the Banach version.

6.1 Modules

We first recall the elementary theory of modules over an algebra.

Definition 6.1.1 *Let A be an algebra. A* left A-module *is a linear space E over $\mathbb{C}$ and a map $(a, x) \mapsto a \cdot x$, $A \times E \to E$, such that:*

(i) $a \cdot (\alpha x + \beta y) = \alpha a \cdot x + \beta a \cdot y$ $(\alpha, \beta \in \mathbb{C},\ a \in A,\ x, y \in E)$;
(ii) $(\alpha a + \beta b) \cdot x = \alpha a \cdot x + \beta b \cdot x$ $(\alpha, \beta \in \mathbb{C},\ a, b \in A,\ x \in E)$;
(iii) $a \cdot (b \cdot x) = ab \cdot x$ $(a, b \in A, x \in E)$.

A right A-module *is a linear space E over $\mathbb{C}$ and a map $(a, x) \mapsto x \cdot a$, $A \times E \to E$, such that:*

(i) $(\alpha x + \beta y) \cdot a = \alpha x \cdot a + \beta y \cdot a$ $(\alpha, \beta \in \mathbb{C},\ a \in A,\ x, y \in E)$;
(ii) $x \cdot (\alpha a + \beta b) = \alpha x \cdot a + \beta x \cdot b$ $(\alpha, \beta \in \mathbb{C},\ a, b \in A,\ x \in E)$;
(iii) $(x \cdot a) \cdot b = x \cdot ab$ $(a, b \in A,\ x \in E)$.

An A-bimodule *is a space E which is a left A-module and a right A-module and which is such that*

$$a \cdot (x \cdot b) = (a \cdot x) \cdot b \quad (a, b \in A,\ x \in E).$$

Suppose that A is commutative and that E is an A-bimodule such that

$$a \cdot x = x \cdot a \quad (a \in A,\ x \in E).$$

Then E is an A-module.

For example, let I be a left ideal in A. Then I and A/I are left A-modules; they are A-bimodules in the case where I is an ideal. Again, let A be commutative, and let $\varphi \in \Phi_A$. Then the linear space $\mathbb{C}$ is an A-module for the map

$$(a, z) \mapsto \varphi(a)z, \quad A \times \mathbb{C} \to \mathbb{C};$$

it is denoted by $\mathbb{C}_\varphi$. Finally, let A be a Banach function algebra on Ω. Then $C(\Omega)$ is a Banach A-module (for the module product equal to the pointwise product).

Let E be a left A-module. Then the map $\rho : A \to \mathcal{L}(E)$ defined by

$$\rho(a)(x) = a \cdot x \quad (a \in A,\ x \in E)$$

is a homomorphism, and every such homomorphism defines a left A-module. The map ρ is called a *representation* of the algebra A on the linear space E. Throughout, one could replace the language of modules by that of representations. This is the approach taken in Palmer (1994, 2001).

Suppose that A is unital. Then a left A-module E is *unital* if

$$e_A \cdot x = x \ (x \in E).$$

Let E be a left A-module. Then we write

$$A \cdot E = \{a \cdot x : a \in E,\ x \in E\}, \quad AE = \operatorname{lin} A \cdot E.$$

We write aE for $\{a\}E$, etc. The module E is *simple* if $A \cdot E \neq \{0\}$ and if $\{0\}$ and E are the only submodules of E. It is easy to check that, for a simple module E, we have $A \cdot x = E$ for each $x \in E \setminus \{0\}$.

Proposition 6.1.2 *Let A be a unital algebra. An ideal I in A is* primitive *if*

$$I = \{a \in A : aE = \{0\}\}$$

for some simple left A-module E.

It is easy to check that I is a primitive if and only if

$$I = \{a \in A : aA \subset M\}$$

for some maximal left ideal M of A, and in this case the simple module E that arises in the definition is isomorphic to A/M as a left A-module. Further, each primitive ideal is the intersection of the maximal left ideals which contain it. In some ways, primitive ideals play the same role for general algebras that the kernels of characters play for commutative algebras.

Let E and F be left A-modules. A *left A-module homomorphism* is a linear map $T : E \to F$ such that

$$T(a \cdot x) = a \cdot Tx \quad (a \in A,\ x \in E).$$

Similarly we define A-bimodule homomorphisms.

Definition 6.1.3 *Let A be a Banach algebra, and let E be a Banach space which is a left A-module. Then E is a* weak Banach left A-module *if the map*

$$\rho(a) : x \mapsto a \cdot x, \quad E \to E,$$

is continuous for each $a \in A$, and E is a Banach left A-module *if the map*

$$(a, x) \mapsto a \cdot x, \quad A \times E \to E,$$

is continuous. Similarly, for right A-modules and A-bimodules.

Thus E is weak Banach if, for each $a \in A$, there exists a constant $C_a > 0$ such that

$$\|a \cdot x\| \leq C_a \|x\| \quad (x \in E),$$

and E is Banach if there exists a constant $C > 0$ such that

$$\|a \cdot x\| \leq C \|a\| \|x\| \quad (a \in A,\ x \in E);$$

in this latter case we may suppose that $C = 1$ by moving to an equivalent norm on E.

Let A be a Banach algebra, and let I be a closed left ideal in A. Then I and A/I are Banach left A-modules. Let A be commutative, and take $\varphi \in \Phi_A$. Then $\mathbb{C}_\varphi$ is a Banach A-module. Let A and B be Banach algebras, and let $\theta : A \to B$ be a homomorphism. Then B is a weak Banach A-bimodule for the maps

$$a \cdot b = \theta(a)b, \quad b \cdot a = b\theta(a) \quad (a \in A,\ b \in B),$$

but we only know that B is a Banach A-bimodule in the case where θ is continuous.

Proposition 6.1.4 *Let A be a unital Banach algebra.*

(i) *Each primitive ideal in A is closed.*

(ii) *Let E be a simple left A-module. Then there is a norm $\|\cdot\|$ on E such that $(E, \|\cdot\|)$ is a Banach left A-module.*

Proof (i) By Proposition 2.2.1(ii), each maximal left ideal is closed, and we know that a primitive ideal is an intersection of maximal left ideals.

(ii) We may suppose that $E = A/M$ as a left A-module for a maximal left ideal M, and then E is a Banach left A-module for the quotient norm $\|\cdot\|$. □

We now come to the important concept of a dual module.

Let E be a Banach space. Then the action of $\lambda \in E'$ on $x \in E$ is denoted by $\langle x, \lambda \rangle$.

Definition 6.1.5 *Let A be a Banach algebra, and let E be a Banach A-bimodule. For $a \in A$ and $\lambda \in E'$, define $a \cdot \lambda$ and $\lambda \cdot a$ by*

$$\langle x, a \cdot \lambda \rangle = \langle x \cdot a, \lambda \rangle, \quad \langle x, \lambda \cdot a \rangle = \langle a \cdot x, \lambda \rangle \quad (x \in E).$$

Then $a \cdot \lambda, \lambda \cdot a \in E'$, and E' is a Banach A-bimodule, called the dual module *to E.*

Check that E' has the stated properties.

For example, take $E = A$. Then A' is a Banach A-bimodule for the operations

$$\langle b, a \cdot \lambda \rangle = \langle ba, \lambda \rangle, \quad \langle b, \lambda \cdot a \rangle = \langle ab, \lambda \rangle \quad (a, b \in A, \lambda \in A').$$

This module is called the *dual module* of A.

6.2 Derivations

Definition 6.2.1 *Let A be an algebra, and let E be an A-bimodule. A linear map $D : A \to E$ is a* derivation *if*

$$D(ab) = a \cdot Db + Da \cdot b \quad (a, b \in A). \tag{6.2.1}$$

Here is an obvious example: let $A = C^{(1)}(\mathbb{I})$, $E = C(\mathbb{I})$, and $D : f \mapsto f'$, $A \to E$. Then E is a Banach A-module for the pointwise product, and D is a continuous derivation.

Equation (6.2.1) is the *derivation identity*. The set of derivations from A into E is denoted by $Z^1(A, E)$; it is a linear subspace of $\mathcal{L}(A, E)$. For example, take $x \in E$, and set

$$\delta_x(a) = a \cdot x - x \cdot a \quad (a \in A).$$

Then, for $a, b \in A$, we have

$$\delta_x(ab) = a \cdot (b \cdot x - x \cdot b) + (a \cdot x - x \cdot a) \cdot b = \delta_x(a) \cdot b + a \cdot \delta_x(b),$$

and so δ_x is a derivation. Derivations of this form are termed *inner derivations*, and an inner derivation δ_x is *implemented by* x; derivations which are not inner derivations are called *outer derivations*. In particular, the map

$$\delta_b : a \mapsto ab - ba, \quad A \to A,$$

is an inner derivation on the algebra A. The set of inner derivations from A to E is a linear subspace $N^1(A, E)$ of $Z^1(A, E)$.

Note that $Z^1(A, \mathbb{C}_\varphi)$ consists of linear functionals $d : A \to \mathbb{C}$ such that

$$d(ab) = \varphi(a)d(b) + d(a)\varphi(b) \quad (a, b \in A).$$

These maps are *point derivations* at φ.

For example, let $A = A(\overline{\mathbb{D}})$, the disc algebra. Then the map $f \mapsto f'(0)$ is a continuous point derivation at the character ε_0 on A.

Let A be a Banach algebra, and let E be a Banach A-bimodule. The space of continuous derivations from A to E is denoted by

$$\mathcal{Z}^1(A, E),$$

and the space of (necessarily continuous) inner derivations is now $\mathcal{N}^1(A, E)$.

A particular case to consider is that of derivations $D : A \to A$ for a Banach algebra A; in this case, we can also define the maps $D^n : A \to A$ for each $n \in \mathbb{N}$. It seems that the range of such a derivation must be 'small' in some sense.

Theorem 6.2.2 (Singer and Wermer) *Let A be a commutative Banach algebra, and let $D : A \to A$ be a continuous derivation. Then $D(A) \subset \operatorname{rad} A$.*

Proof Let $z \in \mathbb{C}$. Then $zD \in \mathcal{Z}^1(A, A)$.

It is easy to check (using Leibniz's identity, Exercise 6.3.4 (iii)) that $\exp(zD)$ is an automorphism in $\mathcal{B}(A)$. For $\varphi \in \Phi_A$ and $a \in A$, define

$$\varphi_z(a) = \varphi((\exp(zD))(a)) = \varphi(a) + \sum_{n=1}^{\infty} \frac{\varphi(D^n a)}{n!} z^n \quad (z \in \mathbb{C}).$$

Then the map

$$F : z \mapsto \varphi_z(a), \quad \mathbb{C} \to \mathbb{C},$$

is an entire function. Also, for each $z \in \mathbb{C}$, we have $\varphi_z \in \Phi_A$, and so it follows that $|\varphi_z(a)| \leq \|a\|$. Thus F is bounded. By Liouville's theorem, F is constant. In particular, $\varphi(Da) = 0$.

This is true for each $\varphi \in \Phi_A$, and so $Da \in \operatorname{rad} A$. □

The following result is the non-commutative generalization of the Singer–Wermer theorem.

Theorem 6.2.3 (Sinclair) *Let A be a Banach algebra, and let $D : A \to A$ be a continuous derivation. Then $D(P) \subset P$ for each primitive ideal P of A.*

Proof Let P be a primitive ideal, say $P = \{a \in A : aE = \{0\}\}$ for a simple left A-module E. Then E is a Banach left A-module.

Assume towards a contradiction that $D(P) \not\subset P$. Then there exists an element $a_0 \in P$ with $Da_0 \cdot x \neq 0$ for some $x \in E$. There exists $b \in A$ with $b \cdot Da_0 \cdot x = x$. But now $D(ba_0) \cdot x = x$ because $a_0 \cdot x = 0$. It follows that $(D(ba_0))^n \cdot x = x \ (n \in \mathbb{N})$.

By a result in the additional notes, below, we have

$$D^n((ba_0)^n) - n!(D(ba_0))^n \in P \quad (n \in \mathbb{N}).$$

Hence $n!x = D^n((ba_0)^n) \cdot x$, and so

$$\|x\|^{1/n} \leq \left(\frac{1}{n!}\right)^{1/n} \|D\| \, \|ba_0\| \, \|x\|^{1/n} \to 0 \quad \text{as } n \to \infty.$$

Thus $x = 0$, a contradiction. The result follows. □

6.3 Exercises

1. Let E and F be linear spaces. Then the tensor product $E \otimes F$ is defined by the following universal property: for each bilinear map $S : E \times F \to G$, there is a linear map $T : E \otimes F \to G$ such that

$$T(x \otimes y) = S(x, y) \quad (x \in E, y \in F).$$

Show that, if E is a left A-module and F is a right A-module, then $E \otimes F$ is an A-bimodule for products satisfying

$$a \cdot (x \otimes y) = a \cdot x \otimes y, \quad (x \otimes y) \cdot a = x \otimes y \cdot a \quad (a \in A, x \in E, y \in F). \tag{6.2.2}$$

2. Let E and F be Banach spaces. Then the *projective norm* on $E \otimes F$ is defined by

$$\|z\|_\pi = \inf \left\{ \sum_{j=1}^{n} \|x_j\| \|y_j\| : z = \sum_{j=1}^{n} x_j \otimes y_j \in E \otimes F \right\}.$$

The completion of $(E \otimes F, \|\cdot\|_\pi)$ is the *projective tensor product* of E and F, denoted by $E \widehat{\otimes} F$. This tensor product has the following universal property: for each continuous bilinear map $S : E \times F \to G$ into a Banach space, there is a continuous linear map $T : E \widehat{\otimes} F \to G$ such that

$$T(x \otimes y) = S(x, y) \quad (x \in E, y \in F).$$

Suppose that E and F are Banach left and right A-modules, respectively. Show that $E \widehat{\otimes} F$ is a Banach A-bimodule for maps satisfying (6.2.2).

In particular, we have the following important example that will return in Chapter 7: the space $A \widehat{\otimes} A$ is a Banach A-bimodule for operations that satisfy the equations

$$a \cdot (b \otimes c) = ab \otimes c, \quad (b \otimes c) \cdot a = b \otimes ca \quad (a, b, c \in A).$$

3. Let A be a Banach algebra.

(i) Let E and F be Banach left A-modules. For $a \in A$ and $T \in \mathcal{B}(E, F)$, define

$$(a \cdot T)(x) = a \cdot Tx, \quad (T \times a)(x) = T(a \cdot x) \quad (x \in E).$$

Show that $\mathcal{B}(E, F)$ is a Banach A-bimodule for the maps

$$(a, T) \mapsto a \cdot T \quad \text{and} \quad (a, T) \mapsto T \times a.$$

(ii) Let E and F be Banach left and right A-modules, respectively. Show that the map $\tau \mapsto T_\tau$ is an isometric A-bimodule isomorphism from the dual A-bimodule $(E \widehat{\otimes} F)'$ onto $\mathcal{B}(E, F')$. Here we are defining $(T_\tau x)(y) = \langle x \otimes y, \tau \rangle$ $(x \in E,\ y \in F)$.

4. Let A be an algebra, let E be an A-module, and let $D \in Z^1(A, E)$.

(i) Show that $p \cdot Dp \cdot p = 0$ for each idempotent p of A.
(ii) If $a \in A$ and $a \cdot Da = Da \cdot a$, then

$$D(a^n) = na^{n-1} \cdot Da \quad (n \geq 2).$$

(iii) In the case where $E = A$, we have Leibniz's identity:

$$D^n(ab) = \sum_{k=0}^{n} \binom{n}{k} D^k a \cdot D^{n-k} b \quad (a, b \in A,\ n \in \mathbb{N}).$$

5. Let A be a unital algebra. Show that $Z^1(A, \mathbb{C}_\varphi)$ can be identified with the space of linear functionals λ on M_φ such that $\lambda \mid M_\varphi^2 = 0$ and $\lambda(e_A) = 0$. What is the analogous result when A is a unital Banach algebra, and we are considering continuous point derivations?

6. (i) Show that all point derivations on the disc algebra $A = A(\overline{\mathbb{D}})$ are continuous. (Use the fact that, if $z \in \mathbb{T}$ and $f \in A$ with $f(z) = 0$, then there exist $g, h \in A$ with $g(z) = h(z) = 0$ and $f = gh$: this follows from Cohen's factorization theorem, described in the notes to Chapter 3.)
(ii) Show that there are many discontinuous point derivations on the Banach algebra $(C^{(n)}(\mathbb{I}), \|\cdot\|_n)$. (See Example 1.2(v).)

6.4 Additional notes

1. The theory of derivations from a Banach algebra A into a Banach A-bimodule E is concerned with the following questions.

(I) When is every derivation from A into a specific Banach A-bimodule automatically continuous? For which Banach algebras A is it true that every derivation from A into an arbitrary Banach A-bimodule is automatically continuous? For an account of this topic, see Part II, Chapter 12.

(II) When is every continuous derivation from A into a class of Banach A-bimodules necessarily an inner derivation? For which commutative Banach algebras A is it true that every continuous derivation from A into a Banach A-module is necessarily zero? Is there a canonical form for an arbitrary continuous derivation from a Banach algebra A into a Banach A-bimodule?

(III) Can an arbitrary derivation be decomposed into the sum of a continuous derivation and a discontinuous derivation of a special type?

Many attractive results are known about these questions, but there remain many challenging open questions. Perhaps the most important is Problem 9.1.13 of Part II; in the notation of Chapter 7, this asks if $\mathcal{H}^1(L^1(G), M(G)) = \{0\}$ for every locally compact group G.

2. Theorem 6.2.2 was proved in Singer and Wermer (1955); the authors conjectured that the result should hold without the assumption that the derivation D be continuous. This was finally proved in the powerful paper (1988) of Thomas, building on an earlier result of Johnson (1969).

3. We used the following algebraic calculation in Theorem 6.2.3.

Let D be a derivation on an algebra A, and let I be an ideal in A. Then, for each $a_1, \dots, a_n \in I$, we have

$$D^n(a_1 \cdots a_n) - n!(Da_1) \cdots (Da_n) \in I .$$

The proof is as follows.

We first make the *claim* that, for each $k \geq 2$, we have $D^j(b_1 \cdots b_k) \in I$ whenever $j \in \{0, \dots, k-1\}$ and $b_1, \dots, b_k \in I$. The claim is true for $k = 2$, for certainly

$$D(b_1 b_2) = b_1 \cdot Db_2 + Db_1 \cdot b_2 \in I$$

whenever $b_1, b_2 \in I$. Assume that the above claim holds for k, and now take elements $b_1, \dots, b_{k+1} \in I$. Then, by Leibniz's identity,

$$D^j(b_1, \dots, b_{k+1}) = \sum_{i=0}^{j} \binom{n+1}{i} D^i(b_1 \cdots b_k) \cdot D^{j-1}(b_{k+1}) \quad (j \in \mathbb{N}),$$

and so $D^j(b_1 \cdots b_{k+1}) \in I$ $(j = 0, \dots, k)$. By induction on k, the claim holds.

We now prove the main result by induction on n. The result is certainly true if $n = 1$. Assume that the result is true for n, and take $a_1, \dots, a_{n+1} \in I$. By Leibniz's identity,

$$D^{n+1}(a_1 \cdots a_{n+1}) = \sum_{j=0}^{n+1} \binom{n+1}{j} D^j(a_1 \cdots a_n) D^{n+1-j}(a_{n+1}) .$$

Each term on the right-hand side of this equality belongs to I, save perhaps for the term $(n+1)D^n(a_1 \cdots a_n)D(a_{n+1})$. By the inductive hypothesis, we have

$$D^n(a_1 \cdots a_n) - n!(Da_1) \cdots (Da_n) \in I ,$$

and so

$$\begin{aligned}&D^{n+1}(a_1\cdots a_{n+1})-(n+1)!(Da_1)\cdots(Da_{n+1})\\&\quad=D^{n+1}(a_1\cdots a_{n+1})-(n+1)D^n(a_1\cdots a_n)(Da_{n+1})\\&\qquad+(n+1)(D^n(a_1\cdots a_n)-n!(Da_1)\cdots(Da_n))(Da_{n+1})\end{aligned}$$

belongs to I. The induction continues.

4. The natural non-commutative analogue of Thomas's general version of the Singer–Wermer theorem would be the following statement.

Let D be a derivation on a Banach algebra A. Then $D(P)\subset P$ for each primitive ideal P of A.

This result has not been proved so far. However Thomas (1993) has proved that $D(P)\subset P$ for all but finitely many primitive ideals P, and that each of these exceptional primitive ideals has finite codimension in A.

A closely related open question is the following. Let D be a derivation on a Banach algebra A. Suppose that $a\in A$ and that $a\cdot Da=Da\cdot a$. Does it follow that Da is a quasi-nilpotent element? The Kleinecke–Shirokov theorem states that this is true in the case where D is continuous. An attractive survey of these questions is given in Mathieu (1994).

5. For many Banach algebras A, all derivations from A into an arbitrary Banach A-bimodule E are continuous. For example, this is true for all $C^\star$-algebras A. However there are discontinuous derivations from the disc algebra $A(\overline{\mathbb{D}})$ into certain Banach $A(\overline{\mathbb{D}})$-modules (despite the fact that all point derivations on $A(\overline{\mathbb{D}})$ are automatically continuous).

6. The algebraic theory of modules and primitive ideals is contained in Dales (2000, §1.4) and Palmer (1994). For Banach modules, see Dales (2000, §2.6). The 'attractive results' about derivations that I know are all contained in Dales (2000, §5.6). Theorem 6.2.3 is from Sinclair (1969).

7

Cohomology

There is a substantial interest nowadays in the cohomology of Banach algebras; it is a development of an earlier, purely algebraic theory. This algebraic theory is now usually presented in the language of homological algebra (exact sequences, functors, Ext, Tor, injective and projective resolutions, ...), and this approach has been adopted by Banach algebraists – see Helemskii (1989). We shall avoid this terminology here, and give an approach that originates with Hochschild.

7.1 Hochschild cohomology

Notation: let A be an algebra, and let E be an A-bimodule. We write $\mathcal{L}^n(A, E)$ for the linear space of all n-linear maps from $A \times \cdots \times A$ (n copies) into E. (Set $\mathcal{L}^0(A, E) = E$.) The formal definition of the Hochschild cohomology groups $H^n(A, E)$ is the following.

Definition 7.1.1 *Let A be an algebra, and let E be an A-bimodule. For each $x \in E$, define*

$$\delta^0(x) : a \mapsto a \cdot x - x \cdot a, \quad A \to E\,,$$

and, for $n \in \mathbb{N}$ and $T \in \mathcal{L}^n(A, E)$, define $\delta^n T \in \mathcal{L}^{n+1}(A, E)$ by

$$\delta^n T(a_1, \ldots, a_{n+1}) = a_1 \cdot T(a_2, \ldots, a_{n+1}) + (-1)^{n+1} T(a_1, \ldots, a_n) \cdot a_{n+1}$$
$$+ \sum_{j=1}^{n} (-1)^j T(a_1, \ldots, a_{j-1}, a_j a_{j+1}, a_{j+2}, \ldots, a_{n+1})\,.$$

Let $n \in \mathbb{Z}^+$. It is clear that δ^n is a linear map from $\mathcal{L}^n(A, E)$ into $\mathcal{L}^{n+1}(A, E)$; these maps are the *connecting maps*. A direct but tedious calculation shows that

$\delta^{n+1} \circ \delta^n = 0$, and so $\operatorname{im} \delta^n \subset \ker \delta^{n+1}$. Indeed we have a complex:

$$\begin{aligned} \mathcal{L}^\bullet(A, E) : 0 \longrightarrow E \xrightarrow{\delta^0} \mathcal{L}(A, E) \xrightarrow{\delta^1} \mathcal{L}^2(A, E) \longrightarrow \cdots \longrightarrow \\ \mathcal{L}^n(A, E) \xrightarrow{\delta^n} \mathcal{L}^{n+1}(A, E) \xrightarrow{\delta^{n+1}} \mathcal{L}^{n+2}(A, E) \longrightarrow \cdots \end{aligned} \quad (7.1.1)$$

of linear spaces and linear maps. For $n \in \mathbb{N}$, the elements of $\ker \delta^n$ and $\operatorname{im} \delta^{n-1}$ are the *n-cocyles* and the *n-coboundaries*, respectively; we set

$$Z^n(A, E) = \ker \delta^n \quad \text{and} \quad N^n(A, E) = \operatorname{im} \delta^{n-1}.$$

Definition 7.1.2 *Let A be an algebra, and let E be an A-bimodule. For $n \in \mathbb{N}$, the* nth cohomology group of A with coefficients in E *is*

$$H^n(A, E) = Z^n(A, E)/N^n(A, E);$$

also, $H^0(A, E) = \ker \delta^0 = \{x \in E : a \cdot x = x \cdot a \ (a \in A)\}$.

In fact, the cohomology groups $H^n(A, E)$ are linear spaces. Despite the impressive generality of our definition of $H^n(A, E)$ for each $n \in \mathbb{N}$, we shall really only consider $H^1(A, E)$ in this chapter (but see the notes); let us see what $H^1(A, E)$ measures.

For $T \in \mathcal{L}(A, E)$, we have $T \in \operatorname{im} \delta^0$ if and only if there exists $x \in E$ with

$$T(a) = a \cdot x - x \cdot a \quad (a, b \in A).$$

Also

$$(\delta^1 T)(a, b) = a \cdot Tb - T(ab) + Ta \cdot b \quad (a, b \in A). \quad (7.1.2)$$

Thus $N^1(A, E)$ and $Z^1(A, E)$ coincide with our previous definitions of this notation in §6.2, and $H^1(A, E)$ is the quotient of the space of all derivations by the space of inner derivations; clearly, $H^1(A, E) = \{0\}$ if and only if every derivation from A into E is inner.

Now let A be a Banach algebra, and let E be a Banach A-bimodule. Then we write $\mathcal{B}^n(A, E)$ for the Banach space of bounded maps in $\mathcal{L}^n(A, E)$, and then define the complex $\mathcal{B}^\bullet(A, E)$ as in (7.1.1), above, with $\mathcal{B}$ replacing $\mathcal{L}$. The elements of $\ker \delta^n$ and $\operatorname{im} \delta^{n-1}$ are now the *continuous n-cocyles* and *continuous n-coboundaries*, respectively, and the spaces thereof are

$$\mathcal{Z}^n(A, E) \quad \text{and} \quad \mathcal{N}^n(A, E).$$

We then define the nth *continuous cohomology group* of A *with coefficients in* E as

$$\mathcal{H}^n(A, E) = \mathcal{Z}^n(A, E)/\mathcal{N}^n(A, E).$$

This is a seminormed space. (In general, $\mathcal{N}^n(A, E)$ is not closed in $\mathcal{Z}^n(A, E)$.) Thus $\mathcal{H}^1(A, E)$ is the quotient of the space of continuous derivations from A to E by the space of inner derivations.

Although we shall not discuss $H^2(A, E)$ in these lectures, we do note that, for each $T \in \mathcal{Z}^2(A, E)$, we have the 2-*cocycle identity*

$$a \cdot T(b, c) - T(ab, c) + T(a, bc) - T(a, b) \cdot c = 0 \quad (a, b, c \in A),$$

and thus $H^2(A, E) = \{0\}$ if and only if each such T has the form $\delta^1 S$ for some element $S \in \mathcal{L}(A, E)$. Even at the level of $n = 2$, direct calculations are not so easy.

7.2 Amenable Banach algebras

Algebraists were interested in characterizing those algebras A such that $H^1(A, E) = \{0\}$ for each A-bimodule E.

Let A be a unital algebra, and let $\pi : A \otimes A \to A$ be the linear map such that $\pi(a \otimes b) = ab$ $(a, b \in A)$. A *diagonal* for A is an element $u \in A \otimes A$ such that $\pi(u) = e_A$ and $a \cdot u = u \cdot a$ $(a \in A)$.

Theorem 7.2.1 *Let A be an algebra. Then the following are equivalent:*

(a) $H^1(A, E) = \{0\}$ *for every A-bimodule E;*
(b) *A is unital and has a diagonal in* $A \otimes A$*;*
(c) *A is semisimple and finite-dimensional.* □

The analogue for Banach algebras is given in the notes.

However, the class of Banach algebras that has proved to be important is that of *amenable* algebras.

Definition 7.2.2 (Johnson) *Let A be a Banach algebra. Then A is* amenable *if* $\mathcal{H}^1(A, E') = \{0\}$ *for every Banach A-bimodule E.*

Thus A is amenable if every continuous derivation into a dual module is inner. We shall explore this notion a little. Throughout A is a unital Banach algebra. It can be checked that, to show A is amenable, it is sufficient to show that $\mathcal{H}^1(A, E') = \{0\}$ for each unital Banach A-bimodule E.

Let E and F be Banach spaces, and let $T \in \mathcal{B}(E, F)$. The *dual* of T is the element $T' \in \mathcal{B}(F', E')$ specified by

$$\langle x, T'\lambda \rangle = \langle Tx, \lambda \rangle \quad (x \in F,\ \lambda \in F');$$

this map is continuous when E' and F' have their weak-$*$ topologies. The second dual of T is $T'' = (T')'$ in $\mathcal{B}(E'', F'')$.

The following generalization of the notion of a diagonal was introduced in Johnson (1972*b*).

Definition 7.2.3 *Let A be a unital Banach algebra. A* virtual diagonal *for A is an element* M *in $(A\widehat{\otimes}A)''$ such that $\pi''(\mathrm{M}) = e_A$ and $a \cdot \mathrm{M} = \mathrm{M} \cdot a$ $(a \in A)$.*

Theorem 7.2.4 *Let A be a unital Banach algebra. Then A is amenable if and only if A has a virtual diagonal.*

Proof Let A have a virtual diagonal M, let E be a unital Banach A-bimodule, and let $D \in \mathcal{Z}^1(A, E')$.

For each $x \in E$, define $\Lambda_x \in (A\widehat{\otimes}A)'$ by

$$\langle a \otimes b,\ \Lambda_x \rangle = \langle x,\ a \cdot Db \rangle = \langle x \cdot a,\ Db \rangle \quad (a, b \in A),$$

and then define $\lambda \in E'$ by

$$\langle x,\ \lambda \rangle = \langle \mathrm{M},\ \Lambda_x \rangle \quad (x \in E).$$

For $a, b, c \in A$ and $x \in E$, we have

$$\langle b \otimes c,\ \Lambda_{a \cdot x - x \cdot a} \rangle = \langle a \cdot x - x \cdot a,\ b \cdot Dc \rangle = \langle x,\ b \cdot Dc \cdot a - ab \cdot Dc \rangle$$

and

$$\begin{aligned} \langle b \otimes c,\ a \cdot \Lambda_x - \Lambda_x \cdot a \rangle &= \langle b \otimes ca - ab \otimes c,\ \Lambda_x \rangle \\ &= \langle x,\ bc \cdot Da + b \cdot Dc \cdot a - ab \cdot Dc \rangle, \end{aligned}$$

and so

$$\langle b \otimes c,\ \Lambda_{a \cdot x - x \cdot a} \rangle = \langle b \otimes c,\ a \cdot \Lambda_x - \Lambda_x \cdot a \rangle - \langle x,\ bc \cdot Da \rangle .$$

Hence, for each $v \in A\widehat{\otimes}A$, it follows that

$$\langle v, \Lambda_{a\cdot x - x\cdot a}\rangle = \langle v, a \cdot \Lambda_x - \Lambda_x \cdot a\rangle - \langle x \cdot \pi(v), Da\rangle \quad (a \in A,\, x \in E). \tag{7.2.1}$$

Let v run through a bounded net $(v_a) \subset A\widehat{\otimes}A$ that converges in the weak-$*$ topology to M. Then $\pi(v_a) \to e_A$ in the weak topology on A; a standard averaging procedure (Mazur's theorem) allows us to suppose that $\pi(v_a) \to e_A$ in the norm-topology of A, and then $x \cdot \pi(v_a) \to x$ in E. Thus, from (7.2.1), we have

$$\begin{aligned}\langle a \cdot x - x \cdot a,\, \lambda\rangle &= \langle a \cdot \mathrm{M} - \mathrm{M} \cdot a,\, \Lambda_x\rangle - \langle x,\, Da\rangle \\ &= -\langle x,\, Da\rangle \quad (a \in A,\, x \in E).\end{aligned}$$

Hence $Da = a \cdot \lambda - \lambda \cdot a$ $(a \in A)$ in E', and D is inner. This shows that A is amenable.

Now suppose that A is amenable. Define $v = e_A \otimes e_A$ in $A\widehat{\otimes}A$. Then

$$\pi''(a \cdot v - v \cdot a) = \pi(a \otimes e_A - e_A \otimes a) = 0,$$

and so $\delta_v \in Z^1(A, \ker \pi'')$. Set

$$X = (A\widehat{\otimes}A)' / \overline{\pi'(A')}.$$

Then X' is identified with $\ker \pi''$ as a Banach A-bimodule. Since A is amenable, there exists $w \in \ker \pi''$ with $\delta_w = \delta_v$. Define $u = v - w$. Then, for each $a \in A$, we have

$$a \cdot u - u \cdot a = (\delta_u - \delta_w)(a) = 0 \quad \text{and} \quad \pi''(u) = \pi''(v) = e_A,$$

and so u is a virtual diagonal. □

Let G be a group, and let $A = \ell^1(G)$. Given $f, g \in A$, we define $f \otimes g$ on $G \otimes G$ by

$$(f \otimes g)(s, t) = f(s)g(t) \quad (s, t \in G).$$

By an earlier remark, there is an isometric isomorphism $A\widehat{\otimes}A \to \ell^1(G \times G)$ that identifies $f \otimes g$ in $A \otimes A$ with $f \otimes g$ as defined above; we thus identify $A\widehat{\otimes}A$ as $\ell^1(G \times G)$.

The dual space of $\ell^1(G)$ is $\ell^\infty(G)$, and the actions of $\delta_s \in \ell^1(G)$ on $f \in \ell^\infty(G)$ are given by

$$(\delta_s \cdot f)(t) = f(ts), \quad (f \cdot \delta_s)(t) = f(st) \quad (t \in G).$$

The dual of $\ell^1(G \times G)$ is $(A\widehat{\otimes}A)' = \ell^\infty(G \times G)$, and this is a Banach A-bimodule: the module operations $\cdot$ satisfy the equations

$$(\delta_s \cdot F)(u, v) = F(u, vs), \quad (F \cdot \delta_s)(u, v) = F(su, v) \quad (u, v \in G),$$

where $s \in G$ and $F \in \ell^\infty(G \times G)$. Next we see that the dual of the induced product map $\pi : A\widehat{\otimes}A \to A$ is $\pi' : \ell^\infty(G) \to \ell^\infty(G \times G)$, where

$$\pi'(f)(u, v) = f(uv) \quad (u, v \in G, \ f \in \ell^\infty(G)).$$

Thus a virtual diagonal is a continuous linear functional M on $\ell^\infty(G \times G)$ such that

$$\langle F \cdot \delta_s, \mathrm{M}\rangle = \langle \delta_s \cdot F, \mathrm{M}\rangle \quad (s \in G, \ F \in \ell^\infty(G \times G)) \tag{7.2.2}$$

and

$$\langle \pi'(f), \mathrm{M}\rangle = f(e_G) \quad (f \in \ell^\infty(G)). \tag{7.2.3}$$

Definition 7.2.5 *Let G be a group. A* mean *on $\ell^\infty(G)$ is a continuous linear functional Λ on $(\ell^\infty(G), |\cdot|_G)$ such that $\Lambda(1) = \|\Lambda\| = 1$. The mean Λ is* left-invariant *if*

$$\langle f, \Lambda\rangle = \langle f \cdot \delta_x, \Lambda\rangle \quad (s \in G, \ f \in \ell^\infty(G)).$$

The group G is amenable *if there is a left-invariant mean on G.*

Thus a mean on G is an element of the state space of the C^*-algebra $\ell^\infty(G)$. It is easy to see that every abelian group and every compact group is amenable, but the free group on two generators $\mathbb{F}_2$ is not amenable.

Here is the theorem that suggested the name 'amenable' for the class of Banach algebras that we are considering.

Theorem 7.2.6 (Johnson) *Let G be a group. Then the Banach algebra $\ell^1(G)$ is amenable if and only if the group G is amenable.*

Proof Suppose that G is amenable, and let Λ be a left-invariant mean on $\ell^\infty(G)$. For $F \in \ell^\infty(G \times G)$, set

$$\widetilde{F}(t) = F(t, t^{-1}) \quad (t \in G),$$

so that $\widetilde{F} \in \ell^\infty(G)$ with $|\widetilde{F}|_G \leq |F|_{G\times G}$, and then define

$$\langle F, \mathrm{M} \rangle = \langle \widetilde{F}, \Lambda \rangle . \tag{7.2.4}$$

Then $\mathrm{M} \in \ell^\infty(G \times G)'$ with $\|\mathrm{M}\| = 1$.

We *claim* that M is a virtual diagonal for $\ell^\infty(G)$. For $F \in \ell^\infty(G \times G)$ and $s, t \in G$, we have

$$\widetilde{F \cdot \delta_s}(t) = (F \cdot \delta_s)(t, t^{-1}) = F(st, t^{-1})$$

and

$$\widetilde{\delta_s \cdot F}(t) = (\delta_s \cdot F)(t, t^{-1}) = F(t, t^{-1}s),$$

and so $\widetilde{\delta_s \cdot F}(st) = \widetilde{F \cdot \delta_s}(t)$. Since Λ is left-invariant, (7.2.2) follows from (7.2.4). Now take $f \in \ell^\infty(G)$, and set $F = \pi'(f) \in \ell^\infty(G \times G)$. Then

$$\widetilde{F}(t) = f(tt^{-1}) = f(e_G) \quad (t \in G),$$

and so $\langle \pi'(f), \mathrm{M} \rangle = f(e_G)$, giving (7.2.3).

Thus M is a virtual diagonal, as claimed, and so $\ell^\infty(G)$ is amenable by Theorem 7.2.4.

For the converse, let M be a virtual diagonal for $\ell^\infty(G)$. For $f \in \ell^\infty(G)$, set

$$\widetilde{f}(u, v) = f(v) \quad (u, v \in G),$$

so that $\widetilde{f} \in \ell^\infty(G \times G)$ and $\left|\widetilde{f}\right|_{G\times G} = |f|_G$, and then define

$$\langle f, \Lambda \rangle = \langle \widetilde{f}, \mathrm{M} \rangle . \tag{7.2.5}$$

Then $\Lambda(1) = 1$. Also, for each $s \in G$ and $f \in \ell^\infty(G)$, we have $\widetilde{f \cdot \delta_s} = \delta_s \cdot \widetilde{f}$ and $\widetilde{f} \cdot \delta_s = \widetilde{f}$, and so

$$\langle f \cdot \delta_s, \Lambda \rangle = \langle \widetilde{f \cdot \delta_s}, \mathrm{M} \rangle = \langle \delta_s \cdot \widetilde{f}, \mathrm{M} \rangle = \langle \widetilde{f} \cdot \delta_s, \mathrm{M} \rangle = \langle \widetilde{f}, \mathrm{M} \rangle = \langle f, \Lambda \rangle .$$

Thus Λ is left-invariant. It may be that $\|\Lambda\| \neq 1$, but this can be fixed by a little trick, and so G is amenable. (See also 11.1.2.) □

Proposition 7.2.7 *Let A and B be Banach algebras, and let $\theta : A \to B$ be a continuous homomorphism with $\overline{\theta(A)} = B$. Suppose that A is amenable. Then B is amenable.*

Proof This is an easy exercise. □

Theorem 7.2.8 *For each compact space Ω, the Banach algebra $C(\Omega)$ is amenable.*

Proof Set $G = C(\Omega, \mathbb{R})$, regarded as an abelian group with respect to addition. The map

$$\theta : \sum_{h \in G} \alpha_h \delta_h \mapsto \sum_{h \in G} \alpha_h \exp(\mathrm{i}h), \quad \ell^1(G) \to C(\Omega),$$

is a continuous homomorphism, where δ_h is the point mass at h. By the Stone–Weierstrass theorem, θ has dense range. Since $\ell^1(G)$ is amenable, so is $C(\Omega)$. □

7.3 Additional notes

1. The algebraic cohomology groups $H^n(A, E)$ were first defined by Hochschild (1945, 1946).
2. We can regard $\mathcal{L}(A, E)$ itself as an A-bimodule for the products $\star$, where

$$(a \star T)(b) = a \cdot Tb, \quad (T \star a)(b) = T(ab) - Ta \cdot b \quad (a, b \in A,\ T \in \mathcal{L}(A, E)).$$

The formal identification of $\mathcal{L}^{n+1}(A, E)$ with $\mathcal{L}^n(A, \mathcal{L}(A, E))$ then leads to the representation of each $\mathcal{L}^n(A, E)$ as an A-bimodule. We then have a useful *reduction-of-dimension* formula: for $k, p \in \mathbb{N}$, we have a linear isomorphism $H^{k+p}(A, E) \simeq H^k(A, \mathcal{L}^p(A, E))$.
3. Let $\mathcal{L}_n(A, E)$ be the space $A \otimes \cdots \otimes A \otimes E$, where there are n copies of A. Then $\mathcal{L}_1(A, E)$ is an A-bimodule for the products given by

$$(a \otimes x) \star b = a \otimes x \cdot b, \quad b \star (a \otimes x) = ba \otimes x - b \otimes a \cdot x \quad (a, b \in A,\ x \in E).$$

The spaces $\mathcal{L}_n(A, E)$ form a complex of A-bimodules for certain connecting maps d_n, and the *homology groups* $H_n(A, E)$ are defined to be

$$\ker d_{n-1} / \operatorname{im} d_n .$$

We have $H_n(A, E) = \{0\}$ if and only if $H^n(A, E^\times) = \{0\}$, where $E^\times$ is the (algebraic) dual space of E.

4. The Banach algebra analogue of Theorem 7.2.1 is the following. Now π is a (continuous) map from $A \widehat{\otimes} A$ onto A.

Theorem 7.3.1 *Let A be a Banach algebra. Then the following are equivalent:*

(a) $\mathcal{H}^1(A, E) = \{0\}$ *for every Banach A-bimodule E;*

(b) *A is unital and has a diagonal in* $A \widehat{\otimes} A$. □

The conditions probably imply that A is semisimple and finite-dimensional, but this has only been proved for various classes of Banach algebras. For a study of this question, see Runde (1998) and Runde (2002).

5. There is a *reduction-of-dimension* theorem for the continuous cohomology groups: for $k, p \in \mathbb{N}$, we have $\mathcal{H}^{k+p}(A, E) \simeq \mathcal{H}^k(A, \mathcal{B}^p(A, E))$. It follows that, for an amenable Banach algebra A, $\mathcal{H}^n(A, E') = \{0\}$ for each $n \in \mathbb{N}$ and each Banach A-bimodule E.

6. A somewhat different proof of Theorem 7.2.6 is given in Bonsall and Duncan (1973, §43).

7. Let G be a locally compact group. Then there is a definition of 'G is amenable' which is essentially the same as Definition 7.2.5, and then it is again true (Johnson 1972*a*) that the group G is amenable if and only if the Banach algebra $L^1(G)$ is amenable; see Dales (2000, §5.6). For a proof related to that of Theorem 7.2.6, see Stokke (2003). For a further discussion of amenable groups, see Part II.

8. There is an industry which seeks to calculate $\mathcal{H}^n(A, E)$ for many special Banach algebras A and Banach A-bimodules E. For example, let A be a von Neumann subalgebra of $\mathcal{B}(H)$. Then $\mathcal{H}^1(A, \mathcal{B}(H))$ is calculated in very many cases in Sinclair and Smith (1996). It is a deep theorem of Haagerup (1983) and others that a C^*-algebra A is amenable if and only if A is what is called *nuclear*. For a new version of this proof, see Runde (2002).

9. Let A be a uniform algebra on Ω. Then A is amenable if and only if $A = C(\Omega)$. This is a theorem of Scheinberg, given in Helemskii (1989).

10. A Banach algebra A is said to be *weakly amenable* if $\mathcal{H}^1(A, A') = \{0\}$. Every C^*-algebra and every group algebra of the form $L^1(G)$ is weakly amenable. A paper by Dales, Ghahramani, and Helemskii (2002) determines when the *measure algebra* $M(G)$ is amenable and when it is weakly amenable. It is an interesting open question whether every natural, weakly amenable uniform algebra on a compact space Ω is necessarily equal to $C(\Omega)$.

11. Let G be an amenable group, so that there is a left-invariant mean on $\ell^\infty(G)$. Then there is another mean on $\ell^\infty(G)$ which is both left- and right-invariant; see Dales (2000, 3.3.49).
12. Let E be an infinite-dimensional Banach space. Then we suspect that $\mathcal{B}(E)$ is never amenable. This has been proved in the case where E is a Hilbert space by surprisingly deep methods from C^*-algebra theory. It has also been proved for the case where E is ℓ^1 by Read (2001). We should like to have a proof that establishes the result at least for each of the spaces ℓ^p (for $1 \leq p \leq \infty$).
13. Let $\mathfrak{A}$ be a Banach algebra with radical $\mathfrak{R}$, and set $A = \mathfrak{A}/\mathfrak{R}$. Then

$$\sum : 0 \to \mathfrak{R} \to \mathfrak{A} \xrightarrow{q} A \to 0$$

is a short exact sequence of Banach algebras which is an extension of A. The sequence $\sum$ is *admissible* in the case where $\mathfrak{R}$ is complemented as a Banach space, and *singular* if $\mathfrak{R}^2 = 0$; in the latter case $\mathfrak{R}$ is a Banach A-bimodule.

The algebra $\mathfrak{A}$ has a *strong Wedderburn decomposition* (SWD) if there is a closed subalgebra $\mathfrak{B}$ of $\mathfrak{A}$ such that $\mathfrak{B} \cong A$ amd $\mathfrak{A} = \mathfrak{B} \oplus \mathfrak{R}$. Then we have a theorem that shows the role of the second cohomology groups $\mathcal{H}^2(A, \cdot)$ of A: in the case where $\sum$ is admissable and singular, $\mathfrak{A}$ has a SWD if and only if $\mathcal{H}^2(A, \mathfrak{R}) = \{0\}$.
14. Algebraic cohomolgy theory is described in Dales (2000, §1.9), and Banach cohomology in Dales (2000, §2.8). A substantial account, from a different perspective, is given in Helemskii (1989). The seminal definition of 'amenable Banach algebra' is from Johnson (1972a), where we also find Theorem 7.2.6. There is an enormous literature on the cohomology of C^*-algebras: for a clear introductory account, see Sinclair and Smith (1995). An attractive new book on amenable Banach algebras is Runde (2002). A classic text on amenability is Paterson (1988); see also Part II, Chapter 11, of the present book. For an essay on Wedderburn decompositions, including references to earlier work, see Bade, Dales, and Lykova (1999).

The literature on Banach algebras

There is an enormous literature on Banach algebras. A good place to start is the fine pair of books by Rudin (1973, 1996). From there, one can progress to more specialized accounts. See also Meise and Vogt (1997).

For the **general theory**, consult Bonsall and Duncan (1973) and Bourbaki (1960). A more recent account at a fairly elementary level is Helemskii (1993). For a work with an emphasis on the cohomology of Banach algebras, see Helemskii (1989). A comprehensive account, with an emphasis on the algebraic side, is in Palmer (1994, 2001). Everything mentioned in this part of the book, and a huge amount more, is contained in Dales (2000).

The classic books on **uniform algebras** are Gamelin (1969) and Stout (1971). It seems that this subject is not so fashionable in recent years.

There are many texts on **harmonic analysis**, which is the study of the Banach algebras $L^1(G)$. For example, see Graham and McGehee (1979), Hewitt and Ross (1963, 1970), Kahane (1970), Katznelson (1976), and Reiter and Stegeman (2000).

For the theory of **amenable groups** and **amenable Banach algebras**, see Paterson (1988) and Runde (2002).

The theory of C^***-algebras** was barely mentioned in the text; this is a world-wide industry. See the following texts, for example: Doran and Belfi (1986), Kadison and Ringrose (1983, 1986), Murphy (1990), Pedersen (1979), and Takesaki (1979, 2002).

For the connection with **abstract algebra** and **set theory** (including the construction of discontinuous homomorphisms from $C(\Omega)$), see Dales and Woodin (1987, 1996).

For more advanced work on spectral theory and the (several-variable) **functional calculus**, see Eschmeier and Putinar (1996).

For a collection of articles on Banach algebra theory, including one on the history of 13 conferences on the subject, see Albrecht and Mathieu (1998).

References

Albrecht, E. and Dales, H. G. (1983). Continuity of homomorphisms from C^*-algebras and other Banach algebras, *Lecture Notes in Mathematics*, **975**, 375–96.

Albrecht, E. and Mathieu, M. (eds.) (1998). *Banach Algebras '97*, Berlin, Walter de Gruyter.

Aupetit, B. (1982). The uniqueness of the complete norm topology in Banach algebras and Banach Jordan algebras, *J. Functional Analysis*, **47**, 1–6.

Bade, W. G. and Curtis, P. C., Jr (1960). Homomorphisms of commutative Banach algebras, *American J. Math.*, **82**, 589–608.

Bade, W. G., Dales, H. G., and Lykova, Z. A. (1999). Algebraic and strong splittings of extensions of Banach algebras, *Memoirs American Math. Soc.*, **656**.

Bhatt, S. J. and Dedania, H. V. (2001). Banach algebras with unique uniform norm II, *Studia Math.*, **147**, 211–35.

Bonsall, F. F. and Duncan, J. (1973). *Complete normed algebras*, Berlin–Heidelberg–New York, Springer-Verlag.

Bourbaki, N. (1960). *Eléments de mathématiques. Théories spectrales*, Paris, Hermann.

Cusack, J. (1977). Automatic continuity and topologically simple radical Banach algebras, *J. London Math. Soc.*, **16**, 493–500.

Dales, H. G. (1979). A discontinuous homomorphism from $C(X)$, *American J. Math.*, **101**, 647–734.

Dales, H. G. (2000). *Banach algebras and automatic continuity*, London Mathematical Society Monographs, **24**, Oxford, Clarendon Press.

Dales, H. G., Ghahramani, F. and Helemskii, A. Ya. (2002). The amenability of the measure algebra, *J. London Math. Soc.*, (2), **66**, 213–26.

Dales, H. G. and Loy, R. J. (1997). Uniqueness of the norm topology for Banach algebras with finite-dimensional radical, *Proc. London Math. Soc.*, (3), **26**, 69–81.

Dales, H. G. and Runde, V. (1997). Discontinuous homomorphisms from non-commutative Banach algebras, *Bull. London Math. Soc.*, **29**, 475–9.

Dales, H. G. and Woodin, W. H. (1987). *An introduction to independence for analysts*, London Mathematical Society Lecture Note Series, **115**, Cambridge University Press.

Dales, H. G. and Woodin, W. H. (1996). *Super-real fields: totally ordered fields with additional structure*, London Mathematical Society Monographs, **14**, Oxford, Clarendon Press.

Dixon, P. G. and Esterle, J. R. (1986). Michael's problem and the Poincaré–Bieberbach phenomenon, *Bull. American Math. Soc.*, **15**, 127–87.

Doran, R. S. and Belfi, V. A. (1986). *Characterizations of C^*-algebras: the Gelfand–Naimark theorems*, New York, Marcel Dekker.

Eidelheit, M. (1940). On isomorphisms of rings of linear operators, *Studia Mathematica*, **9**, 97–105.

Ermert, O. (1996). Continuity of homomorphisms from $AF - C^*$-algebras and other inductive limit C^*-algebras, *J. London Math. Soc.*, (2), **54**, 369–86.

Eschmeier, J. and Putinar, M. (1996). *Spectral decompositions and analytic sheaves*, London Mathematical Society Monographs, **10**, Oxford, Clarendon Press.

Esterle, J. (1978*a*). Semi-norms sur $C(K)$, *Proc. London Math. Soc.*, (2), **36**, 27–45.

Esterle, J. (1978*b*). Injection de semi-groupes divisibles dans des algèbres de convolution et construction d'homomorphismes discontinus de $C(K)$, *Proc. London Math. Soc.*, (2), **36**, 59–85.

Esterle, J. (1980). Theorems of Gel'fand–Mazur type and continuity of homomorphisms from $C(K)$, *J. Functional Analysis*, **36**, 273–86.

Gamelin, T. W. (1969). *Uniform algebras*, Englewood Cliffs, New Jersey, Prentice Hall.

Gelfand, I. M. (1941). Normierte Ringe, *Rec. Math. N. S. Sbornik*, **9**, 43–24.

Gillman, L. and Jerison, M. (1960). *Rings of continuous functions*, Princeton, D. van Nostrand.

Graham, C. C. and McGehee, O. C. (1979). *Essays in commutative harmonic analysis*, New York, Springer-Verlag.

Haagerup, U. (1983). All nuclear C^*-algebras are amenable, *Inventiones Math.*, **74**, 305–19.

Helemskii, A. Ya. (1989). *The homology of Banach and topological algebras*, Dordrecht, Kluwer.

Helemskii, A. Ya. (1993). *Banach and locally convex algebras*, Oxford, Clarendon Press.

Hewitt, E. and Ross, K. A. (1963). *Abstract harmonic analysis, Volume I, Structure of topological groups, integration theory, group representations*, Berlin, Springer-Verlag.

Hewitt, E. and Ross, K. A. (1970). *Abstract harmonic analysis, Volume II, Structure and analysis for compact groups, analysis on locally compact abelian groups*, New York–Berlin, Springer-Verlag.

Hochschild, G. (1945). On the cohomology groups of an associative algebra. *Annals of Math.*, **46**, 58–67.

Hochschild, G. (1946). On the cohomology theory for associative algebras, *Annals of Math.*, **47**, 568–79.

Johnson, B. E. (1967). The uniqueness of the (complete) norm topology, *Bull. American Math. Soc.*, **73**, 537–9.

Johnson, B. E. (1969). Continuity of derivations on commutative algebras, *American J. Math.*, **91**, 1–10.

Johnson, B. E. (1972*a*). Cohomology in Banach algebras, *Memoirs American Math. Soc.*, **127**.

Johnson, B. E. (1972*b*). Approximate diagonals and cohomology of certain annihilator Banach algebras, *American J. Math.*, **94**, 685–98.

Kadison, R. V. and Ringrose, J. R. (1983). *Fundamentals of the theory of operator algebras, Volume I, Elementary theory*, New York–London, Academic Press.

Kadison, R. V. and Ringrose, J. R. (1986). *Fundamentals of the theory of operator algebras, Volume II, Advanced theory*, New York–London, Academic Press.

Kahane, J.-P. (1970). *Séries de Fourier absolument convergentes*, Berlin, Springer-Verlag.

Kaplansky, I. (1949). Normed algebras, *Duke Mathematical J.*, **16**, 399–418.

Katznelson, Y. (1976). *An introduction to harmonic analysis* (2nd edition), New York, Dover.

Mathieu, M. (1994). Where to find the image of a derivation. In J. Zemanek (ed.), *Functional analysis and operator theory*, Warsaw, Banach Center Publications, pp. 237–49.

Murphy, G. J. (1990). *C^*-algebras and operator theory*, San Diego, Academic Press.

Palmer, T. W. (1994). *Banach algebras and the general theory of $*$-algebras, Volume I*, Cambridge University Press.

Palmer, T. W. (2001). *Banach algebras and the general theory of $*$-algebras, Volume II*, Cambridge University Press.

Paterson, A. L. T. (1988). *Amenability*, Providence, Rhode Island, American Mathematical Society.

Pedersen, G. K. (1979). *C^*-algebras and their automorphism group*, London, Academic Press.

Ransford, T. J. (1989). A short proof of Johnson's uniqueness-of-norm theorem, *Bull. London Math. Soc.*, **96**, 309–11.

Read, C. J. (2001). Relative amenability and the non-amenability of $\mathcal{B}(\ell^1)$, preprint.

Reiter, H. and Stegeman, J. D. (2000). *Classical harmonic analysis and locally compact groups*, London Mathematical Society Monographs, **22**, Oxford, Clarendon Press.

Rudin, W. (1973). *Functional analysis*, New York, McGraw-Hill.

Rudin, W. (1996). *Real and complex algebras*, New York, McGraw-Hill.

Runde, V. (1998). The structure of contractible and amenable Banach algebras. In E. Albrecht and M. Mathieu (eds.), *Banach algebras '97*, Berlin, de Gruyter, pp. 415–30.

Runde, V. (2002). *Lectures on amenability*, Lecture Notes in Mathematics, **1774**, Berlin, Springer-Verlag.

Sinclair, A. M. (1969). Continuous derivations on Banach algebras, *Proc. American Math. Soc.*, **20**, 166–70.

Sinclair, A. M. (1974). Homomorphisms from C^*-algebras, *Proc. London Math. Soc.*, (2), **29**, 435–52.

Sinclair, A. M. (1975). Homomorphisms from $C_0(\mathbf{R})$, *J. London Math. Soc.*, **11**, 165–74.

Sinclair, A. M. (1976). *Automatic continuity of linear operators*, London Mathematical Society Lecture Note Series, **21**, Cambridge University Press.

Sinclair, A. M. and Smith, R. (1995). *Hochschild cohomology of von Neumann algebras*, London Mathematical Society Lecture Note Series, **203**, Cambridge University Press.

Singer, I. M. and Wermer, J. (1955). Derivations on commutative normed algebras, *Math. Ann.*, **129**, 260–4.

Stokke, R. (2003). Følner conditions for amenable group and semigroup algebras, preprint.

Stout, E. L. (1971). *The theory of uniform algebras*, Tarrytown-on-Hudson, New York, Bogden and Quigley.

Takesaki, M. (1979). *Theory of operator algebras. Volume I*, Berlin, Springer-Verlag.

Takesaki, M. (2002). Theory of operator algebras, *Volume II*, Berlin, Springer-Verlag.

Thomas, M. P. (1988). The image of a derivation is contained in the radical, *Annals of Math.*, **128**, 435–60.

Thomas, M. P. (1993). Primitive ideals and derivations on non-commutative Banach algebras, *Pacific J. Math.*, **159**, 139–52.

Zame, W. R. (1979). Existence, uniqueness, and continuity of functional calculus homomorphisms, *Proc. London Math. Soc.* (3), **39**, 151–75.

Part II

Harmonic analysis and amenability

GEORGE A. WILLIS
University of Newcastle, New South Wales, Australia

8
Locally compact groups

The set of real numbers, $\mathbb{R}$, plays several fundamental roles in mathematics. It appears as:

- an algebraic object – either an abelian group or a field;
- a topological space – the concepts of continuous function on $\mathbb{R}$ and continuous curve are the most basic in topology;
- a differential manifold – differentiability of functions is defined first for functions on $\mathbb{R}$ and differentiability on higher dimensional manifolds is defined in terms of real parameters; and
- a measure space – the measure of interval length is the basis for theories of integration.

These structures are related to each other and all are important in harmonic, or Fourier, analysis.

Locally compact groups also possess all of these structures. The algebraic and topological structures are hypothesized but differential and measure space properties then follow automatically. The problems and applications of harmonic analysis extend to the more general setting of locally compact groups and the algebraic, topological, differential and integral structures continue to be fundamentally important.

8.1 Definition and examples

The definition of the class of locally compact groups is very simple.

H. G. Dales, P. Aiena, J. Eschmeier, K. B. Laursen, and G. A. Willis, *Introduction to Banach Algebras, Operators, and Harmonic Analysis*. Published by Cambridge University Press.

Definition 8.1.1 *A* topological group *is a group G which is also a Hausdorff topological space so that the maps*

$$(x, y) \mapsto xy : G \times G \to G \quad \text{and} \tag{8.1.1}$$

$$x \mapsto x^{-1} : G \to G \quad (x, y \in G), \tag{8.1.2}$$

are continuous. The topological group G is a locally compact group *if the topology on G is locally compact.*

In these notes group multiplication will be denoted by juxtaposition and the group identity will be denoted by e. The class of locally compact groups includes many important examples.

Examples (a) Any group is locally compact when equipped with the discrete topology. This includes in particular all finite groups. Here are some discrete groups which will be mentioned.

The group of integers is denoted by $\mathbb{Z}$. The cyclic group of order n is denoted by C_n.

The free group on two generators, denoted by $\mathbb{F}_2$, is a frequently used example. Let the generators be a and b. Then a *word* in a, b, a^{-1}, and b^{-1} is a finite sequence $w = l_1 l_2 \cdots l_n$, where $l_j = a, b, a^{-1}$, or b^{-1} for each j. A word w is said to be *reduced* if no cancellation is possible, which means that we do not have a and a^{-1} as adjacent letters in w, and similarly for b and b^{-1}. Thus $abba^{-1}ba$ is reduced and $abbb^{-1}a^{-1}ba$ is not reduced. If $w = l_1 l_2 \cdots l_n$ is reduced, then n is called the *length* of w and denoted by $|w|$.

The *free group on two generators a* and *b* consists of all reduced words in a, b, a^{-1}, and b^{-1}, including the empty word which is denoted by e. Two words are multiplied by concatenating them and then performing any cancellations until a reduced word is achieved. Thus, for example, if $w_1 = aba^{-1}$ and $w_2 = ab^{-1}ab$, then

$$w_1 w_2 = ab(a^{-1}a)b^{-1}ab = a(bb^{-1})ab = aab\,.$$

The empty word is the identity element in $\mathbb{F}_2$ and, if $w = l_1 l_2 \cdots l_n$, then we see that $w^{-1} = l_n^{-1} \cdots l_2^{-1} l_1^{-1}$.

Free groups on any number of generators are defined similarly. The free group on a countably infinite number of generators will be denoted by $\mathbb{F}_\infty$.

(b) The groups $(\mathbb{R}, +)$ and $(\mathbb{C}, +)$ with their usual topology are both locally compact and abelian, as are $(\mathbb{R} \setminus \{0\}, \times)$ and $(\mathbb{C} \setminus \{0\}, \times)$.

Similarly, $(\mathbb{R}^n, +)$ and $(\mathbb{C}^n, +)$ are locally compact groups, and so are all matrix groups over $\mathbb{R}$ and $\mathbb{C}$. Thus for instance $SL(n, \mathbb{R})$, $GL(n, \mathbb{C})$ and $SO(n)$

are locally compact groups with the subspace topology they inherit as subsets of $\mathbb{R}^{n^2}$ or $\mathbb{C}^{n^2}$.

These matrix groups are all examples of Lie groups. A *Lie group* is a group which is also a differential manifold such that the group operations (8.1.1) and (8.1.2) are not just continuous but analytic.

(c) The circle group $\mathbb{T} := \{z \in \mathbb{C} : |z| = 1\}$ with the usual multiplication of complex numbers is a compact Lie group. It is isomorphic to the rotation group $SO(2)$.

Any product of compact groups with the product topology is a compact group, by Tychonov's theorem. Thus $\mathbb{T}^\infty$ is a compact group but is not a Lie group. The infinite product $(C_2)^\infty$ of copies of the cyclic group of order 2 is also a compact group.

(d) The field $\mathbb{Q}_p$ of p-adic numbers is a locally compact field. Hence $(\mathbb{Q}_p, +)$ and $(\mathbb{Q}_p \setminus \{0\}, \times)$ are locally compact abelian groups. The open subgroup $(\mathbb{Z}_p, +)$ of p-adic integers is compact.

In the same way as for the fields $\mathbb{R}$ and $\mathbb{C}$, matrix groups over $\mathbb{Q}_p$ are locally compact groups with the subspace topology as subsets of $\mathbb{Q}_p^{n^2}$. The notion of an analytic map over $\mathbb{Q}_p$ may be defined as well and so may the notion of Lie group over $\mathbb{Q}_p$.

The field $\mathbf{k}((X))$ of Laurent series over the finite field $\mathbf{k}$ is locally compact. The topology on this field is defined by letting the subgroups

$$\mathbf{k}[[X]]_n := \left\{ f = \sum_{j=n}^{\infty} k_j X^j : k_j \in \mathbf{k} \right\}$$

be a base of neighbourhoods of 0. The subring $\mathbf{k}[[X]] = \mathbf{k}[[X]]_0$ of Taylor series over $\mathbf{k}$ is then compact and open, and the topology on $\mathbf{k}[[X]]$ is just the product topology on $\mathbf{k}^\infty$. Matrix groups over $\mathbf{k}((X))$ are also locally compact groups.

(e) Let $\mathcal{G}$ be a locally finite graph. Then the automorphism group $\mathrm{Aut}(\mathcal{G})$ equipped with the topology of uniform convergence on compact sets is a locally compact group.

In particular, let $\mathcal{T}_q$ be the homogeneous tree where every vertex has degree (or valency) $q + 1$. Then $\mathrm{Aut}(\mathcal{T}_q)$ is a locally compact group. A base of neighbourhoods for the identity e consists of the stabilizer subgroups

$$V(\mathcal{F}) := \left\{ x \in \mathrm{Aut}(\mathcal{T}_q) : x(v) = v \;\; (v \in \mathcal{F}) \right\},$$

where $\mathcal{F}$ ranges over all finite subsets of the vertex set V. Then $V(\mathcal{F})$ is a compact open subgroup for each $\mathcal{F}$.

8.2 Structure theory

Here are a couple of often used facts about topological groups whose proofs are left as exercises.

Lemma 8.2.1 *Let G be a locally compact group and V be a neighbourhood of e. Then there is a symmetric neighbourhood W of e with compact closure such that $W^2 \subset V$.* □

Lemma 8.2.2 *Let G be a topological group. Let K and U subsets of G with K compact, U open and $K \subset U$. Then there is an open neighbourhood V of e such that*

$$VK \cup KV \subset U\,.$$ □

The following result is the starting point of the structure theory of locally compact groups.

Theorem 8.2.3 *Let G be a locally compact group. Then the connected component of the identity, called G_e, is a closed normal subgroup, and the quotient group G/G_e is a totally disconnected locally compact group.*

Proof Connected components are always closed. We show that G_e is a normal subgroup.

Let x and y be in G_e. Then e and x are in the same connected component and so $ey = y$ and xy also belong to the same component (because right translation by y is a homeomorphism, see Exercise 8.4.1). Since $y \in G_e$ it follows that $xy \in G_e$. The map $x \mapsto x^{-1}$ is also a homeomorphism and it fixes e. Hence $G_e^{-1} = G_e$ and so G_e is a subgroup.

Next, let $x \in G$. Then the conjugation map $y \mapsto xyx^{-1}$ is a homeomorphism and so xG_ex^{-1} is a connected component of G. Since conjugation fixes e, we have $xG_ex^{-1} = G_e$.

Since G_e is closed, G/G_e is Hausdorff. The image of a compact neighbourhood of e in G will be a compact neighbourhood of the identity in G/G_e. Hence G/G_e is locally compact. The connected components in G/G_e consist of single points, that is, G/G_e is totally disconnected. □

The detailed study of the structure of locally compact groups thus divides into the study of connected and of totally disconnected groups separately and then of extensions.

Among the examples described above, the groups $\mathbb{R}^n$, $SL(n, \mathbb{R})$, $\mathbb{T}$, and $\mathbb{C} \setminus \{0\}$ are connected. Discrete groups are totally disconnected, but some more interesting examples include matrix groups over $\mathbb{Q}_p$ and $\mathbf{k}((X))$, infinite products $(C_n)^\infty$ and automorphism groups of locally finite graphs, such as $\mathrm{Aut}(\mathcal{T}_q)$.

Connected groups are relatively well understood through the following theorem. It was proved in the early 1950s by Gleason (1952), Montgomery and Zippin (1952), and Yamabe (1953), and is the solution of Hilbert's 5th problem. Unified accounts are given in Montgomery and Zippin (1955) and Kaplansky (1971).

Theorem 8.2.4 (Gleason, Montgomery and Zippin, Yamabe) *Let G be a connected locally compact group, and let $\mathcal{O}$ be a neighbourhood of e. Then there is a compact normal subgroup $K \subset \mathcal{O}$ such that G/K is a Lie group.* □

The theorem shows that analytic structure emerges naturally in connected groups, even though only topological structure is assumed. Many results about connected locally compact groups are established by factoring out a small compact normal subgroup and then applying Lie group techniques to the quotient. This approach is known as 'approximation by Lie groups'. The theorem also says that each connected group is a projective limit of Lie groups or, in other words, every connected group is *pro-Lie*.

The following result due to van Dantzig (1936) is the starting point for the structure theory of totally disconnected locally compact groups.

Theorem 8.2.5 (van Dantzig) *Let G be a totally disconnected locally compact group, and let $\mathcal{O}$ be a neighbourhood of e. Then there is a compact open subgroup $U \subset \mathcal{O}$.* □

Note that each totally disconnected locally compact space has a base of neigbourhoods consisting of compact open subsets and therefore has inductive dimension equal to zero. For this reason totally disconnected locally compact groups are also known as *0-dimensional groups*. Van Dantzig's theorem says that there is in fact a base of neighbourhoods of e consisting of compact open *subgroups*.

In the case where G is compact more can be said.

Corollary 8.2.6 *Let G be a totally disconnected compact group. Then G has a base of neighbourhoods of the identity consisting of compact open normal subgroups. Hence G is profinite and is isomorphic to a closed subgroup of a product of finite groups.* □

The theorem does not assert that, for a general totally disconnected locally compact group, there is a base of neighbourhoods of compact open *normal* subgroups however, and groups such as $SL(n, \mathbb{Q}_p)$ and Aut(T_q) have no compact open normal subgroup. Although particular examples of these types are well understood, the general theory of totally disconnected groups did not progress far beyond van Dantzig's theorem for a long time. Relatively recent progress on the structure theory of totally disconnected groups (see Willis 1994, 2001, 2002) determines just how such groups may fail to have compact open normal subgroups. It introduces methods which begin to parallel the Lie techniques used to analyze connected groups.

When G is abelian, the structure of G can be described more fully. The following theorem may be found in Hewitt and Ross (1979), Morris (1997) and Rudin (1962).

Theorem 8.2.7 *Let G be a compactly generated locally compact abelian group. Then G is isomorphic to $\mathbb{R}^m \times \mathbb{Z}^n \times K$, where m and n are non-negative integers and K is a compact group.* □

A compactly generated group is one which is generated by a compact subset. In a locally compact group, any compact neighbourhood of the identity generates an open subgroup. Hence every locally compact abelian group has an open compactly generated subgroup to which Theorem 8.2.7 applies. The requirement that G be compactly generated is essential: the group $(\mathbb{Q}_p, +)$ is not compactly generated and does not satisfy the conclusion of the theorem.

Many classes of locally compact groups have been studied and some also have more-or-less complete structure theories. These classes are given names such as the [SIN]-groups (those which have arbitrarily small neighbourhoods of e invariant under conjugation) and the [Z]-groups (those for which the quotient by the centre is compact). An excellent survey of the current state of knowledge of classes of locally compact groups can be found in Palmer (2001).

8.3 Haar measure

The familiar Lebesgue measure on $\mathbb{R}$ is related to the additive group structure because it is invariant under translation, that is, the measure of a set is equal to the measure of its translate by any real number. Equivalently, the integral of any function on $\mathbb{R}$ is equal to the integral of its translate by any real number. All locally compact groups support a translation-invariant Borel measure. The σ-algebra of Borel subsets of the locally compact group G will be denoted by $\mathfrak{B}(G)$.

Theorem 8.3.1 (Haar) *Let G be a locally compact group. Then there is a positive Borel measure, m_G, on G satisfying the following properties for every $B \in \mathfrak{B}(G)$:*

1. $m_G(xB) = m_G(B)$ *for each* $x \in G$;
2. $m_G(B) = \inf\{m_G(U) : B \subset U,\ U \text{ is open}\}$ *and* $m_G(U) > 0$ *for every non-empty, open set* U;
3. $m_G(B) = \sup\{m_G(K) : K \subset B,\ K \text{ is compact}\}$.

Any measure satisfying these properties is equal to a positive multiple of m_G. □

The measure m_G is called the *Haar measure* on G. Property 1 says that m_G is translation-invariant, while Properties 2 and 3 say that m_G is a regular Borel measure. When there is no ambiguity the Haar measure will be denoted simply by m.

The translation-invariant measure was constructed for compact groups by von Neumann, and he also proved the uniqueness up to a positive multiple. The translation-invariant measure on locally compact groups was constructed by Haar; see Hewitt and Ross (1979).

The Haar measure need not be right-invariant. However, if we fix a Haar measure m then, for each $x \in G$, the measure $m_x : B \mapsto m(Bx),\ \mathfrak{B}(G)) \to \mathfrak{C}$, is left-invariant and so is equal to a positive multiple of m.

Definition 8.3.2 *Let m be a left-invariant Haar measure on G, and, for each $x \in G$, let $\Delta(x)$ be the positive scalar such that $m_x = \Delta(x)m$. Then the map $\Delta : G \to (\mathbb{R}^+, \times)$ is a continuous group homomorphism; Δ is called the* modular function *on G.*

If $\Delta \equiv 1$, then G is a unimodular *group.*

If G is discrete, then the counting measure is translation-invariant and it is conventional to work with this measure.

Proposition 8.3.3 *The locally compact group G is compact if and only if $m_G(G) < \infty$.* □

For compact groups it is conventional to normalize the Haar measure so that $m_G(G)$ equals 1. Finite groups are both discrete and compact and these two conventions are not consistent in this case.

In the case of compact groups there is an alternative way to produce the Haar measure as a limit of convolution powers of a probability measure. The idea of using convolution powers of a probability measure, or equivalently a random walk on a group G, to obtain information about G will be used in some of the results mentioned in later chapters.

Theorem 8.3.4 (Kawada–Itô) *Let G be a compact group, and let μ be a probability measure on G such that the support of μ generates G. Then the sequence of probability measures*

$$\left(\frac{1}{n}\sum_{j=1}^{n}\mu^j\right)_{n=1}^{\infty}$$

converges to the Haar measure on G in the weak-$*$ topology on $M(G)$. □

A. Weil shows how a translation-invariant measure m on a group G may be used to define a topology on G which essentially turns it into a locally compact group and turns m into Haar measure (Halmos 1950; Weil 1951). Thus locally compact groups are the only topological groups which support a translation-invariant measure. This is an important observation because, as will be seen later, the Haar measure is used to produce unitary representations of locally compact groups.

This section concludes with the definitions of some spaces of functions on groups and of the idea of the translate of a function.

Definition 8.3.5 *Let f be a function on the group G. The* left translate *of f by $a \in G$ is the function ${}_a f$ defined by*

$${}_a f(y) = f(a^{-1}y) \quad (y \in G).$$

The right translate *of f by $a \in G$ is the function f_a defined by*

$$f_a(y) = f(ya) \quad (y \in G).$$

Definition 8.3.6 *Let G be a locally compact group.*

(i) *The space of bounded continuous functions on G will be denoted by $C(G)$ and the subspace of functions which converge to zero at infinity by $C_0(G)$. The norm of a function f in $C(G)$ is*

$$\|f\|_\infty := \sup\{|f(x)| : x \in G\}.$$

(ii) *Let $1 \leq p < \infty$. The space of measurable functions (with functions identified if they are equal almost everywhere) on G satisfying $\int_G |f|^p \, dm < \infty$*

will be denoted by $L^p(G)$. The norm of a function f in $L^p(G)$ is

$$\|f\|_p = \left(\int_G |f|^p \mathrm{d}m\right)^{1/p}.$$

(iii) *The space of essentially bounded measurable functions (with functions identified if they are equal locally almost everywhere) on G will be denoted by $L^\infty(G)$. The norm of a function f in $L^\infty(G)$ is*

$$\|f\|_\infty = \text{ess sup } \{|f(x)| : x \in G\}.$$

(iv) *The space of bounded measures on G will be denoted by $M(G)$. The norm of a measure μ in $M(G)$ is*

$$\|\mu\| = \sup\left\{\sum_{j=1}^{n} |\mu(E_j)| : G = \bigcup_{j=1}^{n} E_j \text{ is a Borel partition of } G\right\}.$$

Throughout Part II of this book:

- the dual space of $C_0(G)$ will be identified with $M(G)$ via the pairing

$$\langle f, \mu\rangle = \int_G f \mathrm{d}\mu \quad (f \in C_0(G),\ \mu \in M(G));$$

- the dual space of $L^p(G)$ for $1 \leq p < \infty$ will be identified with $L^q(G)$, where q is the conjugate index to p with $1/p + 1/q = 1$, via the pairing

$$\langle f, g\rangle = \int_G fg \,\mathrm{d}m \quad (f \in L^p(G),\ g \in L^q(G)).$$

Left-invariance of the Haar measure is equivalent to the left-invariance of the integral.

Proposition 8.3.7 *Let f be a measurable function on the locally compact group G. Then*

$$\int_G {}_x f \mathrm{d}m = \int_G f(x^{-1}y)\,\mathrm{d}m(y) = \int_G f(y)\mathrm{d}m(y) = \int_G f \mathrm{d}m \quad (x \in G).$$

□

This identity, when applied to the dual pairing of $L^p(G)$ and $L^q(G)$, implies that the dual of left translation by a is the same as left translation by a^{-1}.

Proposition 8.3.8 *Let $1 \le p < \infty$, and let $f \in L^p(G)$ and $g \in L^q(G)$. Then*

$$\langle {}_a f, g \rangle = \langle f, {}_{a^{-1}} g \rangle \quad (a \in G). \qquad \square$$

8.4 Exercises

1. The *left translation* by $a \in G$ of the group G is the map $x \mapsto ax$, $G \to G$. Show that left translation is a homeomorphism.
2. Prove Lemmas 8.2.1 and 8.2.2.
3. Let $\mathbb{K}$ be a locally compact field. Show that the connected component of 0 is an ideal in $\mathbb{K}$, and deduce that $\mathbb{K}$ is either connected or totally disconnected. (This is the first step in the classifcation of locally compact fields discussed below in the additional notes.)
4. Let M be an $n \times n$ matrix with positive entries such that every row and column sum is equal to 1. Thus $M = (M_{ij})$, where $M_{ij} > 0$, $\sum_{i=1}^n M_{ij} = 1$, and $\sum_{j=1}^n M_{ij} = 1$ for $i, j = 1, 2, \ldots, n$.

   (i) Show that $m := (1, 1, \ldots, 1)$ is an eigenvector for M with eigenvalue 1.
   (ii) Show that $E := \{x \in \mathbb{C}^n : \sum_{j=1}^n x_j = 0\}$ is invariant under M and that $\mathbb{C}^n = \mathbb{C}m \oplus E$.
   (iii) Show that every eigenvector in E for M has an eigenvalue with modulus less than 1.
   (iv) Deduce that the sequence of matrices $(M^n)_{n=1}^\infty$ converges to a projection with range $\mathbb{C}m$ and kernel E.
   (v) Let G be a finite group, and let μ be a probability measure on G with support equal to G. Use the result of part (iv) to show that $(\mu^n)_{n=1}^\infty$ converges to the Haar measure on G.

8.5 Additional notes

1. The structure theory of locally compact fields is more complete than that of locally compact groups. It was shown by D. van Dantzig in the 1930s that there are only two connected locally compact fields, namely $\mathbb{R}$ and $\mathbb{C}$. Shortly afterwards L. Pontryagin completely classified the non-discrete totally disconnected locally compact fields. They are $\mathbb{Q}_p$ and their finite extensions (these are all the characteristic 0 fields) and $\mathbf{k}((X))$ for $\mathbf{k}$ some finite field (these are all the positive characteristic fields).

   The locally compact skew fields were completely classified by N. Jacobson, also in the 1930s. There is one extra connected skew field, namely the quaternions, and the totally disconnected skew fields are all division algebras over $\mathbb{Q}_p$ or $\mathbf{k}((X))$. See Weil (1995) and Jacobson (1980).

Pontryagin (1946) calls this the 'logical necessity' of these fields.

2. Discrete, compact, and abelian groups are all unimodular. An example of a group which is not unimodular is the so-called $ax + b$-group. This is the group of affine motions of the line; it is isomorphic to the matrix group

$$\left\{ \begin{pmatrix} a & b \\ 0 & 1 \end{pmatrix} : a \in \mathbb{R} \setminus \{0\},\ b \in \mathbb{R} \right\}.$$

3. The rate of convergence of the sequence of probability measures in Theorem 8.3.4 is of some interest; see work of Diaconis and Saloff-Coste (1995, 1996) on finite groups. They compute the rate of convergence to the Haar measure of powers of specific probability measures.

9

Group algebras and representations

Several Banach algebras may be associated with each locally compact group. All of them are generalizations to locally compact groups of the algebraic group algebra $\mathbb{C}G$ of the discrete group G.

Abstract harmonic analysis is essentially the study of the structure of these Banach algebras. The development of Banach algebra theory was driven, in part, by work on harmonic analysis.

Throughout this chapter, G is a locally compact group.

9.1 Convolution algebras

Definition 9.1.1 *Let μ and ν belong to $M(G)$. Their* convolution *is the bounded measure $\mu \star \nu$ defined by*

$$\mu \star \nu(B) = (\mu \times \nu)(\{(x, y) \in G \times G : xy \in B\}) \quad (B \in \mathfrak{B}(G)) .$$

For each $\mu \in M(G)$, the adjoint *of μ is the measure μ^* defined by*

$$\mu^*(B) = \overline{\mu(B^{-1})} \quad (B \in \mathfrak{B}(G)) .$$

Proposition 9.1.2 *Convolution $(\mu, \nu) \mapsto \mu \star \nu$ is an associative algebra product on $M(G)$, and satisfies $\|\mu \star \nu\| \leq \|\mu\|\|\nu\|$. The adjoint map $\mu \mapsto \mu^*$ is a conjugate linear involution on $M(G)$, and $(\mu \star \nu)^* = \nu^* \star \mu^*$. Thus $(M(G), \star)$ is a Banach $*$-algebra, called the* measure algebra *of the locally compact group G.* □

The subspace of $M(G)$ consisting of the discrete measures on G will be denoted by $M_d(G)$, the subspace of continuous measures by $M_c(G)$, and the subspace of measures absolutely continuous with respect to Haar measure by $M_a(G)$.

Proposition 9.1.3

(i) *$M_d(G)$ is a subalgebra of $M(G)$, and $M_c(G)$ and $M_a(G)$ are closed two-sided ideals in $M(G)$.*
(ii) *$M(G) = M_c(G) \oplus M_d(G)$, and the quotient algebra $M(G)/M_c(G)$ is isomorphic to $M_d(G)$.* □

Proposition 9.1.4 (Hewitt and Ross 1979, Theorem 20.10) *The space $M_a(G)$ is isometrically isomorphic to the Lebesgue space $L^1(G) := L^1(G, m)$ via*

$$f \in L^1(G) \mapsto f \cdot m \in M_a(G).$$

This isomorphism induces the convolution product $\star$ on $L^1(G)$ specified by

$$f \star g(x) = \int_G f(y)g(y^{-1}x)\, \mathrm{d}m(y) \quad (f, g \in L^1(G),\ a.e.(m)x \in G),$$

and the isometric involution $*$ specified by

$$f^*(x) = \overline{f(x^{-1})}\Delta(x)^{-1} \quad (f \in L^1(G),\ x \in G).$$ □

The convolution formulæ in the proposition agree with those defined in Part I on Banach algebras for the special cases where G is discrete or equal to $\mathbb{R}$.

Definition 9.1.5 *The algebra $(L^1(G), \star, *)$ is the* group algebra *of the locally compact group G. When G is discrete, $L^1(G)$ may be denoted by $\ell^1(G)$.*

The identification of $M_a(G)$ with $L^1(G)$ induces a convolution, $\mu \star f$, for all $\mu \in M(G)$ and $f \in L^1(G)$. Since $M_a(G)$ is an ideal in $M(G)$, $\mu \star f$ belongs to $L^1(G)$.

Proposition 9.1.6 *For each $f \in L^1(G)$, the map $x \mapsto \delta_x \star f = {}_x f$ is continuous with respect to the given topology on G and the norm topology on $L^1(G)$.* □

When G is discrete $M(G) = L^1(G) = \ell^1(G)$ and this algebra has the identity δ_e. When G is not discrete, $L^1(G)$ has a bounded approximate identity.

Definition 9.1.7 *Let A be a Banach algebra. A* left bounded approximate identity *for A is a net $\{u_\lambda\}$ in A such that* $\sup_\lambda \|u_\lambda\| < \infty$ *and*

$$\lim_\lambda \|a - u_\lambda a\| = 0 \quad (a \in A).$$

The notions of right bounded approximate identity *and* two-sided bounded approximate identity *are defined similarly. Let X be a Banach left A-module* (*see* Part I, Chapter 5). *Then a left bounded approximate identity $\{u_\lambda\}$ for A is an* approximate identity for X *if* $u_\lambda \cdot x \to x \ (x \in X)$.

Proposition 9.1.8 *Let $\{\mathcal{E}_\lambda\}$ be a base of compact neighbourhoods of the identity in G. This forms a net directed by inclusion. For each λ, let $u_\lambda = \mathbf{1}_{\mathcal{E}_\lambda}/m(\mathcal{E}_\lambda)$. Then $\|u_\lambda\|_1 = 1$ for every λ and for every $f \in L^1(G)$, and*

$$\|f - f \star u_\lambda\|_1 \xrightarrow{\lambda} 0 \quad \text{and} \quad \|f - u_\lambda \star f\|_1 \xrightarrow{\lambda} 0 . \qquad \square$$

Thus $L^1(G)$ has a two-sided bounded approximate identity. The proof is an exercise. One reason for the significance of bounded approximate identities is the following factorization theorem. The basic version of the theorem is due to P. J. Cohen, but it has many applications and has been strengthened by many authors, such as (Kisyński 2000). See also Bonsall and Duncan (1973, Theorem 11.10) and Dales (2000, Theorem 2.9.24).

Theorem 9.1.9 (Cohen factorization) *Let A be a Banach algebra with a left bounded approximate identity $\{u_\lambda\}$, and let X be a Banach left A-module such that $\{u_\lambda\}$ is an approximate identity for X. Then, for every element $x \in X$, there are $a \in A$ and $y \in X$ such that $x = a \cdot y$. In particular, if I is a closed left ideal in A and (b_n) is a sequence in I with $\|b_n\| \xrightarrow{n} 0$, then there are $a \in A$ and a sequence (c_n) in I with $\|c_n\| \xrightarrow{n} 0$ in I such that $b_n = ac_n \ (n \in \mathbb{N})$.* $\square$

Corollary 9.1.10 *Each f in $L^1(G)$ is a product $f = g \star h$, where we also have $g, h \in L^1(G)$. Moreover, if (f_n) is a sequence in $L^1(G)$ such that $\|f_n\| \xrightarrow{n} 0$, then there are g, h_n in $L^1(G)$ such that $\|h_n\| \xrightarrow{n} 0$ and $f_n = g \star h_n \ (n \in \mathbb{N})$.* $\square$

The group algebra $L^1(G)$ retains all information about G, as was shown by Wendel (1952).

Theorem 9.1.11 (Wendel) *Let $T : L^1(G_1) \to L^1(G_2)$ be an isometric algebra isomorphism. Then there is a continuous group isomorphism $\vartheta : G_1 \to G_2$ and there is a group character χ on G_2 such that*

$$Tf(x) = \chi(x)(f \circ \vartheta^{-1})(x) \quad (f \in L^1(G_1),\ x \in G_2) .$$

Proof Suppose first that G_1 is discrete, so that $L^1(G_1) = \ell^1(G)$. Then the unit ball in $\ell^1(G_1)$ is the convex hull of its extreme points and these are precisely the points of the form $t\delta_x$, where $|t| = 1$ and $x \in G_1$. Since T is an isometry, it follows that the unit ball in $L^1(G_2)$ is the convex hull of its extreme points. Hence the Haar measure on G_2 is the counting measure and it follows that G_2 is discrete.

For each $x \in G_1$, $T(\delta_x)$ is an extreme point of the unit ball in $\ell^1(G_2)$ and so equals $t\delta_y$, where $|t| = 1$ and $y \in G_2$. Define $\vartheta(x) = y$ and $\chi(h) = t$. That ϑ is a group isomorphism and χ is a character follows from the fact that T is an algebra isomorphism.

The proof when G_1 is not discrete is sketched after the next result, which it uses. □

Since $L^1(G)$ is an ideal in $M(G)$, each bounded measure μ on G defines a multiplication operator $M_\mu : f \mapsto \mu \star f$ on $L^1(G)$. Clearly, M_μ satisfies the identity,

$$M_\mu(f \star g) = M_\mu(f) \star g \quad (f, g \in L^1(G)).$$

A linear operator satisfying this identity is called a *left multiplier*. As was first shown by Wendel (1952), every left multiplier arises in this way.

Theorem 9.1.12 (Wendel) *Let M be a left multiplier on $L^1(G)$. Then M is continuous, and there is a unique measure $\mu \in M(G)$ such that $M = M_\mu$.*

Proof Proof Suppose that (f_n) is a sequence in $L^1(G)$ such that $\|f_n\| \xrightarrow{n} 0$. Then, by Corollary 9.1.10, there is $g \in L^1(G)$ and a sequence (h_n) in $L^1(G)$ such that $\|h_n\| \xrightarrow{n} 0$ and $f_n = g \star h_n$ for each n. Hence

$$M(f_n) = M(g \star h_n) = M(g) \star h_n \xrightarrow{n} 0.$$

Therefore M is continuous. (In fact, the continuity of M was hypothesized by Wendel and this hypothesis was removed later by B. E. Johnson.)

Choose a bounded approximate identity $\{u_\lambda\}$ for $L^1(G)$. Then $\{M(u_\lambda)\}$ is a bounded net in $L^1(G)$. Recalling that $L^1(G)$ is a subspace of $M(G)$ and that $M(G)$ is the dual space of $C_0(G)$, it follows that $\{M(u_\lambda)\}$ has a weak-$*$ accumulation point, μ say, in $M(G)$. It may be verified that $M = M_\mu$. □

The proof of the non-discrete case of Theorem 9.1.11 may be completed by lifting the isometric isomorphism T to be an isomorphism of multiplier algebras and hence, following Wendel's Theorem, an isomorphism $M(G_1) \to M(G_2)$. Since the extreme points of the unit ball in $M(G)$ are precisely those measures of the form $t\delta_g$, where $|t| = 1$ and $g \in G$, the argument used in the discrete case applies.

Recall that a *derivation* on a Banach algebra A is a linear map D satisfying the identity $D(ab) = aD(b) + D(a)b$ $(a, b \in A)$. See Chapter 6 for more on derivations. Each measure $\mu \in M(G)$ determines a derivation, D_μ, on $L^1(G)$ defined by

$$D_\mu(f) = f \star \mu - \mu \star f \quad (f \in L^1(G)).$$

At first sight it might seem that it would be as straightforward as the proof of Wendel's Theorem to show that every derivation on $L^1(G)$ arises in this way. This seems to be difficult to prove however.

9.1.13 **Problem** *Let D be a derivation on $L^1(G)$. Does there exist a measure $\mu \in M(G)$ such that $D = D_\mu$?*

This question has been answered affirmatively by B. E. Johnson for [SIN]-groups (and hence for all discrete groups), connected groups, and amenable groups (Johnson 1972, 2001). It has not been solved for all totally disconnected groups, and nor has it been solved for extensions of connected groups by discrete ones. (In fact, a positive solution to this problem has recently been announced by V. Losert.)

9.2 Representations on Hilbert space

There is a correspondence between $*$-representations of $L^1(G)$ on Hilbert space and continuous (with respect to the strong operator topology) unitary representations of G. (A unitary representation of G is a homomorphism from G into the group of unitary operators on some Hilbert space.)

Proposition 9.2.1 *Let $V : G \to \mathcal{B}(\mathfrak{H})$ be a unitary representation of G on the Hilbert space $\mathfrak{H}$ such that the map $x \mapsto V_x v$, $G \to \mathfrak{H}$, is continuous for every $v \in \mathfrak{H}$. For $f \in L^1(G)$, define*

$$\mathcal{R}_V(f) = \int_G f(x) V_x \, dm(x). \tag{9.2.1}$$

Then $\mathcal{R}_V : L^1(G) \to \mathcal{B}(\mathfrak{H})$ is a continuous $*$-representation of $L^1(G)$.

Conversely, every non-degenerate $*$-representation of $L^1(G)$ on Hilbert space is equal to $\mathcal{R}_V$ for some unitary representation V of G.

Proof The proof of the first part is an exercise. For the proof of the second part, note that, if $\mathcal{R}$ is a non-degenerate $*$-representation of $L^1(G)$, then Theorem 9.1.9 implies that, for every $v \in \mathfrak{H}$, there are $f \in L^1(G)$ and $v' \in \mathfrak{H}$ such that $v = \mathcal{R}(f)v'$. A unitary representation V of G on $\mathfrak{H}$ is then well-defined by setting

$$V_g v = \mathcal{R}(\delta_g \star f)v' \quad (g \in G,\ v \in \mathfrak{H}).$$

Then $\mathcal{R}$ may be recovered from V via the definition (9.2.1). □

Locally compact groups and their group algebras have natural representations on the Hilbert space $L^2(G) := L^2(G, m)$. The existence of Haar measure is thus crucial for the existence of unitary representations of such groups.

Definition 9.2.2 *For each $x \in G$, let U_x be the unitary operator on $L^2(G)$ defined by*

$$U_x f(y) = f(x^{-1}y) \quad (f \in L^2(G),\ y \in G).$$

Then $U : x \mapsto U_x$ is a representation of G on $L^2(G)$ and is continuous with respect to the given topology on G and the strong operator topology on $\mathcal{B}(L^2(G))$. The map U is the left regular representation *of G on $L^2(G)$.*

The strong operator closed subalgebra of $\mathcal{B}(L^2(G))$ generated by

$$\{U_x : x \in G\}$$

is the von Neumann algebra *of G; it is denoted by $VN(G)$.*

As shown in Proposition 9.2.1, there is a representation of $L^1(G)$ on $L^2(G)$ corresponding to the left regular representation of G.

Definition 9.2.3 *For each $g \in L^1(G)$, let $\mathcal{R}(g)$ be the operator on $L^2(G)$ defined by*

$$\mathcal{R}(g)f(y) = \int_G g(x)f(x^{-1}y)\,\mathrm{d}m(x) \quad (f \in L^2(G),\ y \in G).$$

Then $\mathcal{R} : L^1(G) \to \mathcal{B}(L^2(G))$ is an injective $*$-homomorphism and $\|\mathcal{R}\| = 1$. The map $\mathcal{R}$ is the left regular representation *of $L^1(G)$ on $L^2(G)$.*

The norm-closed subalgebra of $\mathcal{B}(L^2(G))$ generated by $\mathcal{R}(L^1(G))$ is the reduced C^*-algebra *of G; it is denoted by $C_r^*(G)$.*

The full C^*-algebra *of G, denoted by $C^*(G)$, is the enveloping C^*-algebra of $L^1(G)$ (see* Palmer (2001, Definition 10.1.10) *for the definition of the enveloping C^*-algebra).*

Theorem 9.1.11 of Wendel shows that the group algebra $L^1(G)$ determines G uniquely, but the same is not true for the C^*-algebras $C^*(G)$ and $C_r^*(G)$ and the von Neumann algebra $VN(G)$, as is seen in Exercise 9.4.1.

Among the unitary representations of G the most important are the irreducible ones. Every representation may be decomposed into irreducible representations.

Definition 9.2.4 *The unitary representation V of G on the Hilbert space $\mathfrak{H}$ is* reducible *if $\mathfrak{H}$ is the direct sum $\mathfrak{H} = \mathfrak{H}_1 \oplus \mathfrak{H}_2$ of proper subspaces such that $V(\mathfrak{H}_j) = \mathfrak{H}_j$ for $j = 1, 2$. A unitary representation which is not reducible is* irreducible.

Unitary representations of the group G are studied via the $*$-representations of $C_r^*(G)$ and $C^*(G)$. It is the defining property of the enveloping C^*-algebra that there is a correspondence between $*$-representations of $L^1(G)$ (and therefore, by Proposition 9.2.1, unitary representations of G) and $*$-representations of $C^*(G)$. One of the basic theorems about C^*-algebras asserts that they have sufficiently many irreducible $*$-representations to separate every non-zero point from 0. Since the $*$-homomorphism $\mathcal{R} : L^1(G) \to C_r^*(G)$ is injective and has dense range, it follows that there are sufficiently many irreducible representations of $L^1(G)$ to separate every non-zero point from 0.

Theorem 9.2.5 *Let G be a locally compact group. Then $L^1(G)$ is a semisimple Banach algebra.* □

It also follows that there are sufficiently many unitary representations of G to separate every non-identity element from e.

Theorem 9.2.6 (Gelfand-Raikov) *Let G be a locally compact group, and let $x \neq e$ be in G. Then there is an irreducible, continuous unitary representation V of G on a Hilbert space $\mathfrak{H}$ such that $V(x) \neq I_{\mathfrak{H}}$.* □

The 'sufficiently many' $*$-representations of $L^1(G)$ guaranteed by these theorems are in fact representations of the reduced C^*-algebra $C_r^*(G)$. It may be that there are other irreducible $*$-representations and in this case the full C^*-algebra $C^*(G)$ is not the same as $C_r^*(G)$. This phenomenon is discussed further in Chapter 11; see Proposition 11.2.5.

Definition 9.2.7 *Let V be a unitary representation of the locally compact group G, and let $\mathcal{R}_V$ be the $*$-representation of $L^1(G)$ defined in* (9.2.1). *Then V is* weakly contained *in the regular representation of G if $\|\mathcal{R}_V(f)\| \leq \|\mathcal{R}(f)\|$ for every $f \in L^1(G)$.*

One of the basic problems in harmonic analysis is to describe in some way all the irreducible unitary representations of a given locally compact group G. There are no completely general techniques for doing this and each description relies on particular properties of G. A great deal is known about unitary representations of semisimple Lie goups and nilpotent Lie groups but there is also still much to learn. The class of groups for which the representation theory is most satisfactory is that of abelian groups. Representations of abelian subgroups are also an important tool in the study of representations of other classes of groups.

9.3 The Fourier transform

When G is abelian, $L^1(G)$, $C^*(G)$, and $VN(G)$ are commutative Banach algebras. These algebras, which are discussed in Part I, are best understood with the aid of the *Fourier transform*. In order to define this transform we must first say something about the irreducible representations of G.

Every irreducible unitary representation of a locally compact abelian group G is one-dimensional. The irreducible representations thus correspond to bounded homomorphisms from G to the multiplicative group $\mathbb{T}$ of non-zero complex numbers, that is, to *characters* on G (see §3.4). The set Γ of all continuous characters is itself an abelian group with respect to pointwise multiplication:

$$\langle x, \gamma_1 + \gamma_2 \rangle = \langle x, \gamma_1 \rangle \langle x, \gamma_2 \rangle \quad (x \in G, \gamma_1, \gamma_2 \in \Gamma),$$

where we write $\langle x, \gamma \rangle$ for $\gamma(x)$. When equipped with the topology of uniform convergence on compact subsets of G, Γ is also a locally compact group and is called the *dual group* of G. The dual group of G is sometimes denoted by $\widehat{G}$.

For example, each character on $\mathbb{Z}$ has the form $\gamma_z : n \mapsto z^n$ for some $z \in \mathbb{T}$, and it is easily seen that in fact $\widehat{\mathbb{Z}} \cong \mathbb{T}$ (exercise). With a little more work it may be shown that each continuous character on $\mathbb{T}$ has the form $\gamma_n : z \mapsto z^n$ for some $n \in \mathbb{Z}$, and that $\widehat{\mathbb{T}} \cong \mathbb{Z}$. Similarly, each continuous character on $\mathbb{R}$ has the form $\gamma_y : t \mapsto e^{iyt}$ for some $y \in \mathbb{R}$ and $\widehat{\mathbb{R}} \cong \mathbb{R}$. These basic examples and the structure theory of locally compact abelian groups, Theorem 8.2.7, facilitate the description of the dual group, that is, the set of irreducible representations, of any abelian group.

Let G be a locally compact abelian group with dual group Γ. For $f \in L^1(G)$, the *Fourier transform* of f is the function $\widehat{f}$, defined on Γ by the formula

$$\widehat{f}(\gamma) = \int_G f(s)\langle -s, \gamma \rangle \,\mathrm{d}m(s) \quad (\gamma \in \Gamma). \tag{9.3.1}$$

For example, if $G = \mathbb{R}$, then

$$\widehat{f}(y) = \int_{-\infty}^{\infty} f(t)\mathrm{e}^{-\mathrm{i}yt}\,\mathrm{d}t \quad (y \in \mathbb{R}).$$

You are asked to show in an exercise that the Fourier transform converts the convolution product in $L^1(G)$ to pointwise product of functions on Γ and converts the adjoint in $L^1(G)$ to complex conjugation.

Set

$$A(\Gamma) = \{\widehat{f} : f \in L^1(G)\},$$

the range of the Fourier transform, as in §3.4. The Riemann–Lebesgue lemma (see Rudin 1966, §5.14; or Helson 1983, §1.2) shows that $A(\Gamma)$ is contained in $C_0(\Gamma)$. The Fourier transform is therefore a $*$-homomorphism from $L^1(G)$ into $C_0(\Gamma)$.

Theorem 9.3.1 *Let G be a locally compact abelian group. Then the Fourier transform*

$$\mathcal{F} : f \mapsto \widehat{f}, \quad L^1(G) \to A(\Gamma) \subset C_0(\Gamma),$$

is continuous $*$-homomorphism with $\|\mathcal{F}\| = 1$. □

In fact, the formula (9.3.1) when restricted to $f \in C_{00}(G) \subset L^1(G)$ defines an isometry with respect to the L^2-norm. This isometry extends by continuity to $L^2(G)$. The extended map is also known as the Fourier transform, and is also denoted by $\mathcal{F}$.

Theorem 9.3.2 *Let G be a locally compact abelian group. Then the Fourier transform*

$$\mathcal{F} : g \mapsto \widehat{g}, \quad L^2(G) \to L^2(\Gamma),$$

is an isometric isomorphism. It satisfies the identity

$$\widehat{(\mathcal{R}(f)g)}(\gamma) = \widehat{f}(\gamma)\widehat{g}(\gamma) \quad (f \in L^1(G),\ g \in L^2(G),\ \gamma \in \Gamma). \tag{9.3.2}$$

□

Equation (9.3.2) shows that the Fourier transform converts the regular representation of $L^1(G)$ on $L^2(G)$ to a representation of $A(\Gamma)$ by multiplication operators on $L^2(\Gamma)$. Since $A(\Gamma)$ is self-adjoint and separates points in Γ, the Stone–Weierstrass theorem (Dunford and Schwartz 1958, Theorem IV.6.16) shows that $A(\Gamma)$ is dense in $C_0(\Gamma)$, and it follows that $C_r^*(G)$ is isomorphic to $C_0(\Gamma)$. Since Γ is the set of *all* irreducible representations of G, equation (9.3.2) shows that every irreducible representation is weakly contained in the regular representation and so $C^*(G) = C_r^*(G)$ when G is abelian. (The fact that Γ coincides with the set of all characters on $L^1(G)$ is Theorem 3.4.2 in Part I; there is a full proof in Dales (2000, Theorem 2.5.4).) Further reasoning shows that $VN(G)$ is isomorphic to $L^\infty(\Gamma)$.

The Fourier transform of $L^1(G)$ to $A(\Gamma)$ and the isomorphism of $C^*(G)$ with $C_0(\Gamma)$ are special cases of the *Gelfand transform* of commutative Banach algebras. This transform is discussed in Part I, Chapter 3.

The convolution algebra $M(G)$ is also a commutative Banach algebra. The Fourier transform extends from $L^1(G)$ (identified with the ideal $M_a(G)$) to $M(G)$. This extension is called the *Fourier–Stieltjes* transform, and denoted by $\mu \mapsto \widehat{\mu}$ for $\mu \in M(G)$. Here $\widehat{\mu}$ is defined by

$$\widehat{\mu}(\gamma) = \int_G \gamma(-s)\,\mathrm{d}\mu(s) \quad (\gamma \in \Gamma). \tag{9.3.3}$$

The Fourier–Stieltjes transform is a $*$-homomorphism from $M(G)$ onto an algebra of continuous functions on Γ; the range of the transform is denoted by $B(\Gamma)$ and called the *Fourier–Stieltjes algebra*.

It is no longer the case however that the irreducible (one-dimensional) representations of $M(G)$ correspond to the irreducible representations of G. In the terms introduced in Part I, not every character on $M(G)$ corresponds to a character on G, and the Fourier–Stieltjes transform is not the same as the Gelfand transform on $M(G)$. Indeed, since $M(G)/M_c(G) \cong M_d(G)$ (Proposition 9.1.3) and $M_d(G) \cong \ell^1(G)$, the character space of $M(G)$ contains a copy of the dual group of G with the discrete topology. The Banach algebra $M(G)$ has been studied in some depth when G is abelian and many additional characters have been identified (Taylor 1973). The character space of $M(G)$ even contains analytic discs (Brown and Moran 1976, 1978), which means that $M(G)$ has quotients isomorphic to algebras of analytic functions.

9.4 Exercises

1. Are the group algebras $\ell^1(C_4)$ and $\ell^1(C_2 \times C_2)$ isomorphic? What are $C_r^*(G)$, $C^*(G)$, and $VN(G)$ when G is C_4 and $C_2 \times C_2$?
2. Prove Proposition 9.1.8. Furthermore, show that $\|u_\lambda \star f - f\|_p \xrightarrow{\lambda} 0$ for every $f \in L^p(G)$, where $1 \le p < \infty$, and show that $\|u_\lambda \star f - f\|_\infty \xrightarrow{\lambda} 0$ for every $f \in C_0(G)$.
3. Complete the proof of Proposition 9.1.11 by verifying that ϑ is a homomorphism and χ is a character.
4. (a) For $\nu \in M(G)$ and $\varphi \in C_0(G)$ define

$$\nu \cdot \varphi(x) = \int_G \varphi(xy)\,\mathrm{d}\nu(y).$$

Verify that

$$\langle \varphi, \mu \star \nu \rangle = \langle \nu \cdot \varphi, \mu \rangle \quad (\mu, \nu \in M(G),\ \varphi \in C_0(G)).$$

(b) Let $\{u_\lambda\}$ be the bounded approximate identity for $L^1(G)$ defined in Proposition 9.1.8. Show that

$$\lim_\lambda \|u_\lambda \cdot \varphi - \varphi\|_\infty = 0 \quad (\varphi \in C_0(G)).$$

5. Show that the multiplier M_μ on $L^1(G)$ satisfies $\|M_\mu\| = \|\mu\|$. (*Hint*: if $\{u_\lambda\}$ is a bounded approximate identity for $L^1(G)$, then $\mu \star u_\lambda \overset{w*}{\to} \mu$ in $M(G)$.) Complete the proof of Theorem 9.1.12 by verifying that $M(u_\lambda)$ has a unique weak-$*$ accumulation point μ and that $M = M_\mu$.

6. Verify that the linear operator $\mathcal{R}_V : L^1(G) \to \mathcal{B}(\mathfrak{H})$ defined in Proposition 9.2.1 is a $*$-homomorphism, that is, show that $\mathcal{R}_V(f \star g) = \mathcal{R}_V(f)\mathcal{R}_V(g)$ and $\mathcal{R}_V(f^*) = \mathcal{R}_V(f)^*$.

7. Verify that the map $U : G \to \mathcal{B}(L^2(G))$ defined in 9.2.2 is a unitary group representation, that is, show that $U_{xy} = U_x U_y$ and $U_{x^{-1}} = U_x^*$.

8. Identify the dual group of the finite cyclic group C_n.

9. Show that $\widehat{\mathbb{Z}} \cong \mathbb{T}$.

10. Show that, for all f and g in $L^1(G)$ and γ in Γ, we have

$$\widehat{f \star g}(\gamma) = \widehat{f}(\gamma)\widehat{g}(\gamma) \quad \text{and} \quad \widehat{f^*}(\gamma) = \overline{\widehat{f}(\gamma)}\,.$$

9.5 Additional notes

1. Wendel's theorem (9.1.11) has been improved by N. J. Kalton amd G. V. Wood by weakening the requirement that T be an isometry to require only that it have small norm; see Kalton and Wood (1976) and Wood (1989). Galindo (2000) considers the still weaker condition that T is an isomorphism of topological algebras, but G_1 and G_2 are abelian and torsion-free.
2. Building on earlier work of Cohen (1960), Kepert (1997) has shown that homomorphisms $T : L^1(G_1) \to L^1(G_2)$, where G_1 and G_2 are abelian, have a very precise description and that they automatically have closed range.
3. The factorization theorem 9.1.9 was proved by Cohen (1959) in order to answer the question of whether every element of $L^1(G)$ is a product. This had previously been shown to be the case for certain locally compact groups by Rudin (1957).
4. The theory of the decomposition of unitary representations of groups, or, equivalently, of non-degenerate representations of C^*-algebras, into

irreducible representations is described in Palmer (2001, Chapter 9) and Kadison and Ringrose (1983, 1986).

5. The proofs of Theorems 9.2.5 and 9.2.6 rely on representations of the reduced group C^*-algebra, $C_r^*(G)$, and thus on the left regular representation of G on $L^2(G)$. The fact that locally compact groups have sufficiently many irreducible unitary representations to separate points thus relies ultimately on the existence of Haar measure.

Topological groups which are not locally compact do not have a translation invariant measure, and it can happen that they do not have sufficiently many, or even any, irreducible unitary representations. Unitary representations of non-locally compact groups are the subject of current research with one recent publication being Glöckner and Neeb (2000).

6. Let G be a locally compact abelian group. Then each element x of G determines a character $\iota(x)$ on the dual group $\widehat{G} = \Gamma$ by

$$\langle \gamma, \iota(x) \rangle = \langle x, \gamma \rangle \quad (\gamma \in \Gamma).$$

The map $\iota : G \to \widehat{\widehat{G}}$ is in fact an isomorphism of topological groups. This is known as the *Pontryagin duality theorem*; see Morris (1977) and Rudin (1962).

7. Let Γ be any locally compact group. Define

$$A(\Gamma) := \{f \star g : f, g \in L^2(\Gamma)\}.$$

Then $A(\Gamma)$ is a subalgebra of $C_0(\Gamma)$ and is called the *Fourier algebra* of Γ. It has a norm $\|\cdot\|$ defined by

$$\|h\| = \inf\{\|f\|_2 \|g\|_2 : h = f \star g,\ f, g \in L^2(\Gamma)\},$$

which is in fact a Banach algebra norm on $A(\Gamma)$. When $\Gamma = \widehat{G}$ is abelian, $A(\Gamma)$ is isomorphic to the convolution algebra $L^1(G)$. The Fourier algebra was introduced by Eymard (1964). It carries a great deal of information about Γ; see, for example, Forrest (1992) and Dales (2000, 4.5.30).

10
Convolution operators

The structures of the various group algebras A associated with G, which is a large part of harmonic analysis, is best understood when G is abelian or compact. In these cases the multiplication operators

$$M_a : f \mapsto a \star f \quad (a \in L^1(G),\ f \in A),$$

are decomposable in the sense of Part IV, Definition 21.1.4.

10.1 Compact groups

If G is a compact group, then, for every $a \in L^1(G)$, the convolution operator M_a is a compact operator on $L^1(G)$ (Exercise 10.4.1) and on $L^2(G)$ (Exercise 10.4.2). Therefore, as shown in Chapter 21 on decomposable operators, M_a is decomposable. The decomposability of this operator may be used to analyse the ideal structure of $L^1(G)$. We begin by identifying elements in the centre of $L^1(G)$.

Lemma 10.1.1 *Let G be a compact group, and let $\mathcal{U}$ be an open neigbourhood of e. Then there is a neighbourhood $e \ni \mathcal{V} \subset \mathcal{U}$ such that $x\mathcal{V}x^{-1} = \mathcal{V}$ for every $x \in G$.*

Proof Since $\mathcal{U}$ is open, $G \setminus \mathcal{U}$ is compact. The function

$$(x, y) \mapsto xyx^{-1} : G \times (G \setminus \mathcal{U}) \to G \setminus \{e\}$$

is continuous. Hence its image is compact, and it is also clearly invariant under conjugation by elements of G. Choose $\mathcal{V}$ to be the complement of this image. □

If $\mathcal{V}$ is a neighbourhood of e which is invariant under conjugation, then $\mathcal{V} \cap \mathcal{V}^{-1}$ is invariant and symmetric.

Lemma 10.1.2 *Let $\mathcal{V}$ be a symmetric neighbourhood of e which is invariant under conjugation, and put $u = \mathbf{1}_{\mathcal{V}}/m(\mathcal{V})$. Then u is a self-adjoint element of the centre of $L^1(G)$.*

Proof That u is self-adjoint follows because it is a real-valued function and $\mathcal{V}$ is symmetric.

To see that u is in the centre, let $f \in L^1(G)$. Then, for each $x \in G$, we have

$$\begin{aligned} f \star u(x) &= \frac{1}{m(\mathcal{V})} \int_G f(y) \mathbf{1}_{\mathcal{V}}(y^{-1}x)\, \mathrm{d}m(y) \\ &= \frac{1}{m(\mathcal{V})} \int_G f(y) \mathbf{1}_{\mathcal{V}}(xy^{-1})\, \mathrm{d}m(y), \text{ because } \mathcal{V} \text{ is invariant under conjugation,} \\ &= \frac{1}{m(\mathcal{V})} \int_G f(z^{-1}x) \mathbf{1}_{\mathcal{V}}(z)\, \mathrm{d}m(z), \text{ because } m \text{ is left-invariant and unimodular,} \\ &= u \star f(x), \end{aligned}$$

and so u is in the centre. □

The group algebra $L^1(G)$ may now be analysed as a sum of full matrix algebras.

Theorem 10.1.3 *Let G be a compact group. Then*

$$L^2(G) = \left(\bigoplus_{\sigma \in \Sigma} \mathfrak{H}_\sigma \right)_{\ell^2},$$

where each $\mathfrak{H}_\sigma$ is a finite-dimensional subspace of $L^2(G)$ invariant under the left regular representation U (see Definition 9.2.2*). For each σ, there is a minimal central idempotent z_σ in $L^1(G)$ such that $\mathfrak{H}_\sigma = z_\sigma \star L^2(G)$. Moreover, for each σ, $z_\sigma \star L^1(G) := I_\sigma$ is a minimal two-sided ideal in $L^1(G)$ and is isomorphic to a full matrix algebra. The ideal $\sum_{\sigma \in \Sigma} I_\sigma$ is dense in $L^1(G)$.*

Proof Let $u = \mathbf{1}_{\mathcal{E}}/m(\mathcal{E})$, where $\mathcal{E}$ is a symmetric neighbourhood of e invariant under conjugation. Then M_u is a compact, self-adjoint operator on $L^2(G)$. Hence $L^2(G)$ is the direct sum of the eigenspaces of M_u.

Let $\mathfrak{H}$ be an eigenspace of M_u corresponding to a non-zero eigenvalue. Then there is a projection P in the algebra generated by M_u such that $\mathfrak{H} = P(L^2(G))$.

In fact, there is z in the algebra generated by u such that $P = M_z$. Since u belongs to the centre of $L^1(G)$, by Lemma 10.1.2, so does z. Hence for each $x \in G$ we have

$$U_x(\mathfrak{H}) = \delta_x \star \mathfrak{H} = \delta_x \star z \star \mathfrak{H} = z \star \delta_x \star \mathfrak{H} \subset \mathfrak{H}.$$

The eigenspace is thus G-invariant.

It must be shown that there are sufficiently many eigenspaces corresponding to non-zero eigenvalues for $L^2(G)$ to be their direct sum. For this, consider all central functions of the form $u = \mathbf{1}_{\mathcal{E}}/m(\mathcal{E}) := u_{\mathcal{E}}$. Note that

$$\mathfrak{U} := \{u_{\mathcal{E}} : e \in \mathcal{E} = \mathcal{E}^{-1},\ x\mathcal{E}x^{-1} = \mathcal{E}\ (x \in G)\}$$

is a commuting family of functions, and so the corresponding convolution operators on $L^2(G)$ have common eigenspaces. Suppose that f belongs to the zero eigenspace for all $u_{\mathcal{E}} \in \mathfrak{U}$. Then, since $\mathfrak{U}$ contains an approximate identity for $L^2(G)$ (by Exercise 9.4.2), $f = 0$. Hence $L^2(G)$ is the direct sum of finite-dimensional eigenspaces for $\mathfrak{U}$. For each of these eigenspaces there is a central idempotent in $L^1(G)$ which projects onto it, and so it is invariant under $U(G)$.

Since G is compact, $L^2(G) \subset L^1(G)$. The eigenspaces for $\mathfrak{U}$ are thus eigenspaces in $L^1(G)$ also and, since $L^2(G)$ is dense in $L^1(G)$, the sum of these eigenspaces is also dense in $L^1(G)$. Moreover, if z is a central idempotent projecting onto an eigenspace, then $z \star L^2(G)$ is dense in $z \star L^1(G)$ and so, since the first space is finite-dimensional, $z \star L^2(G) = z \star L^1(G)$. It follows that $z \star L^1(G)$ is a finite-dimensional, two-sided ideal in $L^1(G)$ and that z belongs to this ideal.

Theorem 9.2.5 shows that $L^1(G)$ is semisimple. Hence each ideal $z \star L^1(G)$ is a semisimple, finite-dimensional algebra, and is therefore isomorphic to a direct sum of full matrix algebras. It follows that z is a sum of minimal idempotents which project onto the full matrix algebras. Let $\{z_\sigma : \sigma \in \Sigma\}$ be the collection of all the minimal central idempotents arising in this way and define $\mathfrak{H}_\sigma$ to be $z_\sigma \star L^2(G)$. □

The orthogonal decomposition of $L^2(G)$ may be further refined as follows. Let d_σ be the integer such that the minimal ideal $\mathfrak{H}_\sigma$ is isomorphic to the algebra M_{d_σ} of $d_\sigma \times d_\sigma$ matrices. Then $z_\sigma = \sum_{j=1}^{d_\sigma} e_j$, where e_j are minimal (no longer central) idempotents. Put $\mathfrak{K}_\sigma = \mathfrak{H}_\sigma \star e_1$. Then $\mathfrak{K}_\sigma$ is a d_σ-dimensional Hilbert space and $U(\mathfrak{K}_\sigma) = \mathfrak{K}_\sigma$. Denote the restriction of U to $\mathfrak{K}_\sigma$ by $U^{(\sigma)}$. Then $U^{(\sigma)}$ is an irreducible representation of G.

Choose an orthonormal basis $\{k_\alpha\}$ for $\mathfrak{K}_\sigma$, and define

$$c^\sigma_{\alpha\beta}(x) = \left\langle U^{(\sigma)}_x k_\beta, k_\alpha \right\rangle$$

to be the matrix coefficient of $U_x^{(\sigma)}$ with respect to this basis; see Hewitt and Ross (1979, Theorem 27.5). Then the functions $c_{\alpha\beta}^{\sigma}$ are continuous on G and satisfy the orthogonality relations

$$\int_G c_{\alpha\beta}^{\sigma}\bar{c}_{\gamma\delta}^{\tau}\,\mathrm{d}m = \begin{cases} 1/d_\sigma, & \text{if } \sigma = \tau,\ \ \alpha = \gamma \ \text{ and } \ \beta = \delta\,, \\ 0, & \text{otherwise}\ . \end{cases}$$

See Hewitt and Ross (1979, Theorems 27.15, 27.19). These functions belong to $\mathfrak{H}_\sigma$, and

$$\left\{d_\sigma^{1/2} c_{\alpha\beta}^{\sigma} : \alpha, \beta \in \{1, 2, \ldots, d_\sigma\}\right\}$$

is an orthonormal basis for $\mathfrak{H}_\sigma$. Therefore the set of functions

$$\left\{d_\sigma^{1/2} c_{\alpha\beta}^{\sigma} : \sigma \in \Sigma,\ \alpha, \beta \in \{1, 2, \ldots, d_\sigma\}\right\}$$

is an orthonormal basis for $L^2(G)$.

These functions make the isomorphism of $\mathfrak{H}_\sigma$ with M_{d_σ} explicit because they convolve as scaled matrix units

$$\bar{c}_{\alpha\beta}^{\sigma} * \bar{c}_{\gamma\delta}^{\tau} = \begin{cases} \bar{c}_{\alpha\delta}^{\sigma}/d_\sigma, & \text{if } \ \sigma = \tau \ \text{ and } \ \beta = \gamma\,, \\ 0, & \text{otherwise}\,. \end{cases}$$

See Hewitt and Ross (1979, Theorem 27.20(iii)). The translates of their conjugates can be found by matrix multiplication:

$$\delta_x * \bar{c}_{\alpha\beta}^{\sigma} = \sum_{\gamma=1}^{d_\sigma} c_{\gamma\alpha}^{\sigma}(k)\bar{c}_{\gamma\beta}^{\sigma}\,.$$

See Hewitt and Ross (1979, Theorem 27.20(i)).

We may now see that every irreducible unitary representation of G is contained as a direct summand in the regular representation.

Proposition 10.1.4 *Let G be a compact group. Then every irreducible continuous unitary representation of G is equivalent to $U^{(\sigma)}$ for a unique $\sigma \in \Sigma$.*

Proof Let V be an irreducible unitary representation of G on the Hilbert space $\mathfrak{H}_V$. Then $\mathcal{R}_V(z_\sigma)\mathfrak{H}_V \neq \{0\}$ for at least one $\sigma \in \Sigma$ because otherwise Theorem 10.1.3 would show that $\mathcal{R}_V(L^1(G))\mathfrak{H}_V = \{0\}$. On the other hand, if $\mathcal{R}_V(z_\sigma)\mathfrak{H}_V \neq \{0\}$ for two distinct central idempotents, z_1 and z_2 say, then $\mathfrak{H}_V = \mathcal{R}_V(z_1)\mathfrak{H}_V \oplus \mathcal{R}_V(z_\sigma)\mathfrak{H}_V$ would reduce the representation V. Hence

$\mathcal{R}_V(z_\sigma)\mathfrak{H}_V \neq \{0\}$ for exactly one $\sigma \in \Sigma$. It is left as an exercise to show that V is equivalent to $U^{(\sigma)}$. □

Corollary 10.1.5 *Let G be a compact group. Then $C_r^*(G) = C^*(G)$.* □

10.2 Abelian groups

One of the basic theorems of topology is Urysohn's lemma, which asserts that, if C and D are disjoint closed subsets of a normal space X, then there is a continuous real-valued function f on X such that $0 \leq f(x) \leq 1$ for all $x \in X$, $f(x) = 0$ for all $x \in C$, and $f(x) = 1$ for all $x \in D$. (See Rudin (1963, 2.12) or Engelking (1977, Theorem 15.10)). The class of normal spaces includes all locally compact Hausdorff spaces, and Urysohn's lemma is the basis of all partition of unity arguments. In particular, it implies that multiplication operators on $C_0(X)$ are decomposable.

It was seen in Section 9.3 that, if G is abelian, then the Fourier (or Gelfand) transform of $f \in L^1(G)$ belongs to $C_0(\Gamma)$, where Γ is the dual group. The continuous functions whose existence is guaranteed by Urysohn's lemma need not belong to $A(\Gamma)$, the algebra of Fourier transforms, however. The next result shows that suitable functions for partition of unity arguments can nevertheless be found inside $A(\Gamma)$. See Rudin (1962, Theorem 2.6.1) and Reiter and Stegeman (2000, Proposition 2.1.5).

Lemma 10.2.1 *Let G be a locally compact abelian group. Let $\mathcal{K}$ and $\mathcal{U}$ be subsets of its dual group Γ, with $\mathcal{K}$ compact, $\mathcal{U}$ open, and $\mathcal{K} \subset \mathcal{U}$. Then there is a function $f \in L^1(G)$ such that:*

$$0 \leq \widehat{f}(\gamma) \leq 1 \ (\gamma \in \Gamma); \ \ \widehat{f}(\gamma) = 1 \ (\gamma \in \mathcal{K}); \quad \text{and } \widehat{f}(\gamma) = 0 \ (\gamma \notin \mathcal{U}).$$

Proof Let $\mathcal{V}$ be a neighbourhood of e such that $\mathcal{V}\mathcal{K} \subset \mathcal{U}$ (see Lemma 8.2.2). Let $\mathcal{W}$ be another neighbourhood of e which is symmetric and has compact closure such that $\mathcal{W}^2 \subset \mathcal{V}$ (see Lemma 8.2.1).

Define functions on Γ, $\xi = \mathbf{1}_{\mathcal{W}}/m_\Gamma(\mathcal{W})$ and $\eta = \mathbf{1}_{\mathcal{W}\mathcal{K}}$. Then ξ and η belong to $L^2(\Gamma) \cap L^1(\Gamma)$: $0 \leq \xi \star \eta(\gamma) \leq 1$ for all $\gamma \in \Gamma$; $\xi \star \eta(\gamma) = 1$ for all $\gamma \in \mathcal{K}$; and $\xi \star \eta(\gamma) = 0$ for all $\gamma \notin \mathcal{U}$.

Let h and k be the inverse Fourier transforms of ξ and η, respectively. Then h and k belong to $L^2(G)$, and so hk is in $L^1(G)$. Put $f = hk$. Then

$$\widehat{f} = \widehat{hk} = \widehat{h} \star \widehat{k} = \xi \star \eta\,.$$

Hence $\widehat{f}$ has the desired values on $\mathcal{K}$ and $\mathcal{U}$. □

The norm of f in $L^1(G)$ is usually bigger than 1, but it has the following bound:

$$\|f\|_1 \leq \|h\|_2\|k\|_2 = \|\xi\|_2\|\eta\|_2 = (m_\Gamma(\mathcal{WK})/m_\Gamma(\mathcal{W}))^{1/2}.$$

The lemma is sometimes stated as $L^1(G)$ has *local units*. One consequence is that $L^1(G)$ is a regular Banach algebra; see Exercise 3.6.1.

Local units and regularity are important in the study of *spectral synthesis*. The question of spectral synthesis is as follows. Let $\mathcal{F}$ be a family of functions in $L^1(G)$, and let g be a function such that $\widehat{g}(\gamma) = 0$ whenever $\widehat{f}(\gamma) = 0$ for every $f \in \mathcal{F}$. Can g be synthesized from $\mathcal{F}$ in the sense that it belongs to the closed ideal generated by $\mathcal{F}$? (Note that the stated condition is necessary for g to belong to this ideal.) The first result in this direction was the so-called 'Tauberian theorem' of Wiener (1932), who showed that, if $f \in L^1(\mathbb{R}^n)$ satisfies $\widehat{f}(\gamma) \neq 0$ for every $\gamma \in \mathbb{R}^n$, then f generates $L^1(\mathbb{R}^n)$. Proposition 3.3.2 in Part I is a similar result which applies when G is discrete and abelian. The Banach algebraic proof of this result was an early application of Banach algebra theory.

Wiener's Tauberian theorem has been considerably strengthened by V. A. Ditkin (Rudin 1972) For the statement of this theorem the definitions of the *hull* of an ideal and the *kernel* of a set of maximal ideals must be recalled from Exercise 3.6.1 in Part I.

Theorem 10.2.2 (Ditkin) *Let G be a locally compact abelian group, and let I be a closed ideal in $L^1(G)$. Suppose that the boundary of $\mathfrak{h}(I)$ contains no perfect set. Then $I = \mathfrak{k}(\mathfrak{h}(I))$.* □

(A *perfect set* is an infinite compact set having no isolated points.)

A closed subset X of $\Phi_{L^1(G)}$ such that $\mathfrak{k}(X)$ is the unique ideal with hull X is said to be a *set of synthesis*. Wiener's original theorem asserts that the empty set is a set of synthesis, and Theorem 10.2.2 asserts that any set whose boundary does not contain a perfect set is a set of synthesis. Every finite set is therefore a set of synthesis and so we have the following consequence.

Corollary 10.2.3 *Let G be a locally compact abelian group, and let I be a closed ideal with infinite codimension in $L^1(G)$. Then $\mathfrak{h}(I)$ is infinite.* □

Lemma 10.2.1 implies that, for every $\gamma \in \Gamma$, there is an $f \in L^1(G)$ such that $\widehat{f}$ has compact support and $\widehat{f}(\gamma) \neq 0$. Hence the common zero set of the ideal

$$\mathcal{F} = \{\widehat{f} : \widehat{f} \text{ has compact support}\}$$

is empty. It follows by Wiener's theorem that $L^1(G)$ is the closure of this ideal.

A function algebra such that the functions with compact support are a dense subalgebra is sometimes called a *Wiener algebra.*

Proposition 10.2.4 *Let G be a locally compact abelian group, and let $u \in L^1(G)$. Then the operator $f \mapsto u \star f$, $L^1(G) \to L^1(G)$, is decomposable.*

Proof Let U_1 and U_2 be open subsets of $\mathbb{C}$ with $\mathbb{C} = U_1 \cup U_2$. Without loss of generality, it may be supposed that $0 \notin \mathbb{C} \setminus U_1$. Then $\widehat{u}^{-1}(\mathbb{C} \setminus U_1)$ is a compact subset of G because $\widehat{u} \in C_0(G)$ and $\widehat{u}^{-1}(U_2)$ is an open set with

$$\widehat{u}^{-1}(\mathbb{C} \setminus U_1) \subset \widehat{u}^{-1}(U_2).$$

Choose $g \in L^1(G)$ such that $\widehat{g}(\gamma) = 1$ for every $\gamma \in \widehat{u}^{-1}(\mathbb{C} \setminus U_1)$ and such that $\widehat{g}(\gamma) = 0$ for every $\gamma \notin \widehat{u}^{-1}(U_2)$. Put $X_1 = ((\delta_e - g) \star L^1(G))^-$ and then set $X_2 = (g \star L^1(G))^-$. Then $L^1(G) = X_1 + X_2$, and the spectrum of $M_u|X_j$ is contained in U_j for each j. □

10.3 The free group

In contrast with the algebras of compact and abelian groups, convolution in $\ell^1(\mathbb{F}_2)$ need not be decomposable. This will be shown by the following calculation, which is an improved exposition of an example taken from Willis (1986*b*). Several further consequences will be deduced from this calculation in the next chapter.

Proposition 10.3.1 *Let $\mathbb{F}_2 = \langle a, b \rangle$ be the free group on two generators, and let $t_a, t_b \in \mathbb{C}$ satisfy $|t_a| = 1 = |t_b|$. Then the linear operator T on $c_0(\mathbb{F}_2)$ defined by*

$$T\varphi(w) = \varphi(w) + t_a\varphi(aw) + t_b\varphi(bw) \quad (\varphi \in c_0(\mathbb{F}_2)) \tag{10.3.1}$$

is surjective. More precisely, for each $\varphi \in c_0(\mathbb{F}_2)$, there exists $\psi \in c_0(\mathbb{F}_2)$ with $\|\psi\| \leq \|\varphi\|$ such that $\varphi = T\psi$.

The convolution operator $(\delta_e + t_a\delta_a + t_b\delta_b)\star$ on $\ell^1(\mathbb{F}_2)$ is the dual of T. Hence this operator satisfies

$$\|(\delta_e + t_a\delta_a + t_b\delta_b) \star f\|_1 \geq \|f\|_1 \quad (f \in \ell^1(\mathbb{F}_2))$$

and has closed range.

Proof It is more convenient to carry out the calculation for a related convolution operator on a larger group. This group will be the free product of three copies of the cyclic group of order 2, $C_2 * C_2 * C_2 = \langle u_0, u_1, u_2 \rangle$. The subgroup of

$C_2 * C_2 * C_2$ generated by u_0u_1 and u_0u_2 is just the subgroup consisting of all words of even length and has index 2. This subgroup is isomorphic to $\mathbb{F}_2 = \langle a, b\rangle$, with the isomorphism determined by $a \mapsto u_0u_1$ and $b \mapsto u_0u_2$.

We identify $\mathbb{F}_2$ with the subgroup generated by u_0u_1 and u_0u_2, and so we see that $C_2 * C_2 * C_2$ is the union of cosets

$$C_2 * C_2 * C_2 = \mathbb{F}_2 \cup u_0\mathbb{F}_2 = \{\text{words of even length}\} \cup \{\text{words of odd length}\}.$$

Then $c_0(C_2 * C_2 * C_2)$ is the direct sum $c_0(\mathbb{F}_2) \oplus c_0(u_0\mathbb{F}_2)$. Equation (10.3.1) then defines an operator T on $c_0(C_2 * C_2 * C_2)$, and $c_0(\mathbb{F}_2)$ and $c_0(u_0\mathbb{F}_2)$ are invariant under T. Hence it will suffice to show that T is surjective on the space $c_0(C_2 * C_2 * C_2)$. Since T is the composite $T = SR$, where R is defined by

$$R\varphi(w) = \varphi(u_0w) + t_a\varphi(u_1w) + t_b\varphi(u_2w) \quad (\varphi \in c_0(C_2 * C_2 * C_2))$$

and $S\varphi(w) = \varphi(u_0w)$, and since S is clearly invertible, it will suffice to show that R is surjective. For ease of exposition we will suppose that $t_a = 1 = t_b$.

Let $\varphi \in c_0(C_2 * C_2 * C_2)$. We define $\psi \in c_0(C_2 * C_2 * C_2)$ with $R\psi = \varphi$ by defining $\psi(w)$ inductively on the length of w. Begin by setting $\psi(e) = 0$ and $\psi(w) = \varphi(e)/3$ for $w = u_0$, u_1 or u_2. Then $R\psi(e) = \varphi(e)$.

Assume that $\psi(w)$ has been defined for all words w with $|w| \le n$ such that $R\psi(w) = \varphi(w)$ for all w with $|w| \le n-1$. Let w have length n. Then $|uw| = n-1$ for u equal to one of u_0, u_1, and u_2, and $|uw| = n+1$ for u equal to the other two. Suppose, for instance, that $|u_0w| = n-1$ and that $|u_1w| = n+1 = |u_2w|$. In this case, define $\psi(u_1w) = (\varphi(w) - \psi(u_0w))/2 = \psi(u_2w)$. Then

$$R\psi(w) = \psi(u_0w) + \psi(u_1w) + \psi(u_2w) = \varphi(w).$$

The other cases are treated similarly to define ψ on all words of length up to $n+1$ to achieve $R\psi(w) = \varphi(w)$ for all w of length up to n. Continuing in this way we construct ψ with $R\psi = \varphi$.

It remains to show that $\psi \in c_0(C_2 * C_2 * C_2)$. For this, define for each w the sequence of words $(w_j)_{j=0}^n$, where $n = |w|$, by removing letters one at a time from the left of w. Thus, for example, $w_0 = w$, $w_n = e$, and $|w_j| = n-j$. (If the Cayley graph of $C_2 * C_2 * C_2 = \langle u_0, u_1, u_2\rangle$ is defined so that w is adjacent to u_0w, u_1w, and u_2w, then $(w_j)_{j=0}^n$ is the sequence of words on the path from w to e.) Then it may be shown by induction that

$$\psi(w) = -\sum_{j=1}^{n/2} \left(-\frac{1}{2}\right)^j \varphi(w_{2j-1}) \tag{10.3.2}$$

if n is even, and

$$\psi(w) = \frac{1}{3}\left(-\frac{1}{2}\right)^{(n-1)/2}\varphi(e) - \sum_{j=1}^{(n-1)/2}\left(-\frac{1}{2}\right)^{j}\varphi(w_{2j-1}) \qquad (10.3.3)$$

if n is odd. Since $|\varphi(w)| \to 0$ as $|w| \to \infty$, it follows that $\psi \in c_0(C_2 * C_2 * C_2)$. □

The identities (10.3.2) and (10.3.3) may be used to show that the operator T is surjective on $\ell^p(\mathbb{F}_2)$ for every $p > 2$. By duality, (see Dunford Schwartz 1958, §VI.6), T is injective with closed range on $\ell^p(\mathbb{F}_2)$ for every $p < 2$. The map T is injective and has dense range on $\ell^2(\mathbb{F}_2)$.

Proposition 10.3.2 *The operator T defined in the previous proposition has non-trivial kernel. Hence the operator $f \mapsto (\delta_e + t_a\delta_a + t_b\delta_b) \star f$ is not surjective on $\ell^1(\mathbb{F}_2)$.*

Proof It suffices to show that $(\delta_e + t_a\delta_a + t_b\delta_b)\star$ is not surjective on $\ell^1(\mathbb{F}_2)$ (see Dunford and Schwartz 1958, VI.6). For this, note that, since $\mathbb{F}_2$ is a free group, $a \mapsto t_a^{-1}e^{2\pi i/3}$, $b \mapsto t_b^{-1}e^{-2\pi i/3}$ determines a character on $\mathbb{F}_2$. Then the element $\delta_e + t_a\delta_a + t_b\delta_b$ belongs to the kernel of the corresponding representation of $\ell^1(\mathbb{F}_2)$, which is an ideal. Hence the range of convolution by this element is contained in this proper ideal.

We have shown that T' is not surjective. It follows that T itself is not injective. □

It is not difficult to show that a surjective operator which is not injective does not have the single-valued extension property (SVEP). Hence T does not have property (β) and T' does not have (δ). (For definitions, see 21.2.1 and 21.2.4)

Corollary 10.3.3 *The operator on $\ell^1(\mathbb{F}_2)$ defined by*

$$f \mapsto (\delta_e + t_a\delta_a + t_b\delta_b) \star f$$

is not decomposable. □

A $*$-algebra is *symmetric* if the spectrum of a^*a is contained in $\mathbb{R}^+ \cup \{0\}$ for every $a \in A$ (Dales 2000, §1.10). A locally compact group is *symmetric* if the $*$-algebra $L^1(G)$ is symmetric. Abelian and compact groups are symmetric, but there is no known characterization of symmetric groups. Free groups are not symmetric however.

Proposition 10.3.4 *$\mathbb{F}_2$ is not a symmetric group.*

Proof The element $(\delta_e + \delta_a + \delta_b)^* \star (\delta_e + \delta_a + \delta_b)$ is not invertible, but

$$\|(\delta_e + \delta_a + \delta_b)^* \star (\delta_e + \delta_a + \delta_b) \star f\|_1 \geq \|f\|_1$$

for every $f \in \ell^1(\mathbb{F}_2)$. It follows that the unit disc is contained in its spectrum, and so $\ell^1(\mathbb{F}_2)$ is not symmetric. □

10.4 Exercises

1. Let G be a compact group. Show that the operator $M_a : f \mapsto a \star f$ on $L^1(G)$ is compact when $a \in L^1(G)$ by the following steps.

(i) The function $x \mapsto a \star \delta_x,\ G \to L^1(G)$, is continuous. Hence the set $\{a \star \delta_x : x \in G\}$ is compact (in the norm topology on $L^1(G)$).

(ii) Let $f \in L^1(G)^+$ satisfy $\int_G f\,\mathrm{d}m \leq 1$. Assuming that

$$a \star f = \int_G a \star \delta_x f(x)\,\mathrm{d}m(x)\,,$$

show that $a \star f$ belongs to the closed convex hull of

$$\{a \star \delta_x : x \in G\} \cup \{0\}.$$

(iii) Deduce that $\{a \star f : f \in L^1(G)^+,\ \int_G f\,\mathrm{d}m \leq 1\}$ has compact closure, and hence that $\{a \star f : f \in L^1(G),\ \|f\|_1 \leq 1\}$ has compact closure.

2. (An alternative proof that $f \mapsto a \star f$ is compact.)
Let G be a compact group. We show that for every $a \in L^1(G)$ the operator $f \mapsto a \star f$ is a compact operator on $L^p(G)$ for $1 < p < \infty$ and on $C(G)$. Since the set of compact operators is closed and since $L^1(G) \cap L^\infty(G)$ is dense in $L^1(G)$, it may be supposed that $a \in L^1(G) \cap L^\infty(G)$.

(i) Let $f \in L^p(G)$. Show that $a \star f \in C(G)$ and $\|a \star f\|_\infty \leq \|a\|_q \|f\|_p$, where q is the conjugate index to p.

(ii) By using the fact that the map $x \mapsto \delta_x \star a,\ G \to L^q(G)$, is continuous, show that $\{a \star f : f \in L^p(G),\ \|f\|_p \leq 1\}$ is an equicontinuous and bounded subset of $C(G)$. Conclude (with the aid of the Arzelà–Ascoli theorem; see Dunford and Schwartz (1958, Theorem IV.6.7) that this set is totally bounded.

(iii) Use the fact that the inclusion map $C(G) \to L^p(G)$ is continuous to deduce that $\{a \star f : f \in L^p(G),\ \|f\|_p \leq 1\}$ has compact closure in $L^p(G)$, and hence convolution by a is a compact operator.

3. Describe the functions $c^{\sigma}_{\alpha\beta}$ when G is the circle group $\mathbb{T}$.
4. Describe the functions $c^{\sigma}_{\alpha\beta}$ when G is the permutation group S_3.
5. Complete the proof of Proposition 10.1.4.
6. Let G be a compact group. Show that G is isomorphic to a closed subgroup of a product $\prod_j U(n_j)$, where $U(n)$ denotes the group of $n \times n$ unitary matrices. Deduce that, if $\mathcal{U}$ is a neighbourhood of e, then there is a compact normal subgroup $K \subset \mathcal{U}$ such that G/K is a Lie group.
7. Show that single points are not sets of synthesis for $C^{(n)}(\mathbb{I})$ when $n \geq 1$.

10.5 Additional notes

1. Theorem 10.2.2 suggests that it is not easy to find subsets of $\Gamma = \widehat{G}$ which are not sets of synthesis for $L^1(G)$. However it was shown by Schwartz (1948) that the unit sphere in $\mathbb{R}^3$ is not a set of synthesis for $L^1(\mathbb{R}^3)$. This fact is the starting point of the construction in Grønbaek and Willis (1997).
2. By way of contrast, Herz (1958) showed that the unit circle in $\mathbb{R}^2$ is a set of synthesis for $L^1(\mathbb{R}^2)$. The question of synthesis is thus a very subtle one.
3. If G is compact, then Γ is discrete and so every set is a set of synthesis. If G is not compact, then there is a set of non-synthesis in Γ, as was shown by Malliavin (1959). These ideas are presented very clearly in Rudin (1962).

11

Amenable groups

Many, but certainly not all, results about abelian groups and compact groups have a common extension to the class of amenable groups. Some of these results are discussed in this chapter and it is seen that amenability may be regarded as a type of finiteness condition.

11.1 Definition and examples

Most of the definitions and results in this chapter may be found in the references (Hewitt and Ross 1979, §17; Greenleaf 1969; Reiter and Stegeman 2000, Chapter 8; Paterson 1988; Pier 1984). The first definition repeats 7.2.5.

Definition 11.1.1 *The locally compact group G is* amenable *if there is a left-invariant mean on $L^\infty(G)$. Thus G is amenable if there is a linear functional M on $L^\infty(G)$ such that:*

1. $M(\varphi) \geq 0$ *whenever* $\varphi \geq 0$;
2. $M(\mathbf{1}_G) = 1$; *and*
3. $M({}_x\varphi) = M(\varphi)$ *for every* $\varphi \in L^\infty(G)$ *and* $x \in G$.

Properties 1 and 2 say that M is a *mean*, and Property 3 says that M is invariant under left translation.

It may be shown that $L^\infty(G)$ has a left-invariant mean if and only if it has a right-invariant mean and that this is equivalent to $L^\infty(G)$ having a two-sided invariant mean. It is also equivalent to $L^\infty(G)$ having *topologically invariant mean*, that is, a mean M such that

$$M(f \star \varphi) = \left(\int_G f \, \mathrm{d}m\right) M(\varphi) \quad (f \in L^1(G)).$$

The mean property of M implies that it is continuous (with norm 1), but it is shown in Willis (1988) that the mean property and continuity are both redundant in the definition.

Proposition 11.1.2 *The locally compact group G is amenable if and only if there is a non-zero, left-invariant functional (continuous or discontinuous) on $L^\infty(G)$.* □

Here are some basic examples of amenable groups and of non-amenable groups.

Proposition 11.1.3 *Compact groups and abelian groups are amenable.*

Proof The Haar integral is an invariant mean on $L^\infty(G)$ when G is compact. An indirect proof valid for abelian semigroups is given in Hewitt and Ross (1979, Theorem 17.5). We give a more direct proof for the case where G is the group of integers.

For each $n \in \mathbb{Z}^+$, let $m^{(n)} \in \ell^1(\mathbb{Z})$ be specified by

$$m^{(n)} = \frac{1}{2n+1} \sum_{j=-n}^{n} \delta_j .$$

Then $m^{(n)} \geq 0$ and $\|m^{(n)}\|_1 = 1$. Moreover, when $n > k > 0$, we have

$$m_k^{(n)} - m^{(n)} = \frac{1}{2n+1} \left(\sum_{j=n+1}^{n+k} \delta_j - \sum_{j=-n}^{k-n-1} \delta_j \right),$$

and so $\|m_k^{(n)} - m^{(n)}\|_1 = 2k/(2n+1) \xrightarrow{n} 0$ for every $k \in \mathbb{Z}$. It follows that each weak-$*$ accumulation point of $(m_n)_{n=1}^\infty$ is an invariant mean on $\ell^\infty(\mathbb{Z})$. □

Proposition 11.1.4 *The free group on two generators is not amenable.*

Proof Let $\omega = e^{2\pi i}$. Then the operator $T'' \in \mathcal{B}(\ell^\infty(\mathbb{F}_2))$ given by

$$T''\varphi(x) = \varphi(x) + \omega\varphi(ax) + \omega^2\varphi(bx) \quad (\varphi \in \ell^\infty(\mathbb{F}_2))$$

is the second dual of the operator considered in Proposition 10.3.1, and so is surjective by that proposition and Dunford and Schwartz (1958, VI.6).

Let M be a translation invariant functional on $\ell^\infty(\mathbb{F}_2)$. Then, for every element $\varphi \in \ell^\infty(\mathbb{F}_2)$, we have $\varphi = T''\psi$ for some $\psi \in \ell^\infty(\mathbb{F}_2)$, and so

$$M(\varphi) = M(T''\psi) = M(\psi) + \omega M(\psi_a) + \omega^2 M(\psi_b).$$

Since M is translation-invariant, it follows that

$$M(\varphi) = (1 + \omega + \omega^2)M(\psi) = 0\,.$$

Hence there are no non-zero translation-invariant functionals on $\ell^\infty(\mathbb{F}_2)$, and $\mathbb{F}_2$ is not amenable. □

Proofs of the following stability properties of amenability may be found in Greenleaf (1969) and Paterson (1988).

Theorem 11.1.5

(i) *Let G be a locally compact group, and let N be a closed normal subgroup of G. Then G is amenable if and only if N and G/N are amenable.*
(ii) *Let G be an amenable group, and let $\varphi : G \to H$ be a homomorphism such that $\overline{\varphi(G)} = H$. Then H is amenable.*
(iii) *Let G be an amenable group, and let H be a closed subgroup of G. Then H is amenable.* □

The above theorem enlarges the classes of groups shown to be amenable in Proposition 11.1.3.

Corollary 11.1.6 (i) *Solvable groups are amenable.*
(ii) *Central groups are amenable.* □

Proposition 11.1.4 may also be combined with the theorem to show that many groups are not amenable.

Corollary 11.1.7 *Non-compact, semisimple Lie groups are not amenable.* □

This corollary is proved by showing that non-compact, semisimple Lie groups have closed subgroups isomorphic to $\mathbb{F}_2$. See Reiter and Stegeman (2000, §8.7).

For a long time it was an open question, asked originally by von Neumann, whether every non-amenable, group had a subgroup isomorphic to $\mathbb{F}_2$. This question was answered Ol'shanskiĭ (1980), where a non-amenable group is constructed which has no non-abelian, free subgroup because every one of its proper subgroups is cyclic.

Theorem 11.1.8 *Suppose that $G = \overline{\bigcup_\lambda G_\lambda}$, where each group G_λ is amenable. Then G is amenable.* □

Corollary 11.1.9 *Locally finite groups, that is, discrete groups such that every finite subset generates a finite subgroup, are amenable.* □

The class of groups containing abelian and compact groups and closed under extensions, dense homomorphic images, subgroups, and direct limits is the class of *elementary amenable* groups. For some time these were the only groups which were known to be amenable. However amenable groups which are not elementary were constructed by Grigorchuk (1984, 1988).

11.2 Alternative characterizations of amenability

The amenability condition has many equivalent characterizations which reflect the wide range of applications of amenability. Many of these aspects of amenability show it to be a finiteness condition in some way.

Invariant means By definition, G is amenable if and only if there is a left-translation invariant mean on $L^\infty(G)$. (The name was invented by M. M. Day as a pun.) As already remarked, amenability is equivalent to the existence of right-translation invariant, two-sided translation-invariant, and topologically invariant means on $L^\infty(G)$. It is also equivalent to the existence of a left-invariant mean on $C(G)$, the space of bounded continuous functions on G and on the space of right uniformly continuous functions on G; a function $f \in C(G)$ is *right uniformly continuous* if, for each $\varepsilon > 0$, there exists a neighbourhood U of e such that $|f(x) - f(y)| < \varepsilon$ whenever $x, y \in G$ with $xy^{-1} \in U$.

A characterizing property of compact groups is that their Haar measure is finite, and in this case the Haar measure of the entire group is usually normalized to equal 1. Integration against Haar measure is then an invariant mean on spaces of bounded functions. If G is not compact, bounded functions cannot be integrated generally (because $L^\infty(G, m) \not\subset L^1(G, m)$). The existence of translation-invariant means on $L^\infty(G, m)$ therefore extends some features of compact (and in particular finite) groups to the class of amenable, locally compact groups. Indeed, it will be seen that certain results for finite and compact groups which use the Haar measure in their proof have analogues for amenable groups.

Although the Haar measure is unique, invariant means are not. Rudin (1972) showed that the compact abelian group $\mathbb{T}$ has invariant means other than Haar measure because it has invariant means which are not topological invariant means. The exact number of invariant means on a general locally compact group is found in Chou (1976).

Paradoxical decompositions A *paradoxical decomposition* of the group G consists of a finite partition $G = A_1 \cup \cdots \cup A_k \cup B_1 \cdots \cup B_l$ into measurable subsets and elements $x_1, \ldots, x_k, y_1, \ldots, y_l$ such that $G = x_1A_1 \cup \cdots \cup x_kA_k$ and $G = y_1B_1 \cup \cdots \cup y_lB_l$ are also partitions.

It was shown by Tarski (1938) that G is amenable if and only if it does not have a paradoxical decomposition. Non-amenability of G thus says that G is infinite in a strong sense closely related to the algebraic structure: there is a two-to-one map from G to itself which is a piecewise translation.

Amenability first appeared in connection with paradoxical decompositions. The Banach–Tarski paradox shows that the sphere can be partitioned into finitely many sets which may be rotated and reassembled to form two copies of the original sphere. It follows that there is no finitely additive measure on the family of *all* subsets of the sphere which extends the familiar area measure on Borel sets. This is in contrast to the circle $\mathbb{T}$ or the line $\mathbb{R}$ or the plane $\mathbb{R}^2$, where the familiar Lebesgue measure can be extended to a translation-invariant, finitely additive measure on all subsets. It was observed by von Neumann (1929) that the paradoxical decomposition of the sphere arises from a paradoxical decomposition of a free subgroup of the rotation group of the sphere. The symmetry groups of the line, plane and circle do not contain free subgroups.

The Følner condition This condition says, roughly speaking, that amenable groups have compact subsets almost invariant under translations. For example, if G is compact, then the group itself is invariant under translation. For a second example, the proof in Proposition 11.1.3 that $\mathbb{Z}$ is amenable uses the fact that, for each $n \in \mathbb{N}$, the subset

$$K = \{-n, -n+1, \ldots, n-1, n\}$$

is almost invariant in a sense which we now make precise.

Definition 11.2.1 *The locally compact group G satisfies* Følner's condition *if, for every $\varepsilon > 0$ and every compact subset C of G, there is a compact subset K of G such that*

$$m\left((xK \setminus K) \cup (K \setminus xK)\right)/m(K) < \varepsilon \quad (x \in C).$$

The locally compact group G is amenable if and only if it satisfies Følner's condition.

Almost invariant probabilities Closely related to Følner's condition are several formulations of almost invariance for probability measures on G. Denote the set of non-negative, norm 1 functions in $L^1(G)$ by $P(G)$. These are the probability measures on G which are absolutely continuous with respect to Haar measure.

Proposition 11.2.2 *The locally compact group G is amenable if it satisfies any one of the following three equivalent conditions.*

1. *There is a net $\{f_\lambda\}$ in $P(G)$ such that $\{_x f_\lambda - f_\lambda\}$ converges weakly to 0 for every $x \in G$.*
2. *There is a net $\{f_\lambda\}$ in $P(G)$ such that $\{\|_x f_\lambda - f_\lambda\|_1\}$ converges to 0 for every $x \in G$.*
3. *There is a net $\{f_\lambda\}$ in $P(G)$ such that $\{\|g \star f_\lambda - f_\lambda\|_1\}$ converges to 0 for every $g \in P(G)$.* □

The proof that the first condition implies the second, due to M. M. Day, requires the functional-analytic fact (derived from the Hahn–Banach theorem) that the weak and strong closures of a convex subset of a Banach space are the same. The argument does not show that weak convergence of $\{_x f_\lambda - f_\lambda\}$ to 0 implies norm convergence. It in fact is shown in Rosenblatt and Willis (2001) that there is a net with $\{_x f_\lambda - f_\lambda\}$ weakly convergent but not norm convergent. This paper introduces the notion of *configuration sets*, which could be useful for further work on amenable groups.

The properties in Proposition 11.2.2 are very close to the $p = 1$ case of the following condition introduced by H. Reiter.

Definition 11.2.3 (Property P_p) *Let $1 \le p < \infty$. The locally compact group G satisfies* Property P_p *if, for every $\varepsilon > 0$ and every compact subset C of G, there exists $f \in L^p(G)$ such that $f \ge 0$, $\|p\|_p = 1$, and*

$$\|_x f - f\|_p < \varepsilon \quad (x \in C).$$

Proposition 11.2.4 (Reiter and Stegeman 2000, Theorem 8.3.2) *The conditions P_p for $1 \le p < \infty$ are all equivalent, and the group G is amenable if it satisfies any one of these conditions.* □

Unitary representations Reiter's condition P_2 implies that there is a net $\{f_\lambda\}$ of norm 1 elements in $L^2(G)$ such that $\|_x f_\lambda - f_\lambda\|_2 \xrightarrow{\lambda} 0$. In other words, $\|U_x(f_\lambda) - f_\lambda\|_2 \xrightarrow{\lambda} 0$, where U is the left regular representation of G on $L^2(G)$. The *trivial representation* of G is the one-dimensional representation defined by $V_x = 1$ $(x \in G)$. Condition P_2 therefore says that functions in the regular representation approximate the trivial representation of G. That is the first statement of the next characterization of amenability.

Proposition 11.2.5 *Let G be a locally compact group. Then the following conditions are each equivalent to G being amenable.*

1. *The trivial representation of G is weakly contained in the regular representation.*
2. *Every unitary representation of G is weakly contained in the regular representation.*
3. $C_r^*(G) = C^*(G)$.
4. *For every probability measure in $M(G)$ (or in $M_a(G)$ or $M_d(G)$), the norm of the operator $f \mapsto \mu \star f$, $L^2(G) \to L^2(G)$, is equal to 1.* □

The facts, proved in Chapters 9 and 10, that $C_r^*(G) = C^*(G)$ when G is abelian or compact may now be seen to be consequences of the amenability of abelian groups and compact groups, respectively. Alternatively, they might be viewed as proofs that these groups are amenable.

The unitary representation theory of finite groups shows that, if G is finite, then $\mathbb{C}G$ is isomorphic to a direct sum of full matrix algebras and that every irreducible unitary representation of G is equivalent to a direct summand of the regular representation. In Theorem 10.1.3 the same was seen to hold for compact groups. Condition 2 of the above proposition extends this to amenable groups. It is no longer the case that every irreducible unitary representation is a direct summand in the regular representation; it is only weakly contained in it. *Weak containment* of an irreducible representation V in the regular representation U means that, for every ξ and η in $\mathfrak{H}_V$, there are bounded nets $\{f_\lambda\}$ and $\{g_\lambda\}$ in $L^2(G)$ such that the 'matrix coefficient' $x \mapsto \langle V_x\xi, \eta\rangle$ is approximated in the sense that

$$\langle V_x\xi,\, \eta\rangle = \lim_{\lambda} \langle U_x f_\lambda,\, g_\lambda\rangle\,,$$

where the limit is uniform on compact subsets of G. Amenable groups are precisely those for which we have this weak version of a theorem about finite groups. Amenability thus appears once again as a finiteness condition.

An immediate consequence of Condition 4 in Proposition 11.2.5 is that, if G is not amenable, there are $x_1, \ldots, x_n$ in G and positive numbers $c_1, \ldots, c_n$ with $\sum_{j=1}^n c_j = 1$ such that $|\sum_{j=1}^n (c_j\delta_{x_j}\star)|_{L^2(G)} < 1$. (J. Dieudonné (1960) made a direct calculation in the case of $\mathbb{F}_2$ and showed that $|\delta_e + \delta_a + \delta_b|_{l^2(G)} = \sqrt{8}$.) It is an exercise to deduce the following result from this fact.

Corollary 11.2.6 *Let G be a non-amenable group. Then there is no non-zero, translation-invariant functional on $L^2(G)$.* □

The fixed point property This property means that every continuous and affine action of G on a compact, convex subset C of a locally convex topological linear space has a fixed point. The next characterization was shown by Day and by Rickert (1967).

Proposition 11.2.7 *The locally compact group G is amenable if and only if it has the fixed point property.* □

Cohomology of $L^1(G)$ One of the last formulations of amenability to be obtained concerns the cohomology of $L^1(G)$. Some definitions are required in order to state it.

Definition 11.2.8 *Let A be a Banach algebra.*

(i) *A* Banach A-bimodule *is a Banach space X together with left and right* actions $(a, x) \mapsto a \cdot x$ *and* $(a, x) \mapsto x \cdot a$ *which are bilinear maps satisfying the associative laws*

$$(ab) \cdot x = a \cdot (b \cdot x), \quad (a \cdot x) \cdot b = a \cdot (x \cdot b)$$
$$\text{and } x \cdot (ab) = (x \cdot a) \cdot b \quad (a, b \in A,\ x \in X).$$

These maps must also satisfy the bound condition

$$\|a \cdot x\|_X \leq \|a\|_A \|x\|_X \text{ and } \|x \cdot a\|_X \leq \|a\|_A \|x\|_X \quad (a \in A,\ x \in X).$$

(ii) *A* derivation *from A to a bimodule X is a linear map $D : A \to X$ satisfying the* Leibniz *rule*

$$D(ab) = a \cdot D(b) + D(a) \cdot b \quad (a, b \in A).$$

(iii) *Let x be in X. The map $D_x : A \to X$ is an* inner deriviation *if*

$$D(a) = a \cdot x - x \cdot a \quad (a \in A).$$

(iv) *The set of continuous derivations is a linear subspace of $\mathcal{B}(A, X)$, denoted by $\mathcal{Z}^1(A, X)$. The set of inner derivations is a subspace of $\mathcal{Z}(A, X)$, denoted by $\mathcal{N}^1(A, X)$. The quotient space*

$$\mathcal{H}^1(A, X) := \mathcal{Z}^1(A, X)/\mathcal{N}^1(A, X)$$

is the first continuous cohomology group of A with coefficients in X.

The first and higher cohomology groups of Banach algebras are discussed in Part I, §6.2. For further information, see Dales (2000) and Helemskii (1989).

The dual module of a Banach A-module is defined in Definition 6.1.5. The dual actions of A on X' are defined so that the linear operator $x' \mapsto x' \cdot a$ is the dual of the operator $x \mapsto a \cdot x$ and $x' \mapsto a \cdot x'$ is the dual of the operator $x \mapsto x \cdot a$. As part of his work on derivations on $L^1(G)$, B. E. Johnson proved the following theorem; it gave rise to the definition of an amenable Banach algebra. See Part I and also Johnson (1972, Theorem 2.5) or Dales (2000, Theorem 5.6.2).

Theorem 11.2.9 *Let G be a locally compact group. Then G is amenable if and only if $\mathcal{H}^1(L^1(G), X') = \{0\}$ for every Banach $L^1(G)$-bimodule X.* □

In particular, since $M(G)$ with the convolution action is the dual $L^1(G)$-bimodule of $C_0(G)$ with appropriately defined actions, we have that $\mathcal{H}^1(L^1(G), M(G)) = \{0\}$.

Corollary 11.2.10 *Let G be an amenable group, and let $D : L^1(G) \to L^1(G)$ be a continuous derivation. Then there exists $\mu \in M(G)$ such that*

$$D(f) = f \star \mu - \mu \star f \quad (f \in L^1(G)) .$$ □

A proof of Theorem 11.2.9 in the case where G is discrete was presented in Chapter 7. That this theorem of Johnson's is yet another manifestation of amenability as a finiteness condition can be seen by comparing it with a theorem from purely algebraic cohomology theory. For each algebra A and A-bimodule X, the algebraic cohomology group $H^1(A, X)$ is defined just as in Definition 11.2.8 except that there is no norm; see 7.1.2. Then it may be shown that the algebra A satisfies $H^1(A, X) = \{0\}$ for *every* A-bimodule X if and only if A is finite-dimensional and semisimple; see 7.2.1. It is conjectured that the same holds for Banach algebras and Banach bimodules but all proofs of special cases assume that spaces concerned have the approximation property or compact approximation property at some point, (see Taylor 1972; Selivanov 1976; Runde 1998). Vanishing of all cohomology groups of A with values in dual A-bimodules is thus a weak version of a condition which, in the purely algebraic case at least, implies the finite-dimensionality of A.

11.3 Approximate identities in ideals

Amenability of G is also equivalent to the existence of (bounded) approximate identities in certain ideals of $L^1(G)$. The first result is due to H. Reiter 1968. For this, note that the codimension 1 subspace

$$L_0^1(G) := \left\{ f \in L^1(G) : \int_G f \, dm = 0 \right\}$$

is a closed ideal in $L^1(G)$. It is the kernel of the representation of $L^1(G)$ induced by the trivial representation of G; see equation (7.2.1). It is called the *augmentation ideal.*

Proposition 11.3.1 *The locally compact group G is amenable if and only if $L_0^1(G)$ has a left bounded approximate identity.*

For a full proof, see Dales (2000, Theorem 5.6.42). See the exercises for a direct proof that $\ell_0^1(\mathbb{F}_2)$ does not have an approximate identity. It had previously been shown by W. Rudin that, when G is compact, all ideals in $L^1(G)$ have bounded approximate identities. This was extended to the class of amenable groups by Liu, van Rooij, and Wang (1973).

Proposition 11.3.2 *Let G be an amenable group, and let I be a closed right ideal in $L^1(G)$. Then I has a left bounded approximate identity if and only if it is weakly complemented in $L^1(G)$.*

(A subspace Y of a Banach space X is *weakly complemented* if the annihilator of Y, which is defined by $Y^\perp := \{x' \in X' : \langle y, x' \rangle = 0 \ (y \in Y)\}$, has a Banach space complement in X'.) It is not difficult to see that any ideal which has a bounded approximate identity must be weakly complemented. Amenability is essential for the 'if' direction of the proposition. In fact, the result is proved in Liu, van Rooij, and Wang (1973) for complemented ideals only. The stronger version for weakly complemented ideals may be proved with the aid of the cohomological property shown in Theorem 11.2.9; for this, see Curtis and Loy (1989) and Dales (2000, Theorem 2.9.58).

The previous proposition applies in particular to ideals with finite codimension in $L^1(G)$ because such ideals are complemented. The 'if' direction of Reiter's theorem extends to these ideals (Willis 1982).

Proposition 11.3.3 *Suppose that some closed right ideal of finite codimension in $L^1(G)$ has a left bounded approximate identity. Then G is amenable.* □

These results on approximate identities are related to another characterization of amenability in terms of random walks. Each probability measure on G determines a random walk on G; see for example Kaĭmanovič and Veršik (1983). An important tool in the study of a random walk is its *boundary*. The boundary is a measurable G-space imagined as consisting of points 'at infinity' which the walker approaches as times proceeds.

It is not directly relevant to make this precise here. For the purposes of this chapter, we describe the boundary of the random walk corresponding to the probability measure μ in Banach algebraic terms as follows. First, define

$$ {}_{\mu}J := \overline{\{f - \mu \star f : f \in L^1(G)\}}. $$

Then ${}_{\mu}J$ is a closed right ideal in $L^1(G)$ and has a left bounded approximate identity consisting of functions of the form

$$ \left(\delta_e - \frac{1}{n}\sum_{j=1}^{n}\mu^j\right) \star u_E, $$

where u_E is a function as defined in Proposition 9.1.8. It may be shown that the quotient Banach space $L^1(G)/{}_{\mu}J$ is isometrically isomorphic to $L^1(\Omega, \nu)$, where Ω is the boundary of the random walk; see Willis (1990, Theorem 2.1).

A random walk where the boundary consists of a single point is said to be *ergodic*. It is clear that this occurs if and only if ${}_{\mu}J$ has codimension 1 in $L^1(G)$. Since ${}_{\mu}J \subset L^1_0(G)$, it follows that the random walk is ergodic if and only if ${}_{\mu}J = L^1_0(G)$. As shown in the exercises, ${}_{\mu}J$ has a left bounded approximate identity. Hence, by Proposition 11.3.1, if G supports an ergodic random walk, then G is amenable. The converse was conjectured by H. Furstenberg and proved independently by Rosenblatt (1981) and by Kaĭmanovič and Veršik (1983); also see Willis (1990).

Theorem 11.3.4 *The σ-compact locally compact group G is amenable if and only if there is a probability measure μ on G such that $L^1_0(G) = {}_{\mu}J$.* □

Recall from the Kawada–Itô theorem 8.3.4 that, if μ is a probability measure on the compact group G and if the support of μ generates G, then the sequence of probability measures $\left(\sum_{j=1}^{n}\mu^j\right)_{j=1}^{n}$ converges to Haar measure (in the weak-$*$ topology on $M(G)$). Theorem 11.3.4 is a weaker version of this, where compactness is replaced by amenability, so that once again amenability appears as a weak finiteness condition. There is an important difference however. Theorem 8.3.4 is a universal statement applying to every probability measure μ having large enough support, but Theorem 11.3.4 is an existential statement and μ is quite difficult to construct. A probability measure on an amenable group need not be ergodic even if its support equals G; some examples illustrating how subtle the boundary of a random walk can be are given in Kaĭmanovič (1983).

11.4 Exercises

1. Show that, if G is amenable and N is a closed normal subgroup of G, then G/N is amenable.
2. (An alternative proof that $\mathbb{F}_2$ is not amenable.) Let

$$S_a = \{w \in \mathbb{F}_2 : w = ax \quad \text{as a reduced word}\},$$

and define subsets S_b, $S_{a^{-1}}$, and $S_{b^{-1}}$ of $\mathbb{F}_2$ similarly. Note that

$$\mathbb{F}_2 = \{e\} \cup S_a \cup S_b \cup S_{a^{-1}} \cup S_{b^{-1}}$$

is a partition. Show that $\mathbb{F}_2 = a^{-1}S_a \cup S_{a^{-1}}$ is also a partition of $\mathbb{F}_2$, and deduce that there is no translation invariant mean on $\mathbb{F}_2$.
3. Show that, if G has a paradoxical decomposition, then there is no left-invariant mean on $L^\infty(G)$.
4. Let G be a group which satisfies Følner's condition. Show that G satisfies Property P_1 of Reiter and the second condition of Proposition 11.2.2. Deduce that G has a left-translation invariant mean.
5. Prove Corollary 11.2.6.
6. Show that

$$\left\{\left(\delta_e - \frac{1}{n}\sum_{j=1}^{n}\mu^j\right) \star u_\lambda\right\}_{(n,\lambda)}$$

is a left bounded approximate identity for the ideal J_μ. (See Proposition 9.1.8 for the definition of u_λ.)
7. Show that the distance between the elements

$$(\delta_e + e^{2\pi i/3}\delta_a + e^{-2\pi i/3}\delta_b) \star f \quad \text{and} \quad \delta_e + e^{2\pi i/3}\delta_a + e^{-2\pi i/3}\delta_b$$

is at least 1 for every $f \in \ell_0^1(\mathbb{F}_2)$, and deduce that $\ell_0^1(\mathbb{F}_2)$ does have any (even unbounded) approximate identity.

12

Harmonic analysis and automatic continuity

Knowledge of the structure of $L^1(G)$ can be used to show that derivations from $L^1(G)$ are automatically continuous when G is abelian, compact, or is in one of several other classes. However it is not known whether such derivations are continuous for every locally compact group G. Work on this problem leads to further questions in abstract harmonic analysis.

12.1 Automatic continuity of derivations

It was shown by Ringrose (1972) that every derivation from a C^*-algebra A is automatically continuous. Powerful general techniques for proving automatic continuity results have been developed from this and earlier work; see Part I and the substantial account in Dales (2000). The following general method for proving automatic continuity of derivations was given by Jewell (1977).

Theorem 12.1.1 *Let A be a Banach algebra such that:*

1. *every closed two-sided ideal with finite codimension in A has a bounded approximate identity; and*
2. *if I is a closed two-sided ideal with infinite codimension in A, then there are sequences (a_n) and (b_n) in A such that $b_n a_1 \cdots a_{n-1} \notin I$, but $b_n a_1 \cdots a_n \in I$ for each $n \geq 2$.*

Then every derivation $D : A \to X$, where X is a Banach A-bimodule, is continuous. □

Condition 2 of the theorem is used to show that the continuity ideal (see Part I and Exercise 12.4.1) of any derivation $D : A \to X$ cannot have infinite codimension. A variant of the Main Boundedness Theorem of Bade and Curtis (see Theorem 5.2.2) is used to show this. The existence of a bounded approximate identity for the continuity ideal implies, via an application of Cohen's factorization theorem, that the restriction of D to its continuity ideal is continuous. Since the continuity ideal has finite codimension, it follows that D must be continuous.

Theorem 12.1.1 applies to all C^*-algebras A. The first condition is satisfied because every ideal in a C^*-algebra has a bounded approximate identity. To verify the second condition, it is first observed that, if A/I has an infinite-dimensional commutative $*$-subalgebra C, then its character space Φ_C (see Definition 3.1.1) is infinite, and C may be assumed to be $C_0(\Phi_C)$. Then an infinite sequence of pairwise disjoint subsets of Φ_C may be chosen and Urysohn's lemma used to write down a sequence of non-zero functions (f_n) such that $f_m f_n = 0$ whenever $m \neq n$. These functions show that the second condition is satisfied (exercise). If every $*$-subalgebra of A/I is finite-dimensional, then a separate (and elementary) argument shows that A/I is itself finite-dimensional. (For details, see Dales (2000, Corollary 5.3.7).) No such elementary argument shows that the conditions of Theorem 12.1.1 are satisfied by $L^1(G)$ for general locally compact groups G. However, they can be shown for certain classes of groups, and Condition 1 can be replaced by a weaker hypothesis.

12.2 Finite-codimensional ideals in $L^1(G)$

As we have already seen in Propositions 11.3.2 and 11.3.3, ideals with finite codimension in $L^1(G)$ have a bounded approximate identity if and only if G is amenable. Hence Theorem 12.1.1 can only apply to amenable groups at best. However, Condition 1 of the theorem can be weakened. The bounded approximate identity is needed only so that Cohen's factorization theorem 9.1.9 may be applied. The requisite factoring can be achieved in finite-codimensional ideals of general group algebras with the aid of the following result, which is proved in Willis (2001).

Proposition 12.2.1 *Let G be a σ-compact, locally compact group, and let I be a closed two-sided ideal with finite codimension in $L^1(G)$.*

(i) *There is a closed left ideal $L \subset I$ having a right bounded approximate identity, and a closed right ideal $R \subset I$ having a left bounded approximate identity, such that*

$$I = R + L. \tag{12.2.1}$$

(ii) *For every sequence* (f_n) *in* I *with* $\|f_n\|_1 \xrightarrow{n} 0$, *there are elements* a *and* b *and sequences* (h_n) *and* (k_n) *converging to* 0 *in* I *with*

$$f_n = h_n \star a + b \star k_n \quad (n \in \mathbb{N}). \tag{12.2.2}$$

□

The factoring of null sequences described in Equation (12.2.2) is just what is needed to complete the automatic continuity argument behind Theorem 12.1.1.

It is not possible to give the details of the proof of Proposition 12.2.1 here, but we will compare the construction of the ideals R and L with the construction of the approximate identities in Proposition 11.3.2 in the cases where G is finite, where G is compact, and where G is amenable. It will be clearer to begin with the case where $I = L^1_0(G)$.

When G is a finite group, the element $\delta_e - m_G$ is an identity for the ideal $L^1_0(G)$. When G is compact but not finite, this ideal has a bounded approximate identity of the form

$$\{u_\lambda \star (\delta_e - m_G)\}, \tag{12.2.3}$$

where $\{u_\lambda\}$ is a bounded approximate identity for $L^1(G)$; see Proposition 9.1.8. If G is not compact, the Haar measure m_G is not bounded and (12.2.3) does not make sense, but when G is amenable m_G can be replaced by a net of probability measures $\{f_\nu\}$ as in Proposition 11.2.2(3) or, when G is also σ-compact, by convolution powers of a single probability measure as in Proposition 11.3.4. Then $L^1(G)$ has a bounded approximate identity of the form

$$\{u_\lambda \star (\delta_e - f_\nu)\}_{(\lambda,\nu)} \quad \text{or} \quad \left\{u_\lambda \star \left(\delta_e - \frac{1}{n}\sum_{j=1}^{n} \mu^j\right)\right\}_{(\lambda,n)}. \tag{12.2.4}$$

Random walks, that is, convolution powers of a single probability measure, can also be used in the case where G is not amenable. Define

$$_\mu J := \overline{\{f - \mu \star f : f \in L^1(G)\}} \quad \text{and} \quad J_\mu := \overline{\{f - f \star \mu : f \in L^1(G)\}}.$$

Then, as seen in the last chapter, $_\mu J$ is a closed right ideal with a left bounded approximate identity, and it may be seen in the same way that J_μ is a closed left ideal with a right bounded approximate identity. These are the ideals R and L in (12.2.1) when $I = L^1_0(G)$.

Proposition 12.2.2 (Willis 1990) *Let* G *be a* σ*-compact group, and let* μ *be a probability measure on* G *which is absolutely continuous with respect to Haar* *measure and with support equal to G. (The σ-compactness of G is required only so that such a probability measure exists.) Then*

$$L_0^1(G) =_{\mu} J + J_{\mu}. \qquad \square$$

Note that the assertion holds for any probability measure satisfying the hypothesis; the probability measure is not specially constructed. (This hypothesis can be weakened a little.) The proposition is in this way closer to the Kawada–Itô theorem than to Proposition 11.3.4. It has also been long known that, if G is abelian and the support of the probability measure μ generates G, then $J_\mu = L_0^1(G)$. This is known as the *Choquet–Deny theorem* (Choquet and Deny 1961). Proposition 12.2.2 might therefore be regarded as a common generalization of the Kawada–Itô and Choquet–Deny theorems. The proof represents the quotient $L^1(G)/J_\mu$ as $L^1(\Omega)$ (where Ω is the boundary of the random walk with law μ), and applies the ergodic theorem to a certain operator on this space. A proof in more probabilistic terms has also been given by Kaĭmanovič (1992).

Now we see how R and L are defined when I is any finite-codimensional ideal. The quotient space $L^1(G)/I$ is finite-dimensional and the regular representation of G on $L^1(G)$ induces a representation V of G by isometries on this quotient space. The closure of $V(G)$ in $\mathcal{B}(L^1(G)/I)$ is a compact group, K say, and harmonic analysis on K plays an important part. The representation V factors through K as $V = \widetilde{V} \circ \varphi$, where $\varphi : G \to K$ is a group homomorphism and $\widetilde{V}$ is a representation of K. Then $\widetilde{V}$ is the direct sum of irreducible finite-dimensional representations and there are corresponding minimal central idempotents $z_1, \ldots, z_n$ in $L^1(K)$. When G is finite or compact, I has a bounded approximate identity of the form

$$\left\{ u_\lambda \star \left(\delta_e - \sum_{j=1}^{n} z_j \right) \right\},$$

where $\{u_\lambda\}$ is a bounded approximate identity for $L^1(G)$. Compare this approximate identity with (12.2.3), noting that when $I = L_0^1(G)$, $\widetilde{V}$ is the trivial representation and $\{z_1, \ldots, z_n\} = \{m_G\}$. When G is not compact but is amenable, the functions z_j do not belong to $L^1(G)$, but there are nets $\{z_{j\nu}\}$ of functions in $L^1(G)$ such that I has a bounded approximate identity

$$\left\{ u_\lambda \star \left(\delta_e - \sum_{j=1}^{n} z_{j\nu} \right) \right\}_{(\lambda,\nu)}.$$

Compare this with (12.2.4). The general case also relies on choices of nets of functions in $L^1(G)$ to modify the bounded approximate identities of ${}_\mu J$ and J_μ. This is fairly delicate however, and much of the work in Willis (2001) involves making these choices. The closed ideals R and L are then defined to be the ideals for which these nets are the respective left and right approximate identities.

Proposition 12.2.1 implies that all derivations from $L^1(G)$ are continuous whenever Condition 2 of Theorem 12.1.1 is satisfied.

12.3 Infinite-codimensional ideals in $L^1(G)$

In this section we shall see that Condition 2 of Theorem 12.1.1 is satisfied by $L^1(G)$ when G is abelian or compact. We shall also see that, when G is connected, a similar condition is satisfied by $M(G)$. This condition suffices to show that derivations are continuous.

Abelian groups Suppose first of all that G is an abelian group with dual group Γ, and let I be a closed ideal with infinite codimension in $L^1(G)$. Recall from Exercise 3.6.1 that the hull of I is

$$\mathfrak{h}(I) = \{\gamma \in \Gamma : \widehat{f}(\gamma) = 0 \ (f \in I)\}.$$

Then, by Corollary 10.2.3, $\mathfrak{h}(I)$ is infinite. Since Γ is Hausdorff, there is a sequence (γ_n) of disjoint points in $\mathfrak{h}(I)$ and pairwise disjoint open sets (U_n) with $\gamma_n \in U_n$ for each $n \in \mathbb{N}$. Lemma 10.2.1 assures us that we can choose functions $f_n \in L^1(G)$ such that, for each $n \in \mathbb{N}$, we have $\widehat{f}_n(\gamma_n) = 1$ and $\widehat{f}_n(\gamma) = 0$ for $\gamma \notin U_n$. These functions then satisfy the following for each m and n in $\mathbb{N}$: $f_n^2 \notin I$ because $\widehat{f}_n^2(\gamma_n) = 1$, and $f_m f_n = 0$ whenever $m \neq n$. It follows that $L^1(G)$ satisfies Condition 2 of Theorem 12.1.1 when G is abelian.

Compact groups Suppose next that G is compact, and let I be a closed ideal with infinite codimension in $L^1(G)$. Then infinitely many of the central idempotents z_σ identified in Theorem 10.1.3 do not belong to I. Since they are minimal central idempotents, $z_\sigma z_\tau = z_\sigma$ if $\tau = \sigma$ and $z_\sigma z_\tau = 0$ if $\tau \neq \sigma$. These idempotents show that $L^1(G)$ satisfies Condition 2 of Theorem 12.1.1 when G is compact.

Connected groups Finally, suppose that G is connected, and let I be a closed ideal with infinite codimension in $L^1(G)$. We aim to show that I cannot be the continuity ideal of a derivation. Theorem 8.2.4 tells us that G can be approximated by Lie groups and, after some argument, it may be supposed that G is in fact a connected Lie group. Then G has many 1-parameter subgroups and these are, in particular, abelian.

Let H be a closed abelian subgroup of G. Then $M(H)$ is isomorphic to a subalgebra of $M(G)$, and $L^1(H)$ acts as multipliers on $L^1(G)$ via

$$f \star g(x) = \int_H f(y)g(y^{-1}x)\,\mathrm{d}m_H(y) \quad (a.e. x \in G, f \in L^1(H),\ g \in L^1(G)).$$

Define

$$I_H := \{f \in L^1(H) : f \star g \in I \ (g \in L^1(G))\}.$$

Then I_H is a closed ideal in $L^1(H)$.

Since G is a Lie group, it has abelian subgroups $H_1, \ldots, H_r$ and a compact connected subgroup K such that $G = H_1 \cdots H_r K$; see Iwasawa (1949, Theorem 8). Since K is a Lie group, it has 1-parameter subgroups, say $L_1, \ldots, L_s$, such that $L_1 \cdots L_s$ contains a neighbourhood $\mathcal{O}$ of e. Since K is compact and connected, $\mathcal{O}^n = K$ for some $n \in \mathbb{N}$, and so it may in fact be supposed that there are closed abelian groups $H_1, \ldots, H_t$ such that $G = H_1 \cdots H_t$. (If some H_j is not closed, replace it with its closure. The closure is still an abelian group.) If I_{H_j} had finite codimension in $L^1(H_j)$ for every $j \in \{1, \ldots, t\}$, then it would follow that I had finite codimension in $L^1(G)$ (see Willis 1992) in contradiction to our hypothesis. Hence I_H has infinite codimension in $L^1(H)$ for some closed abelian subgroup H.

Choose a closed abelian subgroup H so that I_H has infinite codimension in $L^1(H)$. Then there are sequences of functions (a_n) and (b_n) in $L^1(H)$ satisfying:

2′: $(b_n a_1 \ldots a_{n-1}) \star L^1(G) \not\subset I$, but $(b_n a_1 \ldots a_n) \star L^1(G) \subset I$ for $n \geq 2$.

These sequences may then be used, exactly as Condition 2 is used, to show that I cannot be the continuity ideal of a derivation.

The above comments show that the continuity ideal of a derivation D from $L^1(G)$ has finite codimension in $L^1(G)$ when G is abelian, compact, or connected. Proposition 12.2.1 may then be used to show that the restriction of D to its continuity ideal is continuous; see Exercise 12.4.3. Hence we have shown the following result.

Proposition 12.3.1 *Let G be a locally compact group which is either abelian, compact or connected, and let X be a Banach $L^1(G)$-bimodule. Then every derivation $D : L^1(G) \to X$ is continuous.* □

It can also be shown that all derivations are continuous for solvable and for locally finite groups and for groups satisfying certain finiteness conditions, such as $[FD]^-$-groups. Proofs of some of these cases can be found in Dales (2000) and

Willis (1986*a*, 1992). Moreover, it can be shown that, if there is a discontinuous derivation from $L^1(G)$ for some locally compact group G, then there is one where G is discrete. Since every discrete group is the quotient of a free group, there would then be a discontinuous derivation from a free-group algebra.

Proposition 12.3.2 *Assume that there is a locally compact group G with a discontinuous derivation $D : L^1(G) \to X$. Then there is a discontinuous derivation*

$$D : \ell^1(\mathbb{F}_\infty) \to \ell^\infty(\mathbb{F}_\infty \times \mathbb{F}_\infty). \qquad \square$$

A partial proof of this reduction is given in Willis (1986).

Apparent progress can be made towards a proof that all derivations from $\ell^1(\mathbb{F}_\infty)$ are continuous. Free groups have many infinite abelian subgroups because every non-identity element generates a subgroup isomorphic to $\mathbb{Z}$. The same argument as used above for abelian groups shows that, if $D : \ell^1(\mathbb{F}_\infty) \to X$ is a derivation, then $\mathcal{I}(D) \cap \ell^1(H)$ has finite codimension in $\ell^1(H)$ for every such abelian subgroup H. The continuity ideal is therefore very large, but it does not follow that it has finite codimension in $\ell^1(\mathbb{F}_\infty)$. This apparent progress is misleading because there are groups for which our knowledge of automatic continuity techniques and of the group algebra are not adequate to show that the continuity ideal is non-zero.

W. Burnside asked in 1902 whether every finitely generated group in which every element satisfies $x^n = e$ for some fixed $n \in \mathbb{N}$ must be finite. A negative solution was announced by P. S. Novikov in 1959 and a proof was published, for odd $n \geq 4381$, by Novikov and S. I. Adian in 1968. Adian subsequently improved the value of the exponent n, and established more results about the counter-examples; see Adian (1979). Let $B(m, n)$ denote the free periodic group with m generators for which $x^n = e$ for every x in the group. Adian showed that $B(m, n)$ is infinite when $m \geq 2$ and $n \geq 665$ is odd. Furthermore, every abelian subgroup of $B(m, n)$ is cyclic of finite order (Adian 1979). Hence the group algebra of each abelian subgroup is finite-dimensional, and automatic continuity techniques give no information about the continuity ideal.

The following problem is a weaker, functional-analytic version of Burnside's original question.

Problem 12.3.3 (i) *Let $f \in \ell^1(B(m, n))$. Is the spectrum of f always totally disconnected?*

(ii) *Let I be an ideal in $\ell^1(\mathbb{F}_\infty)$ such that $\ell^1(H) \cap I$ has finite codimension in $\ell^1(H)$ for every abelian subgroup H of $\mathbb{F}_\infty$. Is the spectrum of $f + I$ totally disconnected for every $f \in \ell^1(G)$?*

A positive answer, or at least a positive answer for sufficiently many f, would show that the continuity ideal of a derivation has finite codimension in $\ell^1(G)$, and hence that every derivation from a group algebra is continuous.

12.4 Exercises

1. Let $D : A \to X$ be a derivation, and let

$$\mathcal{I}(D) := \{f \in A : f \cdot \mathfrak{S}(D) = \{0\}\}$$

be its continuity ideal. Show that $\mathcal{I}(D)$ is a *closed* two-sided ideal in A.

2. Let A be a Banach algebra, and suppose there are elements f_n in A with $f_m f_n = 0$ if $m \neq n$ and $f_n^2 \neq 0$ whenever $m, n \in \mathbb{N}$. Construct sequences of elements (a_n) and (b_n) in A such that $b_n a_1 \cdots a_{n-1} \neq 0$, but $b_n a_1 \cdots a_n = 0$ for $n \geq 2$.
3. Let $D : A \to X$ be a derivation, and let $\mathcal{I}(D)$ be its continuity ideal. Show that, if $\mathcal{I}(D)$ has a bounded approximate identity, then the restriction of D to $\mathcal{I}(D)$ is continuous.

12.5 Additional notes

1. It is easy to see that, if X is an A-bimodule, then $A \oplus X$ is a Banach algebra under the product defined by

$$(a, x)(b, y) = (ab, a \cdot y + x \cdot b) \quad (a, b \in A,\ x, y \in X),$$

and that, if $D : A \to X$ is a derivation, then the map $\Theta : A \to A \oplus X$ defined by $\Theta(a) = (a, D(a))$ is a homomorphism. If D is discontinuous, so is Θ. Hence techniques for showing that homomorphisms from an algebra are continuous can be expected to apply to derivations too.

V. Runde has studied extensively the question of automatic continuity of homomorphisms from group algebras. He has shown that, if G is an $[FIA]^-$-group which for some integer n has an infinite number of inequivalent n-dimensional unitary representations, then (assuming the continuum hypothesis) there is a discontinuous homomorphism from $L^1(G)$ (Runde 1994). Abelian groups fall into this class, and so there are indeed discontinuous homomorphisms from some group algebras. Runde conjectures that the existence of infinitely many inequivalent irreducible representations of G is both a necessary and sufficient condition for there to be discontinuous homomorphisms from $L^1(G)$.

This changes the emphasis of work on the homomorphisms question somewhat. Runde has shown that there are groups for which all homomorphisms are continuous, and has examined whether discontinuous homomorphisms from $L^1(G)$ correspond to discontinuous homomorphisms from its centre; see Runde (1994, 1996, 1997).

2. The question of the automatic continuity of derivations is a particular case of the question of comparison of the continuous and algebraic cohomology theories of Banach algebras. Let A be a Banach algebra, and let M be a Banach A-bimodule. Let $H^q(A, M)$ denote the algebraic cohomology group and $\mathcal{H}^q(A, M)$ the continuous group. Then there is a natural *comparison map*,

$$i^q : \mathcal{H}^q(A, M) \to H^q(A, M).$$

When $q = 1$, this map is injective, and automatic continuity of derivations says that i^1 is surjective as well.

M. Wodzicki (1991) has examined the comparison map for higher values of q. He finds that $i^q = 0$ for every Banach bimodule M provided that A is an amenable Banach algebra and $q \geq n + 3$, where the cardinality of A is equal to $\aleph_n$. Thus there often is no relation between the continuous and algebraic cohomologies for large values of q, in sharp contrast to the case where $q = 1$ when the two cohomology groups are often isomorphic to each other.

References

Adian, S. I. (1979). *The Burnside problem and identities in groups*, Berlin-Heidelberg-New York, Springer-Verlag.

Bonsall, F. F. and Duncan, J. (1973). *Complete normed algebras*, Berlin-Heidelberg-New York, Springer-Verlag.

Brown, G. and Moran, W. (1976). Point derivations on $M(G)$, *Bull. London Math. Soc.*, **8**, 57–64.

Brown, G. and Moran, W. (1978). Analytic discs in the maximal ideal space of $M(G)$, *Pacific J. Math.*, **75**, 45–57.

Choquet, G. and Deny, J. (1961). Sur l'équation de convolution $\mu = \mu \star \sigma$, *Comptes Rendus de l'Academie des Sciences, Paris*, **250**, 794–801.

Chou, C. (1976). The exact cardinality of the set of invariant means on a group, *Proc. American Math. Soc.*, **55**, 103–6.

Cohen, P. J. (1959). Factorization in group algebras, *Duke Math. J.*, **26**, 199–205.

Cohen, P. J. (1960). On homomorphisms in group algebras, *American J. Math.*, **82**, 213–26.

Curtis, P. C., Jr and Loy, R. J. (1989). The structure of amenable Banach algebras, *J. London Math. Soc.* (2), **40**, 89–104.

Dales, H. G. (2000). *Banach algebras and automatic continuity*, London Mathematical Society Monographs, **24**, Oxford, Clarendon Press.

van Dantzig, D. (1936). Zur topologisches Algebra III, Brouwersche and Cantorsche Gruppen, *Compositio Math.*, **3**, 408–26.

Diaconis, P. and Saloff-Coste, L. (1995). Random walks on finite groups: a survey of analytic techniques. In *Probability measures on groups and related structures, XI* (Oberwolfach, 1994), River Edge, NJ, World Scientific Publishing, pp. 44–75.

Diaconis, P. and Saloff-Coste, L. (1996). Walks on generating sets of abelian groups, *Probability Theory and Related Fields*, **105**, 393–421.

Dieudonné, J. Sur une propriété des groupes libres, *Journal für die reine und angewandte Mathematik*, **204** (1960), 30–4.

Dunford, N. and Schwartz, J. T. (1958). *Linear operators I*, New York, John Wiley.

Engelking, R. (1977). *General topology*, Warsaw, Polish Scientific Publishers.

Eymard, P. (1964). L'algèbre de Fourier d'un groupe localement compact, *Bull. Soc. Math. France*, **92**, 181–236.

Forrest, B. (1992). Amenability and ideals in $A(G)$, *J. Australian Math. Soc. (Series A)*, **53**, 143–55.

Galindo, J. (2000). Relations between locally compact abelian groups with isomorphic group algebras, *J. London Math. Soc.* (2), **61**, 110–22.

Gleason, A. M. (1952). Groups without small subgroups, *Annals of Math.*, **56**, 193–212.

Glöckner, H. and Neeb, K.-H. (2000). Minimally almost periodic abelian groups and commutative W^*-algebras, *Research and Exposition in Mathematics*, **24**, 163–85.

Greenleaf, F. P. (1969). *Invariant means on topological groups*, New York, van Nostrand.

Grigorchuk, R. I. (1984). Degrees of growth of finitely generated groups and the theory of invariant means, (Russian) *Izv. Akad. Nauk SSSR Ser. Mat.*, **48**, 939–85.

Grigorchuk, R. I. (1998). An example of a finitely presented amenable group that does not belong to the class EG, (Russian) *Mat. Sb.*, **189**, 79–100; translation in *Sb. Math.*, **189**, 75–95.

Grønbæk, N. and Willis, G. A. (1997). Embedding nilpotent finite-dimensional Banach algebras into amenable Banach algebras, *J. Functional Analysis*, **145**, 99–107.

Halmos, P. R. (1950). *Measure theory*, New York, D. van Nostrand.

Helemskii, A. Ya. (1989). *The homology of Banach and topological algebras*, Dordrecht, Kluwer.

Helson, H. (1983). *Harmonic analysis*, London, Addison-Wesley.

Herz, C. S. (1958). Spectral synthesis for the circle, *Annals of Math.* (2), **68**, 709–12.

Hewitt, E. and Ross, K. A. (1979). *Abstract harmonic analysis, Volume I*, (2nd edition) Berlin–Heidelberg–New York, Springer-Verlag.

Iwasawa, K. (1949). On some types of topological groups, *Annals of Math.*, **50**, 507–58.

Jacobson, N. (1980). *Basic Algebra II*, San Francisco, W. H. Freeman.

Jewell, N. P. (1977). Continuity of module and higher derivations, *Pacific J. Math.*, **68**, 91–8.

Johnson, B. E. (1972). Cohomology in Banach algebras, *Memoirs American Math. Soc.*, **127**.

Johnson, B. E. (2001). The derivation problem for group algebras of connected locally compact groups, *J. London Math. Soc.* (2), **63**, 441–52.

Kadison, R. V. and Ringrose, J. R. (1983, 1986). *Fundamentals of the theory of operator algebras I and II*, Orlando, Academic Press.

Kaĭmanovič, V. A. and Veršik, A. M. (1983). Random walks on discrete groups: boundary and entropy, *Annals of Probability*, **11**, 457–90.

Kaĭmanovič, V. A. (1983). Examples of nonabelian discrete groups with non-trivial exit boundary, (Russian. English summary) *Differential geometry, Lie groups and mechanics, V. Zap. Nauchn. Sem. Leningrad. Otdel. Mat. Inst. Steklov*, **123**, 167–84.

Kaĭmanovič, V. A. (1992). Bi-harmonic functions on groups, *Comptes Rendus de l'Academie des Sciences, Paris*, **314**, 259–64.

Kalton, N. J. and Wood, G. V. (1976). Homomorphisms of group algebras with norm less than $\sqrt{2}$, *Pacific J. Math.*, **62**, 439–60.

Kaplansky, I. (1971). *Lie algebras and locally compact groups*, Chicago and London, University of Chicago Press.

Kawada, Y. and Itô, K. (1940). On the probability distribution on a compact group, *Proc. Phys.-Math. Soc. Japan* (3), **22**, 977–98.

Kepert, A. G. (1997). The range of group algebra homomorphisms, *Canadian Math. Bull.*, **40**, 183–92.

Kisyński, J. (2000). On Cohen's proof of the factorization theorem, *Ann. Polon. Math.*, **75**, 177–92.

Liu, T.-S., van Rooij, A. C. M., and Wang, J.-K. (1973). Projections and approximate identities for ideals in group algebras, *Trans. American Math. Soc.*, **175**, 469–82.

Malliavin, P. (1959). Impossibilité de la synthèse spectrale sur les groupes non compacts, *Inst. Hautes Études Sci. Publ. Math.*, **2**, 61–8.

Montgomery, D. and Zippin, L. (1952). Small subgroups of finite-dimensional groups, *Annals of Math.* (2), **56**, 213–41.

Montgomery, D. and Zippin, L. (1955). *Topological transformation groups*, New York, Interscience.

Morris, S. A. (1977). *Pontryagin duality and the structure of locally compact Abelian Groups*, London Mathematical Society Lecture Notes, **29**, Cambridge University Press.

von Neumann, J. (1929). Zur allgemeinen Theorie des Maßes, *Fundamenta Math.*, **13**, 73–116.

Ol'šanskiĭ, A. Ju. (1980). On the question of the existence of an invariant mean on a group, (Russian) *Uspekhi Mat. Nauk*, **35**, 199–200.

Palmer, T. W. (2001). *Banach algebras and the general theory of ∗-algebras: Volume II, ∗-algebras*, Cambridge University Press.

Paterson, A. L. T. (1988). *Amenability*, Providence, Rhode Island, American Math. Soc.

Pier, J.-P. (1984). *Amenable locally compact groups*, New York, John Wiley.

Pontryagin, L. (1946). *Topological groups*, Princeton University Press.

Reiter, H. (1968). Sur certains idéaux dans $L^1(G)$, *Comptes Rendus de l'Academie des Sciences, Paris*, **267**, 882–5.

Reiter, H. and Stegeman, J. D. (2000). *Classical harmonic analysis and locally compact groups*, London Mathematical Society Monographs, **22**, Oxford, Clarendon Press.

Rickert, N. W. (1967). Amenable groups and groups with the fixed point property, *Trans. American Math. Soc.*, **127**, 221–32.

Ringrose, J. R. (1972). Automatic continuity of derivations of operator algebras, *J. London Math. Soc.* (2), **5**, 432–8.

Rosenblatt, J. M. (1981). Ergodic and mixing random walks on locally compact groups, *Math. Ann.*, **257**, 31–42.

Rosenblatt, J. M. and Willis, G. A. (2001). Weak convergence is not strong convergence for amenable groups, *Canadian Math. Bull.*, **44**, 231–41.

Rudin, W. (1957). Factorization in the group algebra of the real line, *Proc. National Academy Sci. U.S.A.*, **43**, 339–40.

Rudin, W. (1962). *Fourier analysis on groups*, New York, Wiley.

Rudin, W. (1966). *Real and complex analysis* (2nd edn), New York, McGraw-Hill.

Rudin, W. (1972). Invariant means on L^∞, *Studia Mathematica*, **44**, 219–27.

Runde, V. (1994). Homomorphisms from $L^1(G)$ for $G \in [\mathrm{FIA}]^- \cup [\mathrm{Moore}]$, *J. Functional Analysis*, **122**, 25–51.

Runde, V. (1998). The structure of contractible and amenable Banach algebras. In E. Albrecht and M. Mathieu (eds.), *Banach Algebras '97*, Berlin, Walter de Gruyter, 415–30.

Runde, V. (1994). When is there a discontinuous homomorphism from $L^1(G)$?, *Studia Mathematica*, **110**, 97–104.

Runde, V. (1996). Intertwining operators over $L^1(G)$ for $G \in [\mathrm{PG}] \cap [\mathrm{SIN}]$, *Math. Z.*, **221**, 495–506.

Runde, V. (1997). Automatic continuity over Moore groups, *Monatsh. Math.* **123**, 245–52.

Schwartz, L. (1948). Sur une proprieté de synthèse spectrale dans les groupes non compacts, *Comptes Rendus de l'Academie des Sciences, Paris*, **227**, 424–6.

Selivanov, Yu. V. (1976). Banach algebras of small global dimension zero, *Uspekhi Mat. Nauk*, **31**, 227–8.

Tarski, A. (1938). Algebraische Fassung des Maßproblems, *Fundamenta Math.*, **31**, 47–66.

Taylor, J. L. (1972). Homology and cohomology for topological algebras, *Advances in Math.*, **9**, 137–82.

Taylor, J. L. (1973). *Measure algebras*, CBMS Regional Conference Series in Mathematics, **16**, Providence, Rhode Island, American Mathematical Society.

Wagon, S. (1993). *The Banach–Tarski paradox*, Cambridge University Press.

Weil, A. (1951). *L'integration dans les groupes topologiques et ses applications* (2nd edition), Actual Sci. Ind. no. **1145**, Paris, Hermann.

Weil, A. (1955). *Basic number theory*, Berlin–Heidelberg–New York, Springer-Verlag.

Wendel, J. G. (1952). Left centralizers and isomorphisms of group algebras, *Pacific J. Math.*, **2**, 251–61.

Wiener, N. (1932). Tauberian theorems, *Annals of Math.*, **33**, 1–100.

Willis, G. A. (1982). Approximate units in finite codimensional ideals of group algebras, *J. London Math. Soc.* (2), **26**, 143–54.

Willis, G. A. (1986*a*). The continuity of derivations and module homomorphisms, *J. Australian Math. Soc.* (*Series A*), **40**, 299–320.

Willis, G. A. (1986*b*). Translation invariant functionals on $L^p(G)$ when G is not amenable, *J. Australian Math. Soc.* (*Series A*), **41**, 237–50.

Willis, G. A. (1988). Continuity of translation invariant functionals on $C_0(G)$ for certain locally compact groups G, *Monatsh. für Math.*, **105**, 161–4.

Willis, G. A. (1990). Probability measures on groups and some related ideals in group algebras, *J. Functional Analysis*, **92**, 202–63.

Willis, G. A. (1992). The continuity of derivations from group algebras: factorizable and connected groups, *J. Australian Math. Soc.* (*Series A*), **52**, 185–204.

Willis, G. A. (2001). Factorization in finite-codimensional ideals of group algebras, *Proc. London Math. Soc.* (3), **82**, 676–700.

Willis, G. A. (1994). The structure of totally disconnected, locally compact groups, *Math. Ann.*, **300**, 341–63.

Willis, G. A. (2001). Further properties of the scale function on a totally disconnected group, *J. Algebra*, **237**, 142–64.

Willis, G. A. Tidy subgroups for commuting automorphisms of totally disconnected groups: an analogue of simultaneous triangularisation of matrices, preprint, University of Newcastle, NSW.

Wodzicki, M. (1991). Resolution of the cohomology comparison problem for amenable Banach algebras, *Inventiones Math.*, **106**, 541–7.

Wood, G. V. (1989). Small isomorphisms between group algebras, *Glasgow Math. J.*, **33**, 21–8.

Yamabe, H. (1953). A generalization of a theorem of Gleason, *Annals of Math.* (2), **58**, 351–65.

Part III

Invariant subspaces

JÖRG ESCHMEIER
Universität des Saarlandes, Saarbrücken, Germany

13

Compact operators

13.1

Let $T : V \to V$ be a linear operator on a finite-dimensional, complex linear space V. Then T possesses eigenvalues. This means that there is at least one complex number λ such that $\ker(\lambda - T) \neq \{0\}$. Since the latter space is invariant under any linear operator $A : V \to V$ commuting with T, it follows that either $T = \lambda I$ or there is a non-trivial linear subspace of V invariant under any operator in the commutant of T. Here a linear subspace M of V is called non-trivial if $M \neq \{0\}$ and $M \neq V$.

A famous question in operator theory, the *invariant subspace problem*, asks if, more generally, for each continuous linear operator T on an infinite-dimensional complex, separable Banach space (or Hilbert space) X there is a non-trivial, closed linear subspace of X that is invariant under T. This question has guided a considerable amount of research in operator theory. In the beginning of the 1980s first counter-examples were given by C. J. Read and P. Enflo in the case of operators on certain non-reflexive Banach spaces. In the Hilbert-space case, the question seems to be still open at the time of this writing.

Of course, in the infinite-dimensional case, eigenvalues need no longer exist as simple examples show such as the unilateral shift

$$S : \ell^2 \to \ell^2, \quad S(x_k) = (0, x_0, x_1, \dots),$$

or the Volterra operator

$$V : C([0, 1]) \to C([0, 1]), \quad (Vf)(t) = \int_0^t f(s)\,\mathrm{d}s\,,$$

H. G. Dales, P. Aiena, J. Eschmeier, K. B. Laursen, and G. A. Willis, *Introduction to Banach Algebras, Operators, and Harmonic Analysis*. Published by Cambridge University Press.

which is described in Part I, Exercise 1.5.6 and Part IV, Example 22.2.6. A class of operators that is, in many respects, still close to operators on finite-dimensional spaces is given by the compact operators. Any non-zero complex number λ in the spectrum of a compact operator is an eigenvalue for the operator such that the corresponding eigenspace is finite-dimensional (see Part I, Theorem 2.2.5). Hence the invariant subspace problem for compact operators is immediately reduced to the case of quasi-nilpotent compact operators, that is, compact operators T with spectrum $\sigma(T) = \{0\}$ (see Part I, 2.1.2).

In 1954 it was shown by Aronszajn and Smith that each compact operator on an infinite-dimensional Banach space X has a non-trivial, closed invariant subspace. Results of Bernstein and Robinson (1966) and Halmos (1966) show that each bounded operator T on X such that $p(T)$ is compact for some non-zero polynomial p has a non-trivial closed invariant subspace.

The aim of this chapter is to prove an invariant subspace result for compact operators due to Lomonosov (1973) that contains the above results as special cases.

Throughout Chapters 13–20 we shall write $\mathcal{B}(X)$ for the Banach algebra of all bounded linear operators on a (complex) Banach space X. Here the norm on $\mathcal{B}(X)$ is the operator norm. We denote the closed unit ball of X by $X_{[1]}$.

Definition 13.1.1 *A continuous linear operator $K \in \mathcal{B}(X)$ on a Banach space X is* compact *if $K(X_{[1]})$ is a relatively compact subset of X.*

In 1973, Lomonosov proved the existence of non-trivial hyperinvariant subspaces for each operator $T \in \mathcal{B}(X) \setminus \mathbb{C}1_X$ that commutes with a non-zero compact operator; we first describe the elegant proof of this theorem.

Definition 13.1.2 *Let $T \in \mathcal{B}(X)$, and let $\mathcal{M} \subset \mathcal{B}(X)$ be arbitrary.*

(a) *A* closed invariant subspace *for T is a closed linear subspace $Y \subset X$ such that $TY \subset Y$. We write*

$$\mathrm{Lat}(T) = \{Y;\ Y \text{ is a closed invariant subspace for } T\}\,.$$

(b) *More generally, we define*

$$\mathrm{Lat}(\mathcal{M}) = \bigcap\{\mathrm{Lat}(A);\ A \in \mathcal{M}\}\,.$$

(c) *The spaces contained in*

$$\mathrm{Hyp}(T) = \bigcap\{\mathrm{Lat}(A);\ A \in (T)'\}\,,$$

where $(T)' = \{A \in \mathcal{B}(X);\ AT = TA\}$ is the commutant *of T, are called the* hyperinvariant subspaces *for T.*

The following is the result of Lomonosov.

Theorem 13.1.3 (Lomonosov) *Let $T \in \mathcal{B}(X) \setminus \mathbb{C}1_X$. Suppose that T commutes with a non-zero compact operator. Then* Hyp(T) *is non-trivial.*

For the proof of Lomonosov's theorem we shall need a well-known fixed point theorem.

Theorem 13.1.4 (Brouwer) *Let $K \subset \mathbb{R}^n$ be a compact and convex set. Each continuous map $f : K \to K$ has a fixed point, i.e., there is at least one point $x \in K$ with $f(x) = x$.* □

A proof of this result can be found in almost every book on algebraic topology. We are mainly interested in the following infinite-dimensional generalization due to Schauder.

Corollary 13.1.5 (Schauder) *Let X be a normed space. Suppose that $B \subset X$ is convex and that $C \subset B$ is a compact subset. Then each continuous map $f : B \to C$ has a fixed point.*

Proof We reduce Schauder's result to Brouwer's fixed point theorem. To prove the corollary, it suffices to find a sequence (x_k) in B with

$$\|f(x_k) - x_k\| \xrightarrow{k} 0 .$$

Indeed, if y is the limit of a subsequence of $(f(x_k))$, then obviously $f(y) = y$.

Let $\varepsilon > 0$. Since C is compact, there are $c_1, \ldots, c_r \in C$ with

$$C \subset \bigcup_{i=1}^{r} B_\varepsilon(c_i) .$$

Note that the convex hull $K := \langle \{c_1, \ldots, c_r\} \rangle$ is homeomorphic to a convex compact set in $\mathbb{R}^n$. The functions $f_i : C \to [0, \infty)$ and $F : C \to K$ which are defined by $f_i(x) = \max(0, \varepsilon - \|x - c_i\|)$ and

$$F(x) = \sum_{i=1}^{r} \frac{f_i(x)}{\sum_j f_j(x)} c_i$$

are both continuous with $\|F(x) - x\| < \varepsilon$ for all $x \in C$.

To complete the proof, observe that, by Brouwer's fixed point theorem, the map

$$T : K \to K, \quad T(x) = F \circ f(x) ,$$

has a fixed point $z \in K$ and that, for any such point z, we have

$$\|f(z) - z\| = \|f(z) - F(f(z))\| < \varepsilon .$$

Hence f has a fixed point. □

As a consequence of Schauder's fixed point theorem we obtain the so-called Lomonosov lemma.

Lemma 13.1.6 (Lomonosov) *Let $\mathfrak{A} \subset \mathcal{B}(X)$ be a subalgebra with $1 \in \mathfrak{A}$, and suppose that*

$$\mathrm{Lat}(\mathfrak{A}) = \{\{0\}, X\} .$$

Then, for any compact operator $K \in \mathcal{B}(X) \setminus \{0\}$, there are an operator $A \in \mathfrak{A}$ and a vector $x \in X \setminus \{0\}$ such that $KAx = x$.

Proof Choose a vector $y \in X$ such that the compact and convex set $C := \overline{KB_1(y)}$ does not contain the zero vector $0 \in X$. Here we write $B_1(y)$ for the open ball with radius 1 and centre y in X. Since $\mathrm{Lat}(\mathfrak{A})$ is trivial, it follows that $\overline{\mathfrak{A}x} = X$ for all $x \in X \setminus \{0\}$. Hence the compactness of C allows us to choose finitely many operators $A_1, \ldots, A_n \in \mathfrak{A}$ with the property that

$$C \subset \bigcup_{i=1}^{n} A_i^{-1}(B_1(y)) .$$

Fix any continuous map $r : \mathbb{R}_+ \to \mathbb{R}_+$ with $r^{-1}(0) = [1, \infty)$. Then the composition $F : C \xrightarrow{f} B_1(y) \xrightarrow{K} C$, where

$$f(x) = \sum_{i=1}^{n} \frac{r(\|A_i x - y\|)}{\sum_j r(\|A_j x - y\|)} A_i x ,$$

has a fixed point x by Schauder's fixed point theorem. But then we see immediately that $x = F(x) = KAx$ for a suitable operator $A \in \mathfrak{A}$. □

Lomonosov's lemma and elementary linear algebra yield a proof of Lomonosov's invariant subspace theorem.

Proof of Theorem 13.1.3 Under the hypotheses of Theorem 13.1.3, assume towards a contradiction that the algebra $\mathfrak{A} = (T)'$ has no non-trivial invariant subspace. Then, by Lomonosov's lemma, there are $A \in \mathfrak{A}$ and $0 \neq x \in X$ with

$$\ker(1 - KA) \neq \{0\}.$$

Since KA is compact and commutes with T, the space on the left is a finite-dimensional invariant subspace for T. But then T possesses eigenvalues. This is impossible, since any eigenspace for T belongs to $\mathrm{Lat}(\mathfrak{A})$. □

A weaker version of Theorem 13.1.3 can be proved without using a fixed point argument. The spectral radius argument used in the following proof is due to M. Hilden.

Theorem 13.1.7 *Each compact operator $K \in \mathcal{B}(X) \setminus \{0\}$ on an infinite-dimensional Banach space X has a non-trivial hyperinvariant subspace.*

Proof Set $\mathfrak{A} = (K)'$. Assume towards a contradiction that $\mathrm{Lat}(\mathfrak{A})$ is trivial. Choose y, C, and $A_1, \ldots, A_n$ with respect to $\mathfrak{A}$ and the compact operator K exactly as in the proof of Lemma 13.1.6.

Then $Ky \in C$. Hence $A_{i_1}Ky \in B_1(y)$ for some $1 \leq i_1 \leq n$. But then $KA_{i_1}Ky \in C$, and again $A_{i_2}KA_{i_1}Ky \in B_1(y)$ for some $1 \leq i_2 \leq n$. Continuing in this way, we obtain vectors

$$y_N = A_{i_N} \cdots A_{i_1} K^N y \in B_1(y).$$

Define $s = \max(\|A_1\|, \ldots, \|A_n\|)$. Then

$$r \leq \|y_N\| \leq s^N \|K^N\| \|y\| \quad (N \in \mathbb{N})$$

holds with a suitable real number $r > 0$.

We conclude that the spectral radius (see Proposition 2.1.3)

$$\nu(K) = \lim_{N \to \infty} \|K^N\|^{1/N} = \max\{|z|;\ z \in \sigma(K)\}$$

is positive. Since any non-zero element in the spectrum of a compact operator is an eigenvalue, we obtain as before the contradiction that $\mathrm{Lat}(\mathfrak{A})$ is non-trivial. □

13.2 Exercises

1. Let X, Y be Banach spaces, and let $T \in \mathcal{B}(X, Y)$. Show that T is compact if and only if $T' \in \mathcal{B}(Y', X')$ is compact.

2. Let $K \in \mathcal{B}(X)$ be compact. Show that, if λ is a non-zero complex number such that $\lambda - K$ is not bounded below, then λ is an eigenvalue for K. Conclude that each non-zero point $\lambda \in \sigma(K)$ is an eigenvalue for K or for K'. Deduce that on an infinite-dimensional Banach space a compact operator without any non-trivial hyperinvariant subspace is quasi-nilpotent.

3. Let $T \in \mathcal{B}(X)$ be compact. Show that each sequence (λ_n) of pairwise distinct eigenvalues for T converges to zero. Conclude that each point $\lambda \in \sigma(T) \setminus \{0\}$ is isolated in $\sigma(T)$ and is an eigenvalue for T.

4. Use Lomonosov's lemma to show that each norm-closed, unital subalgebra $\mathfrak{A}$ of $\mathcal{B}(X)$ that contains a non-zero compact operator K and possesses no non-trivial closed invariant subspaces contains a non-zero finite-rank operator.

5. Let $T \in \mathcal{B}(X) \setminus \mathbb{C}1_X$ be a bounded operator such that $p(T)$ is a compact operator for some non-zero polynomial $p \in \mathbb{C}[z]$. Show that T has a non-trivial hyperinvariant subspace.

13.3 Additional notes

At the time when Lomonosov's paper appeared, the question whether each bounded operator on a complex and separable infinite-dimensional Banach space possesses a non-trivial closed invariant subspace was still open. Counter-examples on certain non-reflexive Banach spaces were given by C. J. Read (1984) and P. Enflo (1987) (see also Beauzamy 1985). In the Hilbert-space case the question is still open at the time of this writing. Theorem 13.1.3 is from Lomonosov (1973).

Lomonosov's result has been applied and improved in several directions. In 1974, Pearcy and Shields used Lomonosov's result to show that each unital subalgebra $\mathfrak{A}$ of $\mathcal{B}(X)$ without non-trivial closed invariant subspaces that contains a non-zero compact operator has to be strongly dense in $\mathcal{B}(X)$. This gives a positive answer to the so-called transitive algebra conjecture in a special case. In 1975, Kim, Pearcy, and Shields proved that each non-scalar operator $T \in \mathcal{B}(X)$ for which there is a non-zero compact operator K such that $\mathrm{im}(TK - KT)$ is at most one-dimensional has a non-trivial invariant subspace. Other generalizations have been obtained by Fong, Nordgren, Radjabalipour, Radjavi and Rosenthal.

14
Unitary dilations and the H^∞-functional calculus

14.1

In Chapters 14–20 we shall describe and apply a second general method to construct invariant subspaces for certain classes of operators on Banach or Hilbert spaces. This method is due to Scott Brown, who used it in 1978 in his thesis to prove the existence of non-trivial, closed invariant subspaces for subnormal operators. All applications of this so-called *Scott Brown technique* depend on the existence of a suitable H^∞-functional calculus for the operators under consideration. We begin by studying a classical case in which the existence of an H^∞-functional calculus is well-known.

Let H be a complex Hilbert space. Throughout Chapters 14–20, we shall denote by $[x, y]$ the inner product of given vectors $x, y \in H$. For an open set V in $\mathbb{C}$, we write $H(V)$ for the Fréchet space of all complex-valued analytic functions on V equipped with the topology of uniform convergence on all compact subsets of V.

A famous theorem of Sz.-Nagy from 1953 says that every contraction T on H possesses a unitary dilation.

Theorem 14.1.1 *Let $T \in \mathcal{B}(H)$ be a contraction. Then there exists a Hilbert space K containing H and a unitary operator $U \in \mathcal{B}(K)$ such that*

$$T^k = PU^k|H \quad (k \in \mathbb{N}),$$

where P denotes the orthogonal projection from K onto H.

Proof The orthogonal direct sum

$$\mathcal{H} = \bigoplus_{n\in\mathbb{Z}} H = \left\{ (h_n)_{n\in\mathbb{Z}};\ h_n \in H \quad \text{and} \quad \|(h_n)\|^2 = \sum_{n\in\mathbb{Z}} \|h_n\|^2 < \infty \right\}$$

is a Hilbert space relative to the inner product defined by

$$[(h_n),(k_n)] = \sum_{n\in\mathbb{Z}} \lfloor h_n, k_n \rfloor .$$

The operators $D = \sqrt{1 - T^*T}$ and $D_* = \sqrt{1 - TT^*}$ satisfy the relations

$$TD = D_*T \quad \text{and} \quad T^*D_* = DT^* .$$

Indeed, since $T(1 - T^*T) = (1 - TT^*)T$, we obtain that

$$Tp(1 - T^*T) = p(1 - TT^*)T$$

holds for each polynomial p. But then the Stone–Weierstrass theorem implies that

$$Tf(1 - T^*T) = f(1 - TT^*)T$$

for each continuous function $f \in C([0, 1])$. Choosing $f(t) = \sqrt{t}$, we obtain the first intertwining relation. The second follows analogously.

As a consequence we remark that the spaces $\mathcal{D} = \overline{DH}$ and $\mathcal{D}_* = \overline{D_*H}$ satisfy the inclusions

$$T\mathcal{D} \subset \mathcal{D}_* \quad \text{and} \quad T^*\mathcal{D}_* \subset \mathcal{D} .$$

We regard H as a subspace of the Hilbert space

$$K = \{(h_n)_{n\in\mathbb{Z}} \in \mathcal{H};\ h_0 \in H,\ h_n \in \mathcal{D} \text{ and } h_{-n} \in \mathcal{D}_* \text{ for all } n \in \mathbb{N}\}$$

via the canonical isometry $H \to K$ defined by

$$h \mapsto (\ldots, 0, [h], 0, \ldots),$$

where the square bracket indicates the zero-th position.

We claim that the linear operator $U : K \to K$ defined by

$$Uh = (\ldots, h_{-3}, h_{-2}, [D_*h_{-1} + Th_0], -T^*h_{-1} + Dh_0, h_1, h_2, \ldots)$$

is a unitary dilation of T. To see that U is isometric, note that all the terms in

$$\begin{aligned} \|h\|^2 - \|Uh\|^2 = {} & \|h_{-1}\|^2 + \|h_0\|^2 \\ & -[D_*h_{-1} + Th_0,\ D_*h_{-1} + Th_0] \\ & -[-T^*h_{-1} + Dh_0,\ -T^*h_{-1} + Dh_0] \end{aligned}$$

cancel each other. To check that U is surjective, note that the linear operator $V : K \to K$ defined by

$$Vh = (\ldots, h_{-1}, D_* h_0 - T h_1, [T^* h_0 + D h_1], h_2, h_3, \ldots)$$

yields a right inverse for U.

Identifying H with a subspace of K, an elementary induction shows that

$$U^n h = (\ldots, 0, [T^n h], D T^{n-1} h, \ldots, Dh, 0, \ldots).$$

Thus we have shown that U is a unitary dilation of T. □

A unitary dilation $U \in \mathcal{B}(K)$ of a contraction $T \in \mathcal{B}(H)$ is called *minimal* if the only reducing subspace for U containing H is the space K itself or, equivalently, if

$$K = \bigvee_{n \in \mathbb{Z}} U^n H.$$

Remark 14.1.2

(a) It is elementary to check that the unitary dilation constructed in the proof of Theorem 14.1.1 is minimal.

(b) If $U \in \mathcal{B}(K)$ is a unitary dilation of a given contraction $T \in \mathcal{B}(H)$, then the restriction of U to $K_0 = \bigvee_{n \in \mathbb{Z}} U^n H$ is a minimal unitary dilation of T.

(c) If $U \in \mathcal{B}(K)$ and $V \in \mathcal{B}(\widetilde{K})$ are minimal unitary dilations of $T \in \mathcal{B}(H)$, then the unique continuous linear map $\Gamma : K \to \widetilde{K}$ with

$$\Gamma(U^n h) = V^n h \quad (n \in \mathbb{Z},\ h \in H)$$

is a unitary operator with $\Gamma U = V \Gamma$.

As an important application of the dilation theorem, we obtain a continuity property for the polynomial functional calculus of a contraction.

Corollary 14.1.3 (von Neumann's inequality) *Let $T \in \mathcal{B}(H)$ be a contraction. Then*

$$\|p(T)\| \leq |p|_{\overline{\mathbb{D}}}$$

for all polynomials $p \in \mathbb{C}[z]$.

Proof Let U be a unitary dilation of T. Using the fact that the $C(\sigma(U))$–functional calculus of the normal operator U is isometric (see Part I, Additional notes, 4.5.2), we obtain the inequalities

$$\|p(T)\| = \|Pp(U)|H\| \leq \|p(U)\| = |p|_{\sigma(U)} \leq |p|_{\overline{\mathbb{D}}} . \qquad \square$$

Since the restrictions of the complex polynomials to the unit disc $\mathbb{D}$ are uniformly dense in the disc algebra $A(\overline{\mathbb{D}})$ (see Examples 1.2 (iii)), von Neumann's inequality is equivalent to the existence of a contractive disc algebra functional calculus. For more on the functional calculus, see Chapter 4.

Corollary 14.1.4 *Let $T \in \mathcal{B}(H)$ be a contraction. Then there is a unique contractive algebra homomorphism $\Psi : A(\overline{\mathbb{D}}) \to \mathcal{B}(H)$ extending the polynomial functional calculus of the operator T.* □

We want to describe conditions under which the disc algebra functional calculus of T extends to $H^\infty(\mathbb{D})$.

Let $V \subset \mathbb{C}$ be a non-empty, open set. Then $L^\infty(V)$ formed with respect to the planar Lebesgue measure is the dual space of $L^1(V)$, and hence carries a weak-$*$ topology, denoted by σ. Let $H^\infty(V)$ be the Banach algebra of all bounded analytic functions on V (see Examples 1.2 (xi)).

Lemma 14.1.5 *The space $H^\infty(V)$ is a weak-$*$ closed subspace of $L^\infty(V)$.*

Proof Let B be the closed unit ball of $L^\infty(V)$. By the Krein–Smulian theorem (see Schaefer 1966, Theorem IV.6.4), it suffices to show that $H^\infty(V) \cap B \subset B$ is weak-$*$ closed. Since $L^1(V)$ is separable, (B, σ) is metrizable. Let (f_k) be a sequence in $H^\infty(V) \cap B$ with $\sigma - \lim_k f_k = f$. By Montel's theorem, we may suppose, after passing to a subsequence, that $(f_k) \xrightarrow{k} g \in H^\infty(V)$ uniformly on compact subsets. Since $C_{00}(V)$ is dense in $L^1(V)$ (see Lang 1969, Theorem XII.6), the observation that

$$\int_V \varphi g d\lambda = \lim_k \int_V \varphi f_k d\lambda = \int_V \varphi f d\lambda \qquad (\varphi \in C_{00}(V))$$

implies that $g = f$ almost everywhere. □

As a consequence of the previous result we can identify $H^\infty(V)$ with the dual space of the separable Banach space $Q := L^1(V)/{}^\perp H^\infty(V)$.

Lemma 14.1.6 *A sequence (f_k) in $H^\infty(V)$ is a weak-$*$ zero sequence if and only if $\sup_k \|f_k\| < \infty$ and (f_k) converges to zero pointwise (or equivalently, uniformly on compact subsets of V).*

Proof Let (f_k) be a weak-$*$ zero sequence. By the uniform boundedness principle, (f_k) is norm-bounded. By the argument given in the previous proof, each subsequence of (f_k) has a subsequence converging to zero in the Fréchet space $H(V)$. Hence $\lim_k f_k = 0$ in $H(V)$.

Conversely, suppose that (f_k) is norm-bounded and converges to zero pointwise. By the dominated convergence theorem,

$$\int_V \varphi f_k d\lambda \xrightarrow{k} 0 \quad (\varphi \in C_{00}(V)).$$

Since (f_k) is bounded in $L^1(V)'$, this implies that (f_k) is a weak-$*$ zero sequence. □

Let X be a Banach space. Since the space $H^\infty(V)$ has a separable predual, a linear map $\Phi : H^\infty(V) \to X'$ is weak-$*$ continuous if and only if it is sequentially weak-$*$ continuous (Exercise 14.2.2). Thus in particular all point evaluations

$$\mathcal{E}_\lambda : H^\infty(V) \to \mathbb{C}, \quad f \mapsto f(\lambda) \quad (\lambda \in V),$$

are weak-$*$ continuous.

Let H be a Hilbert space. Then $\mathcal{B}(H)$ can be identified isometrically with the dual space of the Banach space $\mathcal{C}^1(H)$ of all trace-class operators on H via the bilinear form

$$\mathcal{C}^1(H) \times \mathcal{B}(H) \to \mathbb{C}, \quad (A, T) \mapsto Tr(AT).$$

For the definition of the trace and trace-class operators the reader is referred to Conway (1991, Chapter 1). The weak-$*$ topology of $\mathcal{B}(H)$ with respect to this duality is usually called the *ultraweak operator topology*. It is given by the seminorms

$$p_{x,y}(T) = \left| \sum_{k=1}^{\infty} [Tx_k, y_k] \right|,$$

where $x = (x_k)$ and $y = (y_k)$ run through all square summable sequences in H (see Stratila and Zsido 1979, Chapter 1). On the bounded subsets of $\mathcal{B}(H)$, the weak operator topology and the weak-$*$ topology coincide. By definition the *weak operator topology* is the locally convex topology on $\mathcal{B}(H)$ given by the seminorms $q_{x,y}(T) = |[Tx, y]|$ where $x, y \in H$ are arbitrary.

Remark 14.1.7 By the previous discussion a map $\Phi : H^\infty(V) \to \mathcal{B}(H)$ is weak-$*$ continuous if and only if, for each weak-$*$ zero sequence (f_k) in $H^\infty(V)$,

$$\lim_k [\Phi(f_k)x, y] = 0$$

for all $x, y \in H$.

Using the same principle, we obtain a useful characterization of those maps $\Psi : A(V) \to \mathcal{B}(H)$ that extend to a weak-$*$ continuous linear map

$$\Phi : H^\infty(V) \to \mathcal{B}(H).$$

Here $A(V)$ denotes the closed subalgebra of $C(\overline{V})$ consisting of all functions $f \in C(\overline{V})$ for which $f|V$ is analytic; it is related to the uniform algebra $A(\overline{V})$ of Examples 1.2(v).

Definition 14.1.8 *Let V be a non-empty, open set in $\mathbb{C}$.*

(i) *A sequence (f_k) in $A(V)$ is a* Montel sequence *if $(f_k|V)$ is a weak-$*$ zero sequence in $H^\infty(V)$.*

(ii) *The space $A(V)$ is* pointwise boundedly dense *in $H^\infty(V)$ if, for each function f in $H^\infty(V)$, there is a sequence (f_k) in $A(V)$ with $\|f_k\| \le \|f\|$ such that (f_k) is weak-$*$ convergent to f.*

Let (A_k) be sequence of operators in $\mathcal{B}(H)$. We shall write

$$\text{WOT} - \lim_{k\to\infty} A_k = A$$

if (A_k) converges to an operator $A \in \mathcal{B}(H)$ in the weak operator topology, that is, if $\lim_{k\to\infty}[A_k x, y] = [Ax, y]$ for all vectors x, y in H.

Lemma 14.1.9 *Suppose that $A(V)$ is pointwise boundedly dense in $H^\infty(V)$. Then a continuous linear operator $\Psi : A(V) \to \mathcal{B}(H)$ extends to a weak-$*$ continuous linear operator $\Phi : H^\infty(V) \to \mathcal{B}(H)$ if and only if*

$$\text{WOT} - \lim_{k\to\infty} \Psi(f_k) = 0$$

for each Montel sequence (f_k) in $A(V)$.

Proof The necessity of the stated condition is obvious.

Suppose that Ψ satisfies the above continuity property. If $f \in H^\infty(V)$ and if x, y are vectors in H, then, for each sequence (f_k) in $A(V)$ with weak-$*$ limit f, the limit

$$f_{x,y} = \lim_k [\Phi(f_k)x, y]$$

exists and is independent of the choice of (f_k). Since we can achieve that $\|f_k\| \leq \|f\|$, this argument gives an extension of Ψ to a continuous linear map $\Phi : H^\infty(V) \to \mathcal{B}(H)$ with $\|\Phi\| = \|\Psi\|$ and

$$[\Phi(f)x, y] = f_{x,y} \quad (f \in H^\infty(V),\ x, y \in H).$$

Using the facts that the unit ball B of $H^\infty(V)$ is metrizable in the weak-$*$ topology and that $B \cap A(V)$ is weak-$*$ dense in B, one can easily show that, for each weak-$*$ zero sequence (f_k) contained in B,

$$\lim_{k\to\infty} [\Phi(f_k)x, y] = 0 \quad (x, y \in H).$$

Therefore the extension Φ of Ψ is weak-$*$ continuous. □

Note that the extension constructed in the above proof satisfies $\|\Phi\| = \|\Psi\|$. It is elementary (Exercise 14.2.3) to show that Φ remains multiplicative if Ψ is supposed to be multiplicative.

A contraction $T \in \mathcal{B}(H)$ is called *completely non-unitary* if there is no non-zero reducing subspace for T such that T restricted to this subspace is a unitary operator. A classical result of Sz.-Nagy and Foiaş shows that each completely non–unitary contraction T possesses a weak-$*$ continuous $H^\infty(\mathbb{D})$-functional calculus. A proof of this result can be found in Sz.-Nagy and Foiaş (1970, Chapter 3).

We indicate briefly an alternative proof, which can be extended to more general settings (see Eschmeier 1997).

Let $\Psi : A(\overline{\mathbb{D}}) \to \mathcal{B}(H)$ be a contractive algebra homomorphism. By the maximum modulus principle we may regard $A(\overline{\mathbb{D}})$ as a closed linear subspace of $C(\mathbb{T})$ (see Part I, Exercise 1.5.4). By the Hahn–Banach and the Riesz representation theorems, there is a family of measures $\mu(x, y) \in M(\mathbb{T})$ $(x, y \in H)$ such that

$$[\Psi(f)x, y] = \int_{\mathbb{T}} f \, \mathrm{d}\mu(x, y) \quad (f \in A(\overline{\mathbb{D}}))$$

and $\|\mu(x, y)\| \leq \|x\| \, \|y\|$. We call $(\mu(x, y))_{x,y\in H}$ a *representing family* for Ψ.

Let m be the normalized linear Lebesgue measure on $\mathbb{T}$. We call a measure μ in $M(\mathbb{T})$ *absolutely continuous* (*singular*) if $\mu \ll m$ ($\mu \perp m$). By the Lebesgue decomposition theorem, each measure $\mu \in M(\mathbb{T})$ has a unique decomposition

$$\mu = \mu_a + \mu_s$$

into an absolutely continuous part μ_a and a singular part μ_s.

Theorem 14.1.10 (F. and M. Riesz theorem) *Let $\mu \in M(\mathbb{T})$ be a measure with*

$$\widehat{\mu}(-n) = \int_{\mathbb{T}} z^n \, \mathrm{d}\mu = 0$$

for all $n \in \mathbb{N}$. Then $\mu \ll m$. In particular, each measure $\mu \in A(\mathbb{D})^\perp$ is absolutely continuous. □

Definition 14.1.11 *A contractive algebra homomorphism $\Psi : A(\mathbb{D}) \to \mathcal{B}(H)$ is* absolutely continuous (singular) *if it possesses a representing family of measures that are absolutely continuous (singular).*

The Lebesgue decomposition for measures yields a corresponding decomposition for disc algebra representations.

Theorem 14.1.12 *Let $\Psi : A(\overline{\mathbb{D}}) \to \mathcal{B}(H)$ be a contractive algebra homomorphism. Then there are contractive algebra homomorphisms*

$$\Psi_a, \Psi_s : A(\overline{\mathbb{D}}) \to \mathcal{B}(H)$$

such that $\Psi = \Psi_a + \Psi_s$, Ψ_a is absolutely continuous, Ψ_s is singular, and

$$\Psi_a(f)\Psi_s(g) = 0 = \Psi_s(g)\Psi_a(f) \quad (f, g \in A(\overline{\mathbb{D}})).$$

Proof Choose a representing family $\mu(x, y)$ of measures for Ψ. Let

$$\mu(x, y) = \mu_a(x, y) + \mu_s(x, y)$$

be the Lebesgue decomposition. Then the map

$$H \times H \to M(\mathbb{T})/A(\overline{\mathbb{D}})^\perp, \quad (x, y) \mapsto [\mu(x, y)],$$

is sesquilinear. By the F. and M. Riesz theorem, the maps

$$H \times H \to M(\mathbb{T})/A(\overline{\mathbb{D}})^\perp, \quad (x, y) \mapsto [\mu_a(x, y)],$$

$$H \times H \to M(\mathbb{T}), \quad (x, y) \mapsto \mu_s(x, y),$$

are sesquilinear. It follows easily that there are unique contractive linear maps $\Psi_a, \Psi_s : A(\overline{\mathbb{D}}) \to \mathcal{B}(H)$ with

$$[\Psi_a(f)x, y] = \int_{\mathbb{T}} f \mathrm{d}\mu_a(x, y), \quad [\Psi_s(f)x, y] = \int_{\mathbb{T}} f \mathrm{d}\mu_s(x, y).$$

For $f, g \in A(\overline{\mathbb{D}})$ and $x, y \in H$, we have

$$\int_{\mathbb{T}} fg \,\mathrm{d}\mu(x, y) = [\Psi(f)x, \Psi(g)^* y] = \int_{\mathbb{T}} f \,\mathrm{d}\mu(x, \Psi(g)^* y).$$

Again by the F. and M. Riesz theorem and the uniqueness of the Lebesgue decomposition, we have

$$g\mu(x, y) - \mu(x, \Psi(g)^* y) = g\mu_a(x, y) - \mu_a(x, \Psi(g)^* y) \in A(\overline{\mathbb{D}})^\perp.$$

It follows that

$$\int_{\mathbb{T}} fg \,\mathrm{d}\mu_a(x, y) = [\Psi_a(f)x, \Psi(g)^* y] = \int_{\mathbb{T}} g \,\mathrm{d}\mu(\Psi_a(f)x, y).$$

As above, we conclude that $\mu_s(\Psi_a(f)x, y) = 0$ and that

$$f\mu_a(x, y) - \mu_a(\Psi_a(f)x, y) \in A(\overline{\mathbb{D}})^\perp .$$

In this way, we obtain the multiplicativity of Ψ_a. Indeed,

$$[\Psi_a(fg)x, y] = \int_{\mathbb{T}} g \,\mathrm{d}\mu_a(\Psi_a(f)x, y) = [\Psi_a(g)\Psi_a(f)x, y].$$

Furthermore, for $f, g \in A(\overline{\mathbb{D}})$ and $x, y \in H$,

$$[\Psi_s(g)\Psi_a(f)x, y] = \int_{\mathbb{T}} g \,\mathrm{d}\mu_s(\Psi_a(f)x, y) = 0.$$

The remaining parts of Theorem 14.1.12 can be obtained in a similar way. □

Definition 14.1.13 *A measure $\mu \in M(\mathbb{T})$ is a* Henkin measure *if*

$$\int_{\mathbb{T}} f_k \,\mathrm{d}\mu \xrightarrow{k} 0$$

for each Montel sequence (f_k) in $A(\overline{\mathbb{D}})$.

Since $\int f \, dm = f(0)$ for all $f \in A(\overline{\mathbb{D}})$, the Lebesgue measure m is a Henkin measure. If $\mu \in M(\mathbb{T})$ is a Henkin measure, then, by Henkin's theorem (see Rudin 1980, Theorem 9.3.1), each measure $\nu \ll \mu$ is a Henkin measure (cf. Exercise 15.2.4).

It can easily be seen that the representations Ψ_a, Ψ_s in Theorem 14.1.12 are uniquely determined by Ψ (Exercise 14.2.4). By the preceding remarks, and by Lemma 14.1.9, the absolutely continuous part Ψ_a of Ψ extends to a weak-$*$ continuous algebra homomorphism $\Phi_a : H^\infty(\mathbb{D}) \to \mathcal{B}(H)$.

Since the family $(\mu_s(x, y))_{x,y \in H}$ depends in a sesquilinear way on x and y and satisfies $\|\mu_s(x, y)\| \le \|x\| \, \|y\|$, we can define a contractive linear extension $\Phi_s : BM(\mathbb{T}) \to \mathcal{B}(H)$ of Ψ_s by the formula

$$[\Phi_s(f)x, y] = \int_{\mathbb{T}} f \, d\mu_s(x, y) \quad (f \in BM(\mathbb{T}),\ x, y \in H).$$

Here $BM(\mathbb{T})$ is the Banach algebra of all bounded measurable functions on $\mathbb{T}$. An argument similar to the one used in the proof of Theorem 14.1.12 shows that Φ_s is multiplicative. Since characteristic functions correspond to orthogonal projections under Φ_s, the map Φ_s is even a C^*-algebra homomorphism (Exercise 14.2.5).

If one denotes by $H_s = \Psi_s(1)H$ the singular part of H, then

$$C(\mathbb{T}) \to \mathcal{B}(H_s), \quad f \mapsto \Phi_s(f)|H_s\,,$$

becomes a unital C^*-algebra homomorphism.

Corollary 14.1.14 (Sz.-Nagy–Foiaş) *Let $T \in \mathcal{B}(H)$ be a completely non-unitary contraction on a Hilbert space H. Then there is a unique weak-$*$ continuous algebra homomorphism $\Phi : H^\infty(\mathbb{D}) \to \mathcal{B}(H)$ with $\|\Phi\| = 1$ and $\Phi(1) = 1_H$, $\Phi(z) = T$.*

Proof If $\Psi : A(\overline{\mathbb{D}}) \to \mathcal{B}(H)$ is the disc algebra functional calculus of T then, with the above notations, $T|H_s = \Phi_s(z)|H_s$ is a unitary operator on the reducing subspace H_s for T. Hence $H_s = \{0\}$ and $\Psi = \Psi_a$ has the claimed extension. □

We call a contraction $T \in \mathcal{B}(H)$ *absolutely continuous* if its disc algebra functional calculus extends to a weak-$*$ continuous algebra homomorphism $\Phi : H^\infty(\mathbb{D}) \to \mathcal{B}(H)$.

14.2 Exercises

1. Show that a minimal unitary dilation of a contraction is uniquely determined in the sense of Remark 14.1.2 (c).
2. Let X, Y be Banach spaces with Y separable. Show that a linear map $S : Y' \to X'$ is weak-$*$ continuous if and only if it is weak-$*$ sequentially continuous.
3. Let $V \subset \mathbb{C}$ be a bounded open set, and let A be a weak-$*$ dense subalgebra of $H^\infty(V)$. Let $\Phi : H^\infty(V) \to \mathcal{B}(H)$ be a weak-$*$ continuous map. Show that Φ is multiplicative if $\Phi|A$ is multiplicative.
4. Let $\Psi : A(\overline{\mathbb{D}}) \to \mathcal{B}(H)$ be a contractive algebra homomorphism. Show that there is at most one pair (Ψ_1, Ψ_2) consisting of contractive algebra homomorphisms $\Psi_i : A(\overline{\mathbb{D}}) \to \mathcal{B}(H)$ with $\Psi = \Psi_1 + \Psi_2$ such that Ψ_1 is absolutely continuous and Ψ_2 is singular.
5. Show that the map Ψ_s constructed in the proof of Theorem 14.1.12 extends to a C^*-algebra homomorphism $\Phi_s : BM(\mathbb{T}) \to \mathcal{B}(H)$.

14.3 Additional notes

The idea of proving the existence of the Nagy–Foiaş functional calculus by decomposing suitable representing measures into an absolutely continuous and a singular part goes back to Mlak (1969). The same idea can be used to decompose representations of the ball algebra $A(\mathbb{B})$ in the multivariable case (see Eschmeier 1997 or Chapter 20). In this case the Lebesgue decomposition theorem has to be replaced by a more general decomposition theorem due to Glicksberg, König, and Seever (see Rudin 1980, Theorem 9.4.4). The classic proof and many other applications can be found in the book of Sz.-Nagy and Foiaş (1970, Chapter 3).

15

Hyperinvariant subspaces

15.1

Let $T \in \mathcal{B}(H)$ be a contraction, and let $\Psi : A(\overline{\mathbb{D}}) \to \mathcal{B}(H)$, $f \mapsto \Psi(f)$, be its disc algebra functional calculus. Recall that T is said to be of type $C_{0\cdot}$ (respectively, $C_{\cdot 0}$) if

$$(T^k) \xrightarrow{\text{SOT}} 0 \quad (\text{respectively, } (T^{*k}) \xrightarrow{\text{SOT}} 0).$$

By a result of Sz.-Nagy and Foiaş, a contraction which is neither of type $C_{0\cdot}$ nor of type $C_{\cdot 0}$ either is a scalar multiple of the identity operator or possesses non-trivial hyperinvariant subspaces. The following observation will help us to extend this result to more general settings.

Lemma 15.1.1 *A contraction T is of type $C_{0\cdot}$ if and only if $(\Psi(f_k)) \xrightarrow{\text{SOT}} 0$ for each Montel sequence (f_k) in $A(\overline{\mathbb{D}})$.*

Proof Fix a Montel sequence (f_k) with $\|f_k\| \leq 1$ and a natural number N, and then use the Taylor expansion of f_k at zero to write

$$f_k = p_k + z^N g_k,$$

where p_k is a polynomial of degree less than N. Then $p_k \to 0$ uniformly on $\mathbb{D}$, and also, by the maximum modulus principle,

$$\|g_k\| \leq \|f_k\| + \|p_k\| \leq 2$$

for k sufficiently large. But then, for any given vector $x \in H$, it follows that

$$\limsup_{k \to \infty} \|\Psi(f_k)x\| \leq 2\|T^N x\|.$$

This shows that the stated condition is necessary for T to be of type $C_{0\cdot}$. The sufficiency of the condition is obvious. □

If T is absolutely continuous, then an analogue of the last lemma for the $H^\infty(\mathbb{D})$-functional calculus follows in the same way. Let V be a bounded open set in $\mathbb{C}$ and let $\Psi : A(V) \to \mathcal{B}(H)$ be a continuous algebra homomorphism. We set $V^* = \{\overline{z};\ z \in V\}$, and we consider the continuous algebra homomorphism

$$\widetilde{\Psi} : A(V^*) \to \mathcal{B}(H), \quad f \mapsto \Psi(\widetilde{f})^* \quad \left(\text{where } \widetilde{f}(z) = \overline{f(\overline{z})}\right).$$

Definition 15.1.2 *A continuous algebra homomorphism $\Psi : A(V) \to \mathcal{B}(H)$ is of* type $C_{0\cdot}$ *if $(\Psi(f_k)) \xrightarrow{\text{SOT}} 0$ for each Montel sequence (f_k) in $A(V)$. We say that Ψ is of type $C_{\cdot 0}$ if $\widetilde{\Psi}$ is of type $C_{0\cdot}$.*

For representations $\Phi : H^\infty(V) \to \mathcal{B}(H)$, we define the $C_{0\cdot}$- and $C_{\cdot 0}$-properties analogously, replacing Montel sequences by arbitrary weak-$*$ zero sequences in $H^\infty(V)$. Our generalization of the Sz.-Nagy–Foiaş result uses the following property of Hankel-type operators.

Theorem 15.1.3 *For each continuous function $g \in C(\overline{V})$, the operator*

$$S_g : A(V) \to C(\overline{V})/A(V), \quad f \mapsto [gf],$$

is compact. □

For a proof of this result, we refer the reader to Cole and Gamelin (1982, Theorem 6.3).

Corollary 15.1.4 *Let $g \in C(\partial V)$. For each Montel sequence (f_k) in $A(V)$, there is a Montel sequence (g_k) in $A(V)$ such that*

$$|f_k g - g_k|_{\partial V} \xrightarrow{k} 0.$$

Proof We may suppose that $g \in C(\overline{V})$. Let (f_k) be a Montel sequence. By Theorem 15.1.3, each subsequence of $([gf_k])_k$ has a convergent subsequence in $C(\overline{V})/A(V)$. Since weak-$*$ $\lim_k f_k = 0$, any such convergent subsequence has limit zero. Hence $\lim_k [gf_k] = 0$ in $C(\overline{V})/A(V)$, and we can choose a sequence (g_k) in $A(V)$ such that $(gf_k - g_k)$ converges to zero uniformly on $\overline{V}$. Clearly (g_k) is a Montel sequence. □

Let $\lambda : \ell^\infty \to \mathbb{C}$ be a continuous linear form of norm 1 such that $\lambda((s_k)) = \lambda((s_{k+1}))$ for each sequence $(s_k) \in \ell^\infty$ and such that $\lambda((s_k)) = \lim_k s_k$ for each convergent sequence of complex numbers. Then, for each bounded sequence (x_k) in H, there is a unique vector $L(x_k)$ in H with

$$\lambda(([x_k, y])) = [L(x_k), y] \quad (y \in H) .$$

The resulting map

$$L : \ell^\infty(H) \to H, \ (x_k) \mapsto L((x_k)) ,$$

is continuous and linear, with norm 1, such that $L((x_k)) = (x_{k+1})$ for each sequence (x_k) in $\ell^\infty(H)$ and $L((x_k)) = x$ whenever (x_k) converges weakly to a vector x in H. The very definition of L implies that $L((Ax_k)) = AL((x_k))$ for each operator $A \in \mathcal{B}(H)$ and each bounded sequence (x_k) in H.

Theorem 15.1.5 *Let $\Psi : A(V) \to \mathcal{B}(H)$ be a continuous algebra homomorphism. If Ψ is not of type $C_{\cdot 0}$, then there is a non-zero bounded linear map $j : C(\partial V) \to H$ with $\Psi(f)j = jM_{(f|\partial V)}$ for all $f \in A(V)$.*

Proof If Ψ is not of type $C_{\cdot 0}$, then there is a Montel sequence (f_k) in $A(V)$ and a vector $x \in H$ such that $\|f_k\| \leq 1$ and such that the limit $\varepsilon = \lim_k \|\Psi(f_k)^* x\|$ exists and is non-zero. Define $x_k = \Psi(f_k)^* x / \|\Psi(f_k)^* x\|$.

Let $g \in C(\partial V)$. By Corollary 15.1.4, there is a Montel sequence (g_k) in $A(V)$ such that $|g f_k - g_k|_{\partial V} \xrightarrow{k} 0$. The mapping

$$j : C(\partial V) \to H, \quad j(g) = L(\Psi(g_k)x_k) ,$$

is well-defined, continuous and linear with $\|j\| \leq \|\Psi\|$. Furthermore, by construction, we have $[j(1), x] = \varepsilon$ and $\Psi(f)j = jM_{f|\partial V}$ for all f in $A(V)$. □

For sufficiently nice domains V in $\mathbb{C}$ (or even $\mathbb{C}^n$), also the converse of the last theorem holds (see Eschmeier 1997, Theorem 2.1).

Let $T \in \mathcal{B}(H)$, and let $j : C(\partial V) \to H$ be a bounded operator intertwining M_z on $C(\partial V)$ and T on H. Denote by $I(j)$ the largest closed ideal in $C(\partial V)$ contained in the kernel of j, and define the support $s(j)$ of j as the common zero set of the functions in $I(j)$. For $f \in C(\partial V)$, let M_f be the multiplication operator

$$M_f : C(\partial V) \to C(\partial V), \quad g \mapsto fg.$$

Using continuous partitions of unity one can prove the next result.

Lemma 15.1.6 *Let j and T be as explained above. Then:*

(i) *if $z \in s(j)$ and if $f \in C(\partial V)$ satisfies $f(z) \neq 0$, then $j \circ M_f \neq 0$;*
(ii) *$s(j \circ M_f) \subset s(j) \cap \operatorname{supp}(f)$ for all $f \in C(\partial V)$;*
(iii) *if $s(j) = \{z\}$, then $\operatorname{im}(j) \in \operatorname{Lat}(T)$ is one-dimensional and $z \in \sigma_p(T)$.* □

We leave the proof as an elementary exercise (Exercise 15.2.1). Using the above observations, we can prove a generalization of the cited result of Sz.-Nagy and Foiaş.

Theorem 15.1.7 *Let $\Psi : A(V) \to \mathcal{B}(H)$ be a continuous algebra homomorphism. If Ψ is neither of type $C_{0\cdot}$ nor of type $C_{\cdot 0}$, then either $T = \Psi(z)$ is a scalar multiple of the identity operator or* $\operatorname{Hyp}(T)$ *is non-trivial.*

Proof The hypothesis that Ψ is neither $C_{\cdot 0}$ nor $C_{0\cdot}$ allows us to choose non-zero bounded linear maps

$$j : C(\partial V) \to H \quad \text{and} \quad k : C(\partial V) \to H'$$

such that j intertwines M_z on $C(\partial V)$ with T on H, and k intertwines M_z on $C(\partial V)$ with the Banach-space adjoint $T' \in \mathcal{B}(H')$ of T. By Lemma 15.1.6, we may suppose that $s(j)$ and $s(k)$ are disjoint. By Tietze's extension theorem, the maps j and k extend to bounded intertwiners

$$J : C(s(j)) \to H \quad \text{and} \quad K : C(s(k)) \to H'.$$

For each operator A commuting with T, the composition

$$X = X(A) : C(s(j)) \xrightarrow{A \circ J} H \xrightarrow{K'} M(s(k))$$

intertwines M_z on $C(s(j))$ and M_z on $M(s(k))$. Since the last two operators have disjoint spectra (their spectra are $s(j)$ and $s(k)$, respectively), a well-known result of Rosenblum (1956) shows that $X = 0$. Hence the space

$$\bigvee \{\operatorname{im} AJ;\ A \in (T)'\} \subset \ker(K')$$

is a non-trivial hyperinvariant subspace for T. □

If $A(V)$ is pointwise boundedly dense in $H^\infty(V)$ and if $\Psi : A(V) \to \mathcal{B}(H)$ is $C_{0\cdot}$ or $C_{\cdot 0}$, then, by Lemma 14.1.9, the map Ψ extends to a weak-$*$ continuous algebra homomorphism $\Phi : H^\infty(V) \to \mathcal{B}(H)$. It is elementary to check that the $C_{0\cdot}$- or $C_{\cdot 0}$-property is inherited by Φ (Exercise 15.2.2).

We end this section with an application of Theorem 15.1.5 to subnormal operators.

Corollary 15.1.8 *Let $S \in \mathcal{B}(H)$ be subnormal, and let $\Psi : A(V) \to \mathcal{B}(H)$ be a continuous unital algebra homomorphism with $\Psi(z) = S$ for some bounded open set V in $\mathbb{C}$. If Ψ is not of type $C_{\cdot 0}$, then there is a non-zero reducing space M for S such that $S|M$ is normal with $\sigma(S|M) \subset \partial V$.*

Proof Let $N \in \mathcal{B}(K)$ be the minimal normal extension of the subnormal operator S, and let $j : C(\partial V) \to H$ be a non-zero bounded linear map intertwining M_z on $C(\partial V)$ and S on H. A Fuglede-type argument (Exercise 15.2.3) shows that j intertwines $M_{\overline{z}}$ on $C(\partial V)$ and N^* on K. Therefore the image of j is contained in the space

$$H_0 = \{x \in H;\ N^{*k}x \in H \text{ for all } k \in \mathbb{Z}^+\},$$

which reduces N, and hence also S. But then $S|H_0 = N|H_0$ is normal with

$$j(C(\partial V)) \subset X_{N|H_0}(\partial V)\,.$$

Clearly the space on the right-hand side can be chosen as M. □

If in Corollary 15.1.8 the operator S is asked to be pure (or to possess no normal part with spectrum in ∂V), then Ψ has to be of type $C_{\cdot 0}$.

15.2 Exercises

1. Prove Lemma 15.1.6.
2. Let $V \subset \mathbb{C}$ be a bounded open set in $\mathbb{C}$ such that $A(V) \subset H^\infty(V)$ is pointwise boundedly dense in $H^\infty(V)$. Show that a weak-$*$ continuous algebra homomorphism $\Phi : H^\infty(V) \to \mathcal{B}(H)$ is of type $C_{0\cdot}$ if $\Phi|A(V)$ has this property.
3. Let $K \subset \mathbb{C}$ be compact and let $N \in \mathcal{B}(H)$ be a normal operator. Suppose that $j : C(K) \to H$ is continuous linear with $j(zf) = Nj(f)$ for all $f \in C(K)$. Show that $j(\overline{z}f) = N^*j(f)$ for all $f \in C(K)$.
4. Let $V \subset \mathbb{C}$ be a bounded open set. Call a measure $\mu \in M(\partial V)$ a *Henkin measure* if $\lim_{k\to\infty} \int_{\partial V} f_k \,\mathrm{d}\mu = 0$ for each Montel sequence (f_k) in $A(V)$. Show that if μ is a Henkin measure, then each measure $\nu \in M(\partial V)$ with $\nu \ll \mu$ is a Henkin measure. *Hint*: use Corollary 15.1.4.

15.3 Additional notes

The classic result of Sz.-Nagy and Foiaş saying that a contraction T which is neither of type $C_{0\cdot}$ nor of type $C_{\cdot 0}$ is either a scalar multiple of the identity or has non-trivial hyperinvariant subspaces can be found in Sz.-Nagy and Foiaş (1970, Theorem II.5.4). Results similar to Theorem 15.1.5 and Theorem 15.1.7 are contained in a paper of Apostol and Chevreau (1982). If in Corollary 15.1.8 the operator S is only supposed to be subdecomposable, then, if Ψ is not of type $C_{\cdot 0}$, it follows that $H_S(\partial V) \neq \{0\}$. Since hyponormal operators are subdecomposable and since a hyponormal operator whose spectrum has planar Lebesgue measure zero is normal, it follows that a hyponormal contraction is completely non-unitary if and only if it is of type $C_{\cdot 0}$.

16

Invariant subspaces for contractions

16.1

Let $T \in \mathcal{B}(H)$ be a contraction. In 1979 Brown, Chevreau, and Pearcy proved that T possesses non-trivial invariant subspaces if its spectrum is rich enough close to the unit circle. It is the aim of this chapter to give a proof of this result and to explain, at the same time, the basic ideas of the Scott Brown factorization technique.

Definition 16.1.1 *Let $V \subset \mathbb{C}$ be open. A set $\sigma \subset \mathbb{C}$ is* dominating *in V if $|f|_V = |f|_{V \cap \sigma}$ for each $f \in H^\infty(V)$.*

With this notion the cited invariant-subspace result is the following.

Theorem 16.1.2 (Brown, Chevreau, and Pearcy) *Let $T \in \mathcal{B}(H)$ be a contraction such that $\sigma(T)$ is dominating in $\mathbb{D}$. Then T has non-trivial invariant subspaces.*

For a subset M of a complex normed space E, we denote by $C(M)$ and $\Gamma(M)$ the convex and absolutely convex hull of M. We write $\overline{C}(M)$ and $\overline{\Gamma}(M)$ for the closures of these sets.

Lemma 16.1.3 *A set σ is dominating in an open set V in $\mathbb{C}$ if and only if*

$$\overline{\Gamma}(\{\mathcal{E}_\lambda;\ \lambda \in \sigma \cap V\}) = \mathcal{Q}_{[1]}\,. \qquad \square$$

This result is a direct consequence of the separation theorem (Exercise 16.2.1). To prove Theorem 16.1.2, we are allowed to make certain reductions. First, according to Corollary 14.1.14 we may suppose that T has a weak-$*$-continuous H^∞-functional calculus $\Phi : H^\infty(\mathbb{D}) \to \mathcal{B}(H)$. The second reduction concerns the type of the spectrum of T.

For an operator $A \in \mathcal{B}(X)$ on a Banach space X, we consider the *left essential spectrum*

$$\sigma_{\ell e}(A) = \{z \in \mathbb{C};\ \text{im}(z - A) \text{ is not closed or dim ker}(z - A) = \infty\}$$

and the *right essential spectrum*

$$\sigma_{re}(A) = \{z \in \mathbb{C};\ \dim(X/(z - A)X) = \infty\}.$$

The *essential spectrum* $\sigma_e(A) = \sigma_{\ell e}(A) \cup \sigma_{re}(A)$ consists of all points $z \in \mathbb{C}$ such that $\dim \ker(z - A) = \infty$ or $\dim(X/\text{im}(z - A)) = \infty$. If we want to prove the existence of a non-trivial invariant subspace for A, we may assume that $\sigma(A) = \sigma_{\ell e}(A) = \sigma_{re}(A)$ (Exercise 16.2.2). For more on the essential spectrum, see Chapter 24.

In view of these remarks we may and shall suppose until the end of the proof of Theorem 16.1.2 that T has a weak-$*$ continuous H^∞-functional calculus and that $\sigma(T) = \sigma_{\ell e}(T)$.

For $x, y \in H$, the linear functional

$$x \otimes y : H^\infty(\mathbb{D}) \to \mathbb{C}, \quad x \otimes y(f) = [\Phi(f)x, y],$$

is weak-$*$-continuous and defines an element in the predual space $Q = L^1(\mathbb{D})/{}^\perp H^\infty(\mathbb{D})$ of $H^\infty(\mathbb{D})$. Note that it suffices to find a point λ in $\mathbb{D}$ and vectors x, y in H with

$$\mathcal{E}_\lambda = x \otimes y.$$

Indeed, in this case the space

$$M = \bigvee \{\Phi(f)x;\ f \in H^\infty(\mathbb{D}) \text{ with } f(\lambda) = 0\}$$

is an invariant subspace for T different from the whole space H. If $M = \{0\}$, then $(\lambda - T)x = 0$. So in any case $\text{Lat}(T)$ is non-trivial.

Let $\mathcal{L}$ be the set of all elements L in Q with the property that, for any given $\varepsilon > 0$ and any choice of vectors $a_1, \ldots, a_s, b_1, \ldots, b_s$ in H (with s arbitrary), there are vectors x, y in the closed unit ball of H with

(i) $\| L - x \otimes y\| < \varepsilon$,

(ii) $\| x \otimes b_i \| < \varepsilon, \quad \|a_i \otimes y\| < \varepsilon \quad (i = 1, \ldots, s)$.

Proposition 16.1.4 *The set $\mathcal{L}$ is norm-closed and absolutely convex.*

Proof Obviously, $\mathcal{L}$ is norm-closed.

Let $L_1, L_2 \in \mathcal{L}$ and $c_1, c_2 \in \mathbb{C}$ with $|c_1| + |c_2| \leq 1$ be given. We indicate only how to find vectors x, y for $L = c_1 L_1 + c_2 L_2$ that satisfy condition (i). For any $r > 0$, there are x_1, x_2, y_1, y_2 in the closed unit ball of H with

$$\|L_i - x_i \otimes y_i\| < r \quad (i = 1, 2)$$

and such that $x_1 \otimes y_2$, $x_2 \otimes y_1$, $x_2 \otimes x_1$, and $y_1 \otimes y_2$ have norm less than r. Factor $c_i = t_i \overline{s_i}$ with complex numbers t_i, s_i such that

$$|t_i| = |s_i| = |c_i|^{1/2} \quad (i = 1, 2).$$

Then $x = t_1 x_1 + t_2 x_2$, $y = s_1 y_1 + s_2 y_2$ satisfy $\max(\|x\|^2, \|y\|^2) \leq 1 + 2r$ as well as

$$\|L - x \otimes y\| < 3r.$$

All that remains is to normalize x and y in a suitable way. □

By Theorem 15.1.7 and the remarks following Theorem 15.1.7, we may suppose that Φ is of type $C_{0\cdot}$ or $C_{\cdot 0}$. The next observation shows how these continuity properties of the Nagy–Foiaş functional calculus can be used. Recall that a sequence (x_k) in H is a weak zero sequence if $\lim_{k\to\infty}[x_k, y] = 0$ for each vector $y \in H$.

Lemma 16.1.5 *If Φ is of type $C_{0\cdot}$ and (y_k) is a weak zero sequence in H, then*

$$\lim_k x \otimes y_k = 0 \quad (x \in H).$$

If Φ is of type $C_{\cdot 0}$ and (x_k) is a weak zero sequence in H, then

$$\lim_k x_k \otimes y = 0 \quad (y \in H).$$

Proof Fix $x \in H$. In the first case, by a normal family argument, the set

$$K = \{\Phi(f)x;\ f \in H^\infty(\mathbb{D})_{[1]}\} \subset H$$

is compact. Hence, for each weak zero sequence (y_k) in H,

$$\|x \otimes y_k\| = \sup\{|[z, y_k]|;\ z \in K\} \xrightarrow{k} 0.$$

In the second case we use the same argument for $\widetilde{\Phi}$ instead of Φ. □

A complex number z belongs to $\sigma_{\ell e}(T)$ if and only if there is an orthonormal sequence (x_k) in H with $((z-T)x_k) \xrightarrow{k} 0$ (Exercise 16.2.3). The next result shows why it is useful to have rich left essential spectrum.

Lemma 16.1.6 *If (x_k) is an orthonormal sequence with $\lim_k(z-T)x_k = 0$, then*

$$\lim_{k\to\infty} x_k \otimes y = 0 \quad (y \in H) \quad \text{and} \quad \lim_{k\to\infty} (\mathcal{E}_z - x_k \otimes x_k) = 0\,.$$

Proof Let $f \in H^\infty(\mathbb{D})_{[1]}$, and let $\lambda \in \mathbb{D}$. Then the unique function $g \in H(\mathbb{D})$ with $(z-\lambda)g(z) = f(z) - f(\lambda)$ for $z \in \mathbb{D}$ is bounded with $\|g\|_\infty \leq 2/(1-|\lambda|)$ (Exercise 16.2.4). For $f \in H^\infty(\mathbb{D})_{[1]}$, choose g as above, and note that

$$|x_k \otimes y(f)| \leq |[\Phi(g)(T-\lambda)x_k, y]| + |[x_k, y]| \quad (k \in \mathbb{N})\,.$$

This proves the first part of the assertion. Similarly,

$$|(x_k \otimes x_k - \mathcal{E}_\lambda)(f)| = |[\Phi(f)x_k, x_k] - f(\lambda)| = |[\Phi(g)(T-\lambda)x_k, x_k]| \xrightarrow{k} 0$$

uniformly for f in the unit ball of $H^\infty(\mathbb{D})$. □

Since $\mathrm{Lat}(T)$ is non-trivial if and only if $\mathrm{Lat}(T^*)$ is non-trivial, we may suppose that Φ is of type $C_{0\cdot}$. Since $\sigma_{\ell e}(T)$ is dominating in $\mathbb{D}$, the previous results imply that the closed unit ball of Q is contained in $\mathcal{L}$. Now a standard procedure, generally called the *Scott Brown technique*, can be used to factor all elements in Q.

Lemma 16.1.7 *For $L \in Q$, $x_0, y_0 \in H$, and $\varepsilon > 0$, there are $x, y \in H$ with $\|L - x \otimes y\| < \varepsilon$ and $\|x - x_0\|, \|y - y_0\| \leq \|L - x_0 \otimes y_0\|^{1/2}$.*

Proof Let $d = \|L - x_0 \otimes y_0\| > 0$. By the above remarks, for each $r > 0$ there are vectors a, b in the closed unit ball of H with

$$\|(1/d)(L - x_0 \otimes y_0) - a \otimes b\| < r$$

and $\|x_0 \otimes b\| < r$, $\|a \otimes y_0\| < r$. In view of the estimate

$$\|L - (x_0 + d^{1/2}a) \otimes (y_0 + d^{1/2}b)\| \leq \|L - x_0 \otimes y_0 - d(a \otimes b)\| + 2d^{1/2}r\,,$$

it is clear that $x = x_0 + d^{1/2}a$ and $y = y_0 + d^{1/2}b$ will satisfy all assertions, provided that r is small enough. □

Theorem 16.1.8 *For $L \in Q$, $x_0, y_0 \in H$, and $\varepsilon > 0$, there are $x, y \in H$ with $L = x \otimes y$ and $\|x - x_0\|, \|y - y_0\| < \|L - x_0 \otimes y_0\|^{1/2} + \varepsilon$.*

Proof Define $d_0 = \|L - x_0 \otimes y_0\|$ and choose $d_k > 0$ $(k \geq 1)$ with

$$\sum_{k=1}^{\infty} d_k^{1/2} < \varepsilon\,.$$

Inductively, using in each step Lemma 16.1.7, we can choose sequences (x_k), (y_k) in H with $\|L - x_k \otimes y_k\| \leq d_k$ and

$$\|x_{k+1} - x_k\|,\ \|y_{k+1} - y_k\| \leq d_k^{1/2} \quad (k \geq 0).$$

Clearly, x and y can be chosen as $x = \lim_k x_k$ and $y = \lim_k y_k$. □

Applying the last result to the particular case that L is a point evaluation, we obtain the existence of non-trivial invariant subspaces for T. Thus the proof of Theorem 16.1.2 is complete.

16.2 Exercises

1. Let M be a subset of a Banach space X. Show that $\overline{\Gamma}(M)$ is the closed unit ball in X if and only if $\sup\{|u(x)|;\ x \in M\} = \|u\|$ for each $u \in X'$.
2. Let $T \in \mathcal{B}(X)$ (X a Banach space with $\dim X \geq 2$). Show that $\mathrm{Lat}(T)$ is non-trivial if $\sigma(T) \neq \sigma_{\ell e}(T)$ or $\sigma(T) \neq \sigma_{re}(T)$.
3. Let X be a Banach space, and let $T \in \mathcal{B}(X)$. Show that a complex number λ belongs to $\sigma_{\ell e}(T)$ if and only if each closed subspace M of finite codimension in X contains a sequence (x_k) of unit vectors such that $\lim_{k\to\infty}(\lambda - T)x_k = 0$.
4. Let $V \subset \mathbb{C}$ be a non-empty bounded open set, and let $\lambda \in V$. Show that the map $Q : H^\infty(V) \to H^\infty(V)$ associating with each function $f \in H^\infty(V)$ the unique function $f_\lambda \in H^\infty(V)$ with $(z - \lambda)f_\lambda(z) = f(z) - f(\lambda)$ for $z \in V$ is weak-$*$ continuous, maps polynomials to polynomials and satisfies the inequality $\|Q\| \leq 2/\mathrm{dist}(\lambda, \partial V)$.
5. Suppose that a compact set $\sigma \subset \mathbb{C}$ is dominating in $\mathbb{D}$. Show that $\partial\mathbb{D} \subset \sigma$.

16.3 Additional notes

The invariant subspace result stated as Theorem 16.1.2 is from Brown, Chevreau and Pearcy (1979). If T is a contraction such that $\sigma(T)$ is dominating in $\mathbb{D}$, then $\sigma(T)$ contains the unit circle (Exercise 16.2.5). It was proved by Brown, Chevreau and Pearcy (1988) that the latter condition suffices to guarantee the existence of a non-trivial invariant subspace. A very readable proof of this result is contained in Bercovici (1990). An extension to the case of polynomially bounded operators on Banach spaces is given in Ambrozie and Müller (2003). Results of Bercovici, Foiaş and Pearcy show that the H^∞-functional calculus of a completely non–unitary contraction whose essential spectrum $\sigma_e(T)$ is dominating in $\mathbb{D}$ allows the factorization of infinite matrices with coefficients in the predual Q of $H^\infty(\mathbb{D})$. It follows that contractions of this type possess an extremely rich invariant subspace lattice. For details on this class of contractions, and many other results on the structure of the invariant subspace lattice of contractions obtained with the Scott Brown technique, the reader is referred to Bercovici, Foiaş and Pearcy (1985).

17

Invariant subspaces for subnormal operators

17.1

In 1978 Scott Brown proved the following general invariant-subspace result.

Theorem 17.1.1 (Brown) *Each subnormal operator S on a Hilbert space H of dimension at least 2 has a non-trivial invariant subspace.*

The original proof of Brown (1978) was quite involved, and it was the starting point for a whole new area in operator theory. Later J. E. Thomson (1986) gave a simpler proof of the original Scott Brown result. We begin by explaining Thomson's proof of Theorem 17.1.1.

Fix a non-zero vector x in H. To prove Theorem 17.1.1, we may suppose that

$$H = \bigvee \{S^k x;\ k \in \mathbb{Z}^+\}.$$

But then there is a positive measure μ on a compact set K in $\mathbb{C}$ such that S is unitarily equivalent to $M_z \in \mathcal{B}(P^2(\mu))$ (Exercise 17.2.4). Therefore we only consider the case where $S = M_z$ on $P^2(\mu)$ as above. Clearly we may suppose that $P^2(\mu) \neq L^2(\mu)$. As before we denote by λ the two-dimensional Lebesgue measure on $\mathbb{C}$.

Lemma 17.1.2 *Let $g \in L^q(\mu) \setminus \{0\}$ for some real number $1 < q < 2$. Then there is a point $w \in \mathbb{C}$ such that*

$$\text{(i)}\ \ g/(z-w) \in L^q(\mu) \quad \text{and} \quad \text{(ii)}\ \ \int_K \frac{g(z)}{z-w}\,\mathrm{d}\mu(z) \neq 0\,.$$

Proof Fix $R > 0$ with $|z| \leq R$ for all $z \in K$. The function h defined on $K \times K$ by

$$h(z,w) = |g(z)|^q / |(z-w)|^q \quad (= 0 \text{ for } z = w)$$

is measurable. Since, for each z in K, we have

$$\int_K |z-w|^{-q} \mathrm{d}w \le \int_{D_R(z)} |w|^{-q} \mathrm{d}w \le \int_{D_{2R}(0)} |w|^{-q} \mathrm{d}w < \infty ,$$

Fubini's theorem shows that the function h is $(\lambda \times \mu)$-integrable and that

$$\int_K \left(\int_K h(z,w)\, \mathrm{d}\mu(z) \right) \mathrm{d}w < \infty ,$$

where the inner integral is finite for λ-almost every w in $\mathbb{C}$.

Let $\varphi \in C^1_{00}(\mathbb{C})$. In a similar way Fubini's theorem shows that the function $(g/(z-w))\overline{\partial}\varphi(w)$ is μ-integrable for λ-almost every w in $\mathbb{C}$ and that

$$\int_{\mathbb{C}} \left(\int_K \frac{g(z)}{z-w} \overline{\partial}\varphi(w)\, \mathrm{d}\mu(z) \right) \mathrm{d}w = \int_K g(z) \left(\int_{\mathbb{C}} \frac{\overline{\partial}\varphi(w)}{z-w} \mathrm{d}w \right) \mathrm{d}\mu(z)$$
$$= \pi \int_K g(z)\varphi(z)\, \mathrm{d}\mu(z) .$$

The last equality follows from the complex version of Green's formula. By Stone–Weierstrass, $C^1_{00}(\mathbb{C})|K$ is dense in $C(K)$. Hence the last integral is non-zero for some function φ as above.

In particular, on a set of positive Lebesgue measure the integral in condition (ii) exists and is non-zero. Hence there are complex numbers w satisfying both conditions. □

In the case where $g = 1$, the above proof shows that the *Cauchy transform*

$$\widetilde{\mu}(w) = \int_K \frac{1}{z-w} \mathrm{d}\mu(z)$$

of μ exists λ-almost everywhere on $\mathbb{C}$ and is locally λ-integrable with

$$\overline{\partial}\widetilde{\mu} = -\pi\mu \quad \text{(in the distributional sense).}$$

For μ as above and $1 \le p < \infty$, we denote by $P^p(\mu)$ the closure of the set of all polynomials in $L^p(\mu)$.

Theorem 17.1.3 (Brennan) *Let $p > 2$ be a real number with $P^p(\mu) \neq L^p(\mu)$. Then there is a point w in $\mathbb{C}$ such that, for some positive constant $c > 0$,*

$$|f(w)| \le c\|f\|_p \quad (f \in \mathbb{C}[z]) .$$

Proof Choose $g \in L^q(\mu) \setminus \{0\}$ with $g \perp P^p(\mu)$, where $q \in (1,2)$ is the conjugate exponent of p. By the previous lemma, there is a point w in $\mathbb{C}$ with $g/(z-w) \in L^q(\mu)$ and $(g\mu)\tilde{}(w) \neq 0$. But then from

$$0 = \int_K \frac{f(z) - f(w)}{z - w} g(z) \, d\mu(z) \quad (f \in \mathbb{C}[z])$$

we deduce that the map

$$f \mapsto f(w) = \frac{1}{(g\mu)\tilde{}(w)} \int_K f(z) \frac{g(z)}{z - w} \, d\mu(z)$$

on $\mathbb{C}[z]$ extends to a continuous linear functional on $P^p(\mu)$. □

To prove the existence of an invariant subspace for $S = M_z$ on $P^2(\mu)$, where we require that $P^2(\mu) \neq L^2(\mu)$, it suffices to find $x, y \in P^2(\mu)$ and $w \in \mathbb{C}$ with

$$[p(S)x, y] = p(w)$$

for all polynomials p. Note that the hypothesis that $P^2(\mu) \neq L^2(\mu)$ implies that we have $P^p(\mu) \neq L^p(\mu)$ for all $p \geq 2$.

Proof of Theorem 17.1.1 Fix S and μ as above. Set $p = 3$ and $q = 3/2$. According to Theorem 17.1.3, there is a point w in $\mathbb{C}$ such that

$$\mathbb{C}[z] \to \mathbb{C}, \quad p \mapsto p(w),$$

extends to a continuous linear functional L on $P^p(\mu)$. By Hahn–Banach, there is a function h in $L^q(\mu)$ with $\|h\|_q = \|L\|$ such that h represents L on $P^p(\mu)$. Since $P^p(\mu)$ is reflexive, there is a function r of norm 1 in $P^p(\mu)$ with $L(r) = \|L\|$. Because we have equality in Hölder's inequality

$$\|h\|_q = \|L\| = L(r) = \int_K rh \, d\mu \leq \|r\|_p \|h\|_q ,$$

there is a constant $c > 0$ (see Hewitt and Stromberg 1965, p. 190) with

$$|r|^p = c^{3/2} |h|^q ,$$

or equivalently $|r|^2 = c|h|$ μ-almost everywhere.

Since $u = h/r$ ($= 0$ where $r = 0$) belongs to $L^2(\mu)$ and since

$$[p(S)r, \overline{u}]_{L^2(\mu)} = \int_K pru \, d\mu = \int_K ph \, d\mu = p(w)$$

for each polynomial p, it follows that

$$\bigvee\{p(S)r;\ p \in \mathbb{C}[z] \text{ with } p(w) = 0\} \in \mathrm{Lat}(S)$$

is non-trivial or $w \in \sigma_p(S)$. □

As a strengthening of the above result, we can prove (see Thomson 1986; or Martin and Putinar 1989):

Theorem 17.1.4 *Each rationally cyclic subnormal operator S on a Hilbert space of dimension at least 2 has a non-trivial hyperinvariant subspace.* □

It is not known whether or not each subnormal operator, which is not a multiple of the identity has a non-trivial hyperinvariant subspace.

A point w as in Theorem 17.1.3 is called a *bounded point evaluation* for $P^p(\mu)$. More recently, it was shown by Thomson (1991) that Theorem 17.1.3 also holds in the case where $p = 2$.

17.2 Exercises

1. Let $N \in \mathcal{B}(H)$ be a normal operator on a Hilbert space H, and let $x \in H$ be a $*$-cyclic vector for N, that is, suppose that

$$H = \bigvee\{N^k N^{*j} x;\ k, j \in \mathbb{Z}^+\}.$$

Show that there is a positive measure $\mu \in M(\sigma(N))$ on the spectrum of N and a unitary operator $U : H \to L^2(\mu)$ such that $UN = M_z U$.

2. A normal extension $N \in \mathcal{B}(K)$ of a subnormal operator $S \in \mathcal{B}(H)$ is called *minimal* if K is the only reducing subspace for N that contains H. Show the following:

   (a) a normal extension N is minimal if and only if $K = \bigvee\{N^{*k} H;\ k \geq 0\}$;
   (b) if $S_1, S_2 \in \mathcal{B}(H)$ are subnormal operators with minimal normal extensions $N_i \in \mathcal{B}(K_i)$ $(i = 1, 2)$, and if $U \in \mathcal{B}(H)$ is a unitary operator with $US_1 = S_2 U$, then there is a unitary operator $V \in \mathcal{B}(K_1, K_2)$ with $VN_1 = N_2 V$ and $V|H = U$;
   (c) any two minimal normal extensions of a subnormal operator are unitarily equivalent.

3. Let $N \in \mathcal{B}(K)$ be the minimal normal extension of a subnormal operator $S \in \mathcal{B}(H)$. Show that $\partial\sigma(S) \subset \sigma(N) \subset \sigma(S)$ and that a bounded connected component C of $\mathbb{C} \setminus \sigma(N)$ either is contained in $\sigma(S)$ or is disjoint to $\sigma(S)$.

4. Let $S \in \mathcal{B}(H)$ be a subnormal operator with a cyclic vector. Show that there is a positive measure $\mu \in M(\sigma(S))$ such that S is unitarily equivalent to M_z on $P^2(\mu)$ (= closure of $\mathbb{C}[z]$ in $L^2(\mu)$).

17.3 Additional notes

Using methods similar to the above, K. Yan (1988) proved that each subnormal tuple, that is, each commuting tuple $S = (S_1, \ldots, S_n) \in \mathcal{B}(H)^n$ for which there is a commuting tuple $N = (N_1, \ldots, N_n) \in \mathcal{B}(K)^n$ of normal operators on a larger Hilbert space $K \supset H$ with $N_i H \subset H$ and $S_i = N_i | H$, has a non-trivial joint invariant subspace.

Let $T \in \mathcal{B}(H)$ be arbitrary, and let $M \geq 1$ be a constant. A compact set K in $\mathbb{C}$ containing $\sigma(T)$ is called *M-spectral* for T if $\|r(T)\| \leq M|r|_K$ for each rational function r with poles off K. For a subnormal operator S, its spectrum is a *spectral* (= 1-spectral) *set* (Exercise 18.2.2). It was shown by Stampfli (1980) that each operator $T \in \mathcal{B}(H)$ for which the spectrum $\sigma(T)$ is an M-spectral set possesses a non-trivial invariant subspace. Extensions of this result to the case of operators on a Banach space are given in Prunaru (1996).

18

Invariant subspaces for subdecomposable operators

18.1

Subnormal operators are examples of hyponormal operators. An operator $T \in \mathcal{B}(H)$ is *hyponormal* if $\|T^*x\| \leq \|Tx\|$ for all $x \in H$. The following generalization of Theorem 17.1.1 was obtained in Scott Brown (1987).

Theorem 18.1.1 (Brown) *Each hyponormal operator with thick spectrum has a non-trivial invariant subspace.* □

Here a compact set K in $\mathbb{C}$ is called *thick* if there is a bounded open set V in $\mathbb{C}$ such that K is dominating in V. If a compact set K in $\mathbb{C}$ is not thick, then $R(K) = C(K)$ (see Brown 1987). The uniform algebra $R(K)$ is described in Part I, Examples 1.2(v). In particular, each subnormal operator with non-thick spectrum is normal (Exercise 18.2.3).

To prove the above theorem, Scott Brown used a result of Putinar showing that each hyponormal operator is the restriction of an operator with a $C^\infty(\mathbb{C})$-functional calculus (see Eschmeier and Putinar 1996, Chapter 6.4). Since operators of the latter type are decomposable, each hyponormal operator is subdecomposable, that is, it is the restriction of a decomposable operator. Therefore the next result extends Theorem 18.1.1. For the definition and the theory of decomposable operators, see Part IV.

Theorem 18.1.2 (Eschmeier and Prunaru) *Let $S \in \mathcal{B}(X)$ be a subdecomposable operator on a complex Banach space X.*

(i) *If $\sigma(S)$ is thick, then* Lat(S) *is non-trivial.*
(ii) *If $\sigma_e(S)$ is thick, then* Lat(S) *is rich.*

Here Lat(S) is called *rich* if it contains a sublattice which is isomorphic (as a lattice) to the lattice of all closed linear subspaces of some infinite-dimensional Banach space.

One of the main difficulties in proving Theorem 18.1.2 is to extend the Scott Brown technique to the Banach-space case. Below we indicate some of the main ideas. First note that it suffices to prove the second part. Indeed, if $\sigma(S) \neq \sigma_e(S)$ and if $\dim(X) > 1$, then Lat(S) is non-trivial.

To prove part (ii), it is enough to find a complex number z and sequences (x_j) in X, (y_j) in X' such that

$$\delta_{j,k} p(z) = \langle p(S)x_j, y_k \rangle$$

for all natural numbers j, k and all polynomials p. In this case

$$M = \bigvee \{p(S)x_j;\ p \in \mathbb{C}[z] \text{ and } j \in \mathbb{N}\},$$
$$N = \bigvee \{p(S)x_j;\ p \in \mathbb{C}[z] \text{ with } p(z) = 0 \text{ and } j \in \mathbb{N}\}$$

are invariant for S such that $N \subset M$, $(z - S)M \subset N$, and $\dim(M/N) = \infty$. If $\pi : M \to M/N$ is the quotient map, then

$$\mathrm{Lat}(M/N) \to \mathrm{Lat}(S), \quad L \mapsto \pi^{-1}(L),$$

defines a lattice embedding (Exercise 18.2.4).

Since S is subdecomposable, S' is the quotient of a decomposable operator. More precisely, there is a decomposable operator $T \in \mathcal{B}(Z)$ on some Banach space Z and a surjective, continuous linear map $q : Z \to X'$ with $qT = S'q$.

Let $\sigma_e(S)$ be dominating in V. For a given compact set K in $\mathbb{C}$, we denote by $Z_T(K)$ the local analytic subspace of the decomposable operator $T \in \mathcal{B}(Z)$ associated with K (see Chapter 22). The space

$$Z_T(V) = \bigcup \{Z_T(K);\ K \subset V \text{ compact}\}$$

becomes an $H(V)$-module via

$$fz = f(T|Z_T(K))z$$

whenever $K \subset V$ is a compact set with $z \in Z_T(K)$. For $x \in X$ and $z \in Z_T(V)$, the map

$$x \otimes z : H^\infty(V) \to \mathbb{C}, \quad f \mapsto \langle x, q(fz) \rangle,$$

defines a weak-$*$ continuous linear functional.

The Scott Brown technique works best for the left essential spectrum. Therefore the following observation is useful.

Lemma 18.1.3 *Let $S \in \mathcal{B}(X)$ be subdecomposable. Then*

$$\sigma_e(S) = \sigma_{re}(S) = \sigma_{\ell e}(S') .$$

Proof The second equality follows from standard duality theory (Exercise 18.2.5). If there were a point z in $\sigma_e(S) \setminus \sigma_{re}(S)$, then

$$\dim(X/(z - S)X) < \infty = \dim \ker(z - S) .$$

In this case S would not even possess the single-valued extension property (see Finch 1975, proof of Corollary 11). □

By Lemma 18.1.3, the left essential spectrum of S' is dominating in V. Let $z \in \sigma_{\ell e}(S')$. Then each closed finite-codimensional subspace $M \subset X'$ contains a sequence (y_k) of unit vectors such that $\lim_{k \to \infty} (z - S')y_k = 0$.

Question *If (y_k) is as above, is there a sequence (z_k) in Z with $q(z_k) = y_k$ and* $\lim_k (z - T)z_k = 0$?

In Eschmeier and Prunaru (1990), essential use of a lifting property of the above type for *approximate eigensequences* is made. However, simple examples show that the answer to the above question is in general negative. For instance, the bilateral shift

$$U : \ell^2(\mathbb{Z}) \to \ell^2(\mathbb{Z}), \quad U((x_k)_{k \in \mathbb{Z}}) = ((x_{k+1})_{k \in \mathbb{Z}}) ,$$

is a decomposable lifting of the backward shift

$$S : \ell^2 \to \ell^2, \quad S((x_k)_{k \in \mathbb{N}}) = ((x_{k+1})_{k \in \mathbb{N}}) .$$

Since $\sigma_p(S) = \mathbb{D}$ and $\sigma(U) = \mathbb{T}$, there is no way to solve the above lifting problem in this concrete case.

Fortunately it is possible to show that each operator that possesses a decomposable lifting at all has a canonical decomposable lifting which allows the lifting of approximate eigensequences in the above sense (see Eschmeier and Prunaru 1990, Proposition 1.4).

As in Chapter 16, define $\mathcal{L}$ as the set of all elements L in the quotient space $Q = L^1(V)/^{\perp}H^{\infty}(V)$ such that, for any given finitely many vectors $a_1, \ldots, a_s \in X$, $b_1, \ldots, b_s \in Z_T(V)$ and any $\varepsilon > 0$, there are vectors $x \in X$ and $z \in Z_T(V)$ with $\|x\|, \|qz\| \leq 1$ and

$$\text{(i) } \|L - x \otimes z\| < \varepsilon ,$$

$$\text{(ii) } \|x \otimes b_i\| < \varepsilon, \quad \|a_i \otimes z\| < \varepsilon \quad (i = 1, \ldots, s) .$$

It suffices to find a positive constant c such that

$$\{L \in Q;\ \|L\| \leq c\} \subset \mathcal{L}.$$

Then a standard factorization method due to Apostol, Bercovici, Foiaş, Pearcy (see Bercovici, Foiaş and Pearcy 1988) allows one to factor infinite matrices with coefficients in Q (see Eschmeier and Putinar 1996, Chapter 6.5) in the following sense. For each matrix $L = (L_{jk}) \in M(\mathbb{N}, Q)$, we can prove the existence of vectors x_N in X^N, z_N in $Z_T(V)^N$ ($N \in \mathbb{N}$) such that the limits

$$x(j) = \lim_{N\to\infty} x_N(j) \in X, \quad y(j) = \lim_{N\to\infty} q\, z_N(j) \in X'$$

exist for each $j \in \mathbb{N}$ and such that, for all $j, k \in \mathbb{N}$, we have

$$L_{jk} = \lim_{N\to\infty} x_N(j) \otimes z_N(k).$$

Here $x_N(j)$ and $z_N(k)$ are the components of the N-tuples x_N and z_N. By applying this result to the special case of a diagonal matrix of the special form $L = (\delta_{j,k}\mathcal{E}_\lambda)_{j,k\in\mathbb{N}}$ with λ in V arbitrary, one obtains the relations

$$\delta_{j,k}\, p(\lambda) = \lim_{N\to\infty} \langle x_N(j), q\, p(T) z_N(k)\rangle = \langle p(S)x(j),\ y(k)\rangle$$

for all polynomials p and all natural numbers j, k.

Since $\mathcal{L}$ is norm-closed, it suffices to find a constant $d > 0$ such that, for each absolutely convex combination $L = \sum_{i=1}^r c_i \mathcal{E}_{\lambda_i}$ of point evaluations at points $\lambda_1, \ldots, \lambda_r$ in $\sigma_{\ell e}(S') \cap V$, the factorization problems (i) and (ii) can be solved with factors $x \in X$ and $z \in Z_T(V)$ bounded by $\|x\|$, $\|qz\| \leq d$. A straightforward decomposition into real and imaginary, positive and negative, parts reduces the problem to the case of convex combinations. To indicate a possible solution of the remaining problem we need two results from the geometry of normed spaces.

Denote by $\mathcal{F}(X)$ the set of all finite-dimensional subspaces of X and by $\mathrm{Cof}(X)$ the set of all closed, finite-codimensional subspaces.

Lemma 18.1.4 *Let $\delta > 0$. For each space M in $\mathcal{F}(X)$, there is a space N in* $\mathrm{Cof}(X)$ *with $\|m + n\| \geq (1 - \delta)\|m\|$ for $m \in M$ and $n \in N$.* □

Lemma 18.1.5 (Zenger) *Let $x_1, \ldots, x_r \in X$ be linearly independent, and let $c_1, \ldots, c_r$ be non-negative real numbers with $\sum_i c_i = 1$. Then there are a linear combination $\sum_{i=1}^r \mu_i x_i$ in the closed unit ball of X and a functional y in the closed unit ball of X' with $\langle \mu_i x_i, y\rangle = c_i$ ($i = 1, \ldots, r$).* □

Lemma 18.1.4 is proved in Singer (1981, Lemma III.1.1). A proof of the second result can be found in Bonsall and Duncan (1973, p. 20).

Exercise 18.1.6 Let I be an index set. For $i \in I$, let P_i be a property that a unit vector in X might have. Suppose that, for each $i \in I$, each space $N \in \mathrm{Cof}(X)$ contains unit vectors with property P_i. Then, for each M in $\mathcal{F}(X)$ and each finite set of indices $i_1, \ldots, i_r$, there are unit vectors $x_1, \ldots, x_r$ such that x_ν has property P_{i_ν} and such that, with $L = \mathrm{lin}\{x_1, \ldots, x_r\}$, we have

(i) the projection from $M + L$ onto L along M has norm less than 3,
(ii) $\max_{1 \le i \le r} |\alpha_i| \le 3 \left\| \sum_{i=1}^r \alpha_i x_i \right\|$ for all $\alpha_1, \ldots, \alpha_r \in \mathbb{C}$.

By Zenger's lemma, for $c_1, \ldots, c_r \ge 0$ with $\sum_{i=1}^r c_i = 1$, there are a linear combination $\sum_{i=1}^r \mu_i x_i$ in the closed unit ball of X and a functional y on X with norm less than 3 such that $\langle \mu_i x_i, y \rangle = c_i$ for all i and such that $y|M = 0$.

Fix $\delta > 0$. Let $K \subset V$ be compact. For λ in K, we say that a unit vector $y \in X'$ *has property* P_λ if there are $z \in Z$ and $w \in Z_T(K)$ with

$$y = qz, \quad \|(\lambda - T)z\| < \delta, \quad \|z - w\| < \delta.$$

Since we are supposing that approximate eigensequences for S' can be lifted and since $\sigma(T/Z_T(K)) \cap \mathrm{int}(K) = \emptyset$, it follows that, for each λ in $\sigma_{\ell e}(S') \cap \mathrm{int}(K)$, each space $N \in \mathrm{Cof}(X')$ contains unit vectors with property P_λ.

Hence, for $\lambda_1, \ldots, \lambda_r$ in $\sigma_{\ell e}(S') \cap \mathrm{int}(K)$, $M \in \mathcal{F}(X')$ and $c_1, \ldots, c_r \ge 0$ adding up to 1, there are unit vectors $y_1, \ldots, y_r$ in X' and associated vectors z_i, w_i as above such that there is a linear combination $\Sigma \mu_i y_i$ in the unit ball of X' and a vector x in ${}^\perp M$ of norm less than 3 with $|\mu_i| \le 3$ and

$$\langle x, \mu_i y_i \rangle = c_i \quad (i = 1, \ldots, r).$$

For $f \in H^\infty(V)$ and $\lambda \in V$, let $f_\lambda \in H^\infty(V)$ be the function with

$$(z - \lambda) f_\lambda(z) = f(z) - f(\lambda)$$

for $z \in V$. Define

$$z = \sum_{i=1}^r \mu_i w_i .$$

Then $\|qz\| \le 1 + 3r\|q\|\delta$ and, when δ approaches zero, the right-hand side of

$$\left| x \otimes z(f) - \sum_{i=1}^{r} c_i f(\lambda_i) \right| \le \left| \sum_{i=1}^{r} \mu_i \langle x, q f_{\lambda_i}(T|Z_T(K))(T - \lambda_i) w_i \rangle \right| + \left| \sum_{i=1}^{r} \mu_i \langle x, q f(\lambda_i)(w_i - z_i) \rangle \right|$$

tends to zero uniformly for f in the unit ball of $H^\infty(V)$. Choosing M and N appropriately one can achieve that x and z also satisfy the remaining conditions in the following result.

Theorem 18.1.7 *Let $c_1, \dots, c_r \ge 0$ with $\sum_{i=1}^{r} c_i = 1$, and let*

$$\lambda_1, \dots, \lambda_r \in \sigma_{\ell e}(S') \cap V.$$

If $a_1, \dots, a_s \in X$, $b_1, \dots, b_s \in Z_T(V)$ and $\varepsilon > 0$ are arbitrary, then there are vectors $x \in X$, $z \in Z_T(V)$ with $\|x\| \le 3$, with $\|qz\| \le 2$, and with
(i) $\| \sum_{i=1}^{r} c_i \mathcal{E}_{\lambda_i} - x \otimes z \| < \varepsilon$,
(ii) $\|x \otimes b_i\| < \varepsilon, \quad \|a_i \otimes z\| < \varepsilon \quad (i = 1, \dots, s)$. □

Thus we have summarized the main steps in a possible proof of Theorem 18.1.2. A more detailed proof, including also the case of quotients of decomposable operators, can be found in Eschmeier and Prunaru (1990).

18.2 Exercises

1. Show that each subnormal operator is hyponormal.
2. Let $S \in \mathcal{B}(H)$ be subnormal. Show that, for each function $f \in H(\sigma(S))$, $\|f(S)\| \le |f|_{\sigma(S)}$ (using Exercise 17.2.3, as well as the spectral radius and the analytic spectral mapping theorem, one can even show that equality holds here). Conclude that S possesses a contractive functional calculus $\Psi : R(\sigma(S)) \to \mathcal{B}(H)$.
3. Let $S \in \mathcal{B}(H)$ be subnormal with $R(\sigma(S)) = C(\sigma(S))$. Show that S is normal *Hint*: Exercise 15.2.3.
4. Let $T \in \mathcal{B}(X)$ be a bounded operator on a Banach space. Suppose that, for some point $z \in \mathbb{C}$, there are sequences (x_j) in X, (y_k) in X' with

$$\delta_{j,k} p(z) = \langle p(T) x_j, y_k \rangle \quad (j, k \in \mathbb{Z}^+).$$

Show that $\mathrm{Lat}(T)$ is rich.

5. For $T \in \mathcal{B}(X)$ (X a Banach space) show that $\sigma_{re}(T) = \sigma_{\ell e}(T')$.
6. Solve Exercise 18.1.6 as formulated in the main text.

18.3 Additional notes

The first applications of the Scott Brown technique to Banach-space operators were given by Apostol (1981). Theorem 18.1.2 is from Eschmeier and Prunaru (1990). Special cases of Theorem 18.1.2 for operators acting on quotients of closed subspaces of ℓ^p ($1 < p < \infty$) have been obtained by Albrecht and Chevreau (1987). A multivariable version of Theorem 18.1.2 is contained in Eschmeier and Putinar (1996). Using duality methods one can prove an analogue of Theorem 18.1.2 for quotients of decomposable operators (see Eschmeier and Prunaru 1990, Theorem 2.1). Intrinsic characterizations of subdecomposable operators and of quotients of decomposable operators have been given in Albrecht and Eschmeier (1997). For instance, a continuous linear operator $S \in \mathcal{B}(X)$ on a Banach space X is subdecomposable if and only if it satisfies Bishop's property (β), that is, all multiplication operators

$$H(U, X) \to H(U, X), \quad f \mapsto (z - S)f \quad (U \subset \mathbb{C} \text{ open}),$$

are continuous with closed range. For a discussion of property (β), see Chapter 23. A proof of the first part of Theorem 18.1.2 which does not make use of canonical extensions and liftings was given more recently in Eschmeier and Prunaru (2002).

19
Reflexivity of operator algebras

19.1

A subalgebra $\mathfrak{A} \subset \mathcal{B}(H)$ is called *reflexive* if $\mathfrak{A} = \text{Alg Lat}(\mathfrak{A})$, where the space on the right-hand side consists of all operators $C \in \mathcal{B}(H)$ with $\text{Lat}(\mathfrak{A}) \subset \text{Lat}(C)$. A single operator $T \in \mathcal{B}(H)$, or more generally, an n-tuple $T = (T_1, \ldots, T_n)$ in $\mathcal{B}(H)^n$ is called *reflexive* if the smallest WOT-closed subalgebra $\mathcal{W}_T$ of $\mathcal{B}(H)$ containing $T_1, \ldots, T_n$ and the identity operator is reflexive. Equivalently, one can ask that

$$\text{Alg Lat}(T) = \mathcal{W}_T,$$

where Alg Lat(T) consists of all operators $C \in \mathcal{B}(H)$ with $\text{Lat}(T) \subset \text{Lat}(C)$.

Sarason proved in 1966 that each normal operator and each analytic Toeplitz operator on $H^2(\mathbb{T})$ is reflexive. In 1971 it was shown by Deddens that all isometries are reflexive. Since all these operators are subnormal, the following result of Olin and Thomson (1980) contains the above observations as special cases.

Theorem 19.1.1 (Olin and Thomson) *Every subnormal operator is reflexive.*

In the following we indicate some of the main steps leading to a proof of the Olin and Thomson result. For simplicity, we shall restrict ourselves to the case of pure subnormal operators.

Let $S \in \mathcal{B}(H)$ be subnormal. It is easy to see that the set

$$V = \bigcup \{U \subset \mathbb{C} \text{ bounded open; } \sigma(S) \text{ is dominating in } U\}$$

is the largest bounded open set in $\mathbb{C}$ in which $\sigma(S)$ is dominating. For a compact set K in $\mathbb{C}$, let $R(K)$ and $A(K)$ be as above and as in Chapter 1. As usual $R(K)$

is called a *Dirichlet algebra* if

$$\{\mathrm{Re}(f|\partial K);\ f \in R(K)\} \subset C_{\mathbb{R}}(\partial K)$$

is uniformly dense (see Conway 1991, Chapter V.14).

If V is empty, then one can show (Brown 1987) (this result is non-trivial!) that $R(\sigma(S)) = C(\sigma(S))$. In this case, the operator S is normal (Exercise 18.2.3). If the set V is non-empty, then it possesses very nice properties.

Theorem 19.1.2 *Let $\sigma \subset \mathbb{C}$ be compact. Suppose that the largest bounded open set V in $\mathbb{C}$ in which $\sigma(S)$ is dominating is non-empty. Then:*

(i) $\partial V = \partial \overline{V}$ *and* $V = \mathrm{int}(\overline{V}) \subset \widehat{\sigma}$;
(ii) $R(\overline{V})$ *is Dirichlet,* $R(\overline{V}) = A(\overline{V}) = A(V)$*, and the components of V are simply connected;*
(iii) $R(\overline{V}) = A(V)$ *is pointwise boundedly dense in* $H^\infty(V)$*;*
(iv) *the polynomials are weak-$*$ dense in* $H^\infty(V)$*.* □

A proof of Theorem 19.1.2 is given in the appendix to Part III.

In the following let $S \in \mathcal{B}(H)$ be subnormal such that the largest bounded open set V in $\mathbb{C}$ in which $\sigma(S)$ is dominating is non-empty.

Lemma 19.1.3 *If $\sigma(S)$ is not contained in $\overline{V}$, then S has a non-trivial normal part.*

Proof Let D be a closed disc with $D \cap \overline{V} = \emptyset$ and centre $\lambda \in \sigma(S)$. Then $D \cap \sigma(S)$ is dominating in no open set anymore. Thus by the above remarks it follows that $R(D \cap \sigma(S)) = C(D \cap \sigma(S))$. By a result of Clancey and Putnam (1972), the operator S has a non-trivial normal part. □

The results of Chapter 15 can be used to obtain an $H^\infty(V)$-functional calculus for pure subnormal operators.

Theorem 19.1.4 *Let $S \in \mathcal{B}(H)$ be a pure subnormal operator, and let V be the largest bounded open set in $\mathbb{C}$ in which $\sigma(S)$ is dominating. Then S has a unique contractive weak-$*$ continuous functional calculus*

$$\Phi : H^\infty(V) \to \mathcal{B}(H).$$

This functional calculus is isometric and of type $C_{\cdot 0}$.

Proof The uniqueness follows from Theorem 19.1.2(iii).

Since $\sigma(S) \subset \overline{V}$, the subnormal operator S possesses a contractive functional calculus over $R(\overline{V}) = A(V)$. By Corollary 15.1.8, this functional calculus is of type $C_{\cdot 0}$. By Theorem 19.1.2(iii) and the remarks following Theorem 15.1.7, it extends to a contractive $H^\infty(V)$-functional calculus of type $C_{\cdot 0}$. To see that Φ is isometric, note that

$$f(V \cap \sigma(S)) \subset \sigma(\Phi(f))$$

for each $f \in H^\infty(V)$. This spectral inclusion follows from the identity

$$(S - \lambda)\Phi\Big(\frac{f - f(\lambda)}{z - \lambda}\Big) = \Phi(f) - f(\lambda),$$

which is valid for all $f \in H^\infty(V)$ and $\lambda \in V$. □

Since Φ is isometric and weak-$*$ continuous, its range is weak-$*$ closed. By Theorem 19.1.2(iv), the range of Φ coincides with the smallest unital, weak-$*$ closed subalgebra $\mathfrak{A}_S$ of $\mathcal{B}(H)$ containing S. Note that $\Phi : H^\infty(V) \to \mathfrak{A}_S$ is an isometric isomorphism and a weak-$*$ homeomorphism if $H^\infty(V)$ and $\mathfrak{A}_S$ are regarded as the dual spaces of Q and $\mathcal{C}^1(H)/{}^\perp\mathfrak{A}_S$, respectively. In this situation, Φ is called a *dual algebra isomorphism.*

Write V as the disjoint union of its connected components

$$V = \bigcup\{V_k;\ 0 \le k < N\} \quad (N \in \mathbb{Z}^+ \cup \{\infty\}).$$

Let $e_k \in H^\infty(V)$ be the characteristic function of V_k. The operators $P_k = \Phi(e_k)$ are orthogonal projections such that the spaces $H_k = P_k H$ are pairwise orthogonal reducing subspaces for $\mathrm{im}(\Phi) = \mathfrak{A}_S$. Furthermore,

$$\mathrm{SOT} - \sum_k P_k = 1_H .$$

For $f \in H^\infty(V_k)$, let $\widetilde{f}$ be the trivial extension of f onto all of V. Then

$$\Phi_k : H^\infty(V_k) \to \mathcal{B}(H_k), \quad f \mapsto \Phi(\widetilde{f})|H_k ,$$

is an isometric weak-$*$ continuous functional calculus for $S_k = S|H_k$ of type $C_{\cdot 0}$. Since the polynomials are weak-$*$ dense in $H^\infty(V_k)$, the algebra homomorphisms $\Phi_k : H^\infty(V_k) \to \mathfrak{A}_{S_k}$ are dual algebra isomorphisms again. Note that

$$\mathfrak{A}_S = \{C \in \mathcal{B}(H);\ H_k \in \mathrm{Lat}(C) \text{ and } C|H_k \in \mathfrak{A}_{S_k} \text{ for all } k\}$$

(Exercise 19.2.1). We abbreviate the space on the right by $\bigoplus_{0 \le k < N} \mathfrak{A}_{S_k}$.

Fix k and a conformal map $\varphi : V_k \to \mathbb{D}$. The composition

$$\Phi^{(k)} : H^\infty(\mathbb{D}) \longrightarrow H^\infty(V_k) \xrightarrow{\Phi_k} \mathfrak{A}_{S_k},$$

where the first map is given by substituting the argument by φ, is a dual algebra isomorphism (of type $C_{\cdot 0}$). A standard argument (Exercise 19.2.3) shows that each operator in $\mathfrak{A}_{S_k}$ is subnormal again. Hence $\Phi^{(k)}$ is an isometric weak-$*$ continuous H^∞-functional calculus of the subnormal operator $T_k = \Phi_k(\varphi)$. In particular, we have $\mathfrak{A}_{T_k} = \mathfrak{A}_{S_k}$.

Thus we have proved the following decomposition theorem.

Theorem 19.1.5 (Conway and Olin) *Let S be a pure subnormal operator. Then there is an orthogonal decomposition $H = \bigoplus_{0 \le k < N} H_k$ into countably many reducing spaces for S such that, with $S_k = S|H_k$,*

$$\mathfrak{A}_S = \bigoplus_{0 \le k < N} \mathfrak{A}_{S_k}$$

and such that $H^\infty(\mathbb{D}) \cong \mathfrak{A}_{S_k}$ as dual algebras for all k. □

To prove that a pure subnormal operator is reflexive it therefore suffices to consider the following special case.

Theorem 19.1.6 *Let $T \in \mathcal{B}(H)$ be subnormal with a weak-$*$ continuous isometric functional calculus $\Phi : H^\infty(\mathbb{D}) \to \mathcal{B}(H)$. Then T is reflexive and* Alg Lat$(T) = \mathfrak{A}_T$. □

Indeed, suppose that Theorem 19.1.6 holds and that $C \in$ Alg Lat(S) for some pure subnormal operator S. Then, with the notations fixed above,

$$C|H_k \in \text{Alg Lat}(S_k) = \text{Alg Lat}(T_k) = \mathfrak{A}_{T_k} = \mathfrak{A}_{S_k}$$

for each k. Hence $C \in \mathfrak{A}_S$, and the reflexivity of S is proved.

Before we indicate some of the ideas leading to a proof of Theorem 19.1.6, we consider an elementary, but typical, example.

Example 19.1.7 Let $T = M_z$ on $H = H^2(\mathbb{D})$. Since H is a Hilbert space of analytic functions, it possesses a reproducing kernel. More precisely, there is a conjugate analytic function $k : \mathbb{D} \to H$ with

$$[f, k(\lambda)] = f(\lambda) \quad (f \in H,\ \lambda \in \mathbb{D})$$

such that H is the closed linear span of all the k_λ. It follows that

$$(\lambda - T)^* k(\lambda) \equiv 0 .$$

If $C \in \text{Alg Lat}(T)$, then $C^* \in \text{Alg Lat}(T^*)$ acts on the eigenspaces of T^* as a scalar multiple of the identity (Exercise 19.2.4). Hence $C^* k(\lambda) = \overline{g(\lambda)} k(\lambda)$ for $\lambda \in \mathbb{D}$ with a bounded function g. Since

$$[Cf, k(\lambda)] = g(\lambda)[f, k(\lambda)] = (gf)(\lambda),$$

we conclude that $C = \Phi(g) \in \mathfrak{A}_T$, where Φ is the H^∞-functional calculus of T.

The above example motivates us to give the next definition.

Definition 19.1.8 *Let $T \in \mathcal{B}(H)$. A space $M \in \text{Lat}(T)$ is an* analytic *(or* $\mathbb{D}$-analytic*) invariant subspace for T if there is a non-zero conjugate analytic function $k : \mathbb{D} \to M$ with $(\lambda - T|M)^* k(\lambda) \equiv 0$.*

In the situation of Theorem 19.1.6 one cannot expect the whole space H to be an analytic invariant subspace for T (as in Example 19.1.7). Nevertheless the proof of Theorem 19.1.6 depends on the observation that there is a dense set of vectors x in H such that the spaces

$$M_x = \bigvee \{T^k x;\ k \geq 0\}$$

are analytic invariant subspaces for T. As before, for $x, y \in H$, regard

$$x \otimes y : H^\infty(\mathbb{D}) \to \mathbb{C}, \quad x \otimes y(f) = [\Phi(f)x, y],$$

as an element in the predual Q of $H^\infty(\mathbb{D})$. For $k \in \mathbb{Z}^+$, define functionals $\mathcal{E}^{(k)}$ in Q by

$$\langle \mathcal{E}^{(k)}, f \rangle = f^{(k)}(0)/k! \quad (f \in H^\infty(\mathbb{D})).$$

Suppose that x, y_k are vectors in H such that $x \otimes y_k = \mathcal{E}^{(k)}$ for all $k \in \mathbb{Z}^+$ and such that $k(\lambda) = \sum_j y_j \overline{\lambda}^j$ converges on $\mathbb{D}$. Then $\mathcal{E}_\lambda = x \otimes k(\lambda)$ for $\lambda \in \mathbb{D}$, and the function

$$\mathbb{D} \to M_x, \quad \lambda \mapsto P_{M_x} k(\lambda),$$

turns M_x into an analytic invariant subspace for T.

Thus to prove that there is a dense set of vectors x generating an analytic invariant subspace for T, it suffices to prove the second part of the next result.

Theorem 19.1.9 *Let T be as in Theorem 19.1.6.*

(i) *There is a constant $r > 0$ such that, for each $L \in Q$ and any given vectors $a, b \in H$, there are $x, y \in H$ with $L = x \otimes y$ and*

$$\|x - a\| \leq r\|L - a \otimes b\|^{1/2}, \quad \|y\| \leq r(\|L - a \otimes b\|^{1/2} + \|b\|) .$$

(ii) *For each $\varepsilon > 0$, there is a constant $c = c(\varepsilon) > 0$ such that, for each sequence $(L_k)_{k\geq 1}$ in Q and each vector $a \in H$, there are vectors $x, y_k \in H$ with $\|x - a\| < \varepsilon$ and such that, for all $k \in \mathbb{N}$,*

$$L_k = x \otimes y_k, \quad \|y_k\| \leq ck^{12}\|L_k\| . \qquad \square$$

For a proof of Theorem 19.1.9 that even works in the multivariable case (for subnormal n-tuples with isometric H^∞-functional calculus over the Euclidean unit ball in $\mathbb{C}^n$), the reader is referred to Eschmeier (1999, Theorem 1.10 and Theorem 2.3).

Let M_x be an analytic invariant subspace for T via a conjugate analytic function $k : \mathbb{D} \to M_x$. Then, as in Example 19.1.7, it follows (Exercise 19.2.5) that, for each operator $C \in \text{Alg Lat}(T)$, there is a function $g = g_{C,x}$ in $H^\infty(\mathbb{D})$ with

$$(C|M_x)^* k(\lambda) = \overline{g(\lambda)} k(\lambda)$$

for $\lambda \in \mathbb{D}$. Since by Theorem 19.1.9 the set

$$\mathcal{C} = \{x \in H;\ M_x \text{ is a } \mathbb{D}\text{–analytic invariant subspace for } T\}$$

is dense in H, to complete the proof of Theorem 19.1.6, it suffices to show that, for each operator $C \in \text{Alg Lat}(T)$ and each vector $x \in \mathcal{C}$, we have

$$\Phi(g_{C,x})|M_x = C|M_x$$

and that all the functions $g_{C,x}$ $(x \in \mathcal{C})$ coincide.

For a complete proof of Theorem 19.1.6 along these lines, we refer the reader to Section 3 in Eschmeier (1999, Theorem 3.6 and Theorem 3.7).

19.2 Exercises

Throughout Exercises 19.2.1 to 19.2.3, let $S \in \mathcal{B}(H)$ be a subnormal operator with minimal normal extension $N \in \mathcal{B}(K)$.

1. Let $\Phi : H^\infty(V) \to \mathfrak{A}_S$ be a dual algebra isomorphism. For $k = 0, \ldots, N-1$, denote by $e_k \in H^\infty(V)$ the characteristic functions of the connected components of V. Set $H_k = \operatorname{im}\Phi(e_k)$, $S_k = S|H_k$. Show:

(a) $\mathfrak{A}_S = \{A \in \mathcal{B}(H);\ H_k \in \operatorname{Lat}(A) \text{ and } A|H_k \in \mathfrak{A}_{S_k} \text{ for all } k\}$;

(b) $\mathfrak{A}_S \to \bigoplus_{\ell^\infty} \mathfrak{A}_{S_k}$, $A \mapsto (A|H_k)_k$, is an isometric isomorphism;

(c) if $L : \mathfrak{A}_S \to \mathbb{C}$ is weak-$*$ continuous, then $\|L\| = \sum_k \|L|\mathfrak{A}_{S_k}\|$.

2. Suppose that $f : \sigma(N) \to \mathbb{C}$ is bounded and measurable with $f(N)H \subset H$. Show that $\sigma(f(N))$ is equal to the spectrum of the minimal normal extension of $f(N)|H$. Deduce that $\|f(N)\| = \|f(N)|H\|$.

3. Let μ be a scalar spectral measure for N, and let $\Phi : L^\infty(\mu) \to W^*(N)$ be the associated isomorphism of von Neumann algebras (= weak-$*$ continuous $*$-isomorphism). Show that Φ induces a dual algebra isomorphism

$$\Psi : P^\infty(\mu) \to \mathfrak{A}_S, \quad f \mapsto \Phi(f)|H\,.$$

Hint: use Exercise 19.2.2 to show that Ψ is isometric.

4. Let $T \in \mathcal{B}(H)$ and let $C \in \operatorname{Alg}\operatorname{Lat}(T)$. Show that C restricted to the eigenspaces of T acts as a scalar multiple of the identity operator.

5. Let M be an analytic invariant subspace for T via a conjugate analytic function $k : \mathbb{D} \to M$. Show that, for each $C \in \operatorname{Alg}\operatorname{Lat}(T)$, there is a unique bounded analytic function $g : \mathbb{D} \to \mathbb{C}$ with $(C|M)^*k(\lambda) = \overline{g(\lambda)}k(\lambda)$ for $\lambda \in \mathbb{D}$.

A weak-$*$ closed subalgebra $\mathfrak{A} \subset \mathcal{B}(H)$ is said to possess *property* $(\mathbb{A}_1(r))$, where $r \geq 1$, if, for any weak-$*$ continuous linear functional $L : \mathfrak{A} \to \mathbb{C}$ and any $s > r$, there are $x, y \in H$ with $L = [x \otimes y]$ in $\mathcal{C}^1(H)/{}^\perp\mathfrak{A}$ and with $\|x\|\,\|y\| \leq s\|L\|$.

6. Let $N \in \mathcal{B}(H)$ be normal. Show that $\mathfrak{A}_N$ has property $(\mathbb{A}_1(1))$.

7. Let $T \in \mathcal{B}(H)$ be such that $\mathfrak{A}_T$ has property $(\mathbb{A}_1(r))$. Show that $\mathfrak{A}_T = W_T$ and that the WOT-topology and the weak-$*$ topology coincide on this algebra.

8. Show that the dual algebra generated by a pure subnormal operator has property $(\mathbb{A}_1(r))$ for some $r \geq 1$. *Hint*: use Theorem 19.1.5 to reduce the assertion to Theorem 19.1.9(i).

9. Suppose that the reflexivity of normal and pure subnormal operators has been shown. Prove that each subnormal operator is reflexive.

19.3 Additional notes

Let T be a contraction mapping with a weak-$*$ continuous functional calculus map $\Phi : H^\infty(\mathbb{D}) \to \mathcal{B}(H)$. If $\sigma(T)$ is dominating in $\mathbb{D}$, then Φ is isometric (*cf.* the proof of Theorem 19.1.4). Brown and Chevreau (1988) proved that each contraction T with an isometric weak-$*$ continuous H^∞-functional calculus is reflexive. Using this result Conway and Dudziak (1990) (and independently, Prunaru) proved that each operator $T \in \mathcal{B}(H)$ whose spectrum $\sigma(T)$ is a spectral set for T is reflexive. In the multivariable case it was shown by Bercovici (1994) that each commuting family of isometries is reflexive. Theorem 19.1.6 was generalized in Eschmeier (1999) to subnormal n-tuples with a weak-$*$ continuous H^∞-functional calculus over the unit ball in $\mathbb{C}^n$.

20

Invariant subspaces for commuting contractions

20.1

A tuple $T = (T_1, \ldots, T_n) \in \mathcal{B}(H)^n$ of commuting bounded operators on a Hilbert space H is called a *spherical contraction* if $\sum_{i=1}^n T_i^* T_i \leq 1$. We know from Chapter 16 that a contraction $T \in \mathcal{B}(H)$ with dominating spectrum in the open unit disc has non-trivial invariant subspaces. Therefore it seems natural to ask whether, for every spherical contraction $T \in \mathcal{B}(H)^n$ with dominating *joint spectrum* in the open Euclidean unit ball $\mathbb{B} = \{z \in \mathbb{C}^n;\ |z| < 1\}$, the *joint invariant subspace lattice* of T

$$\mathrm{Lat}(T) = \bigcap_{i=1}^{n} \mathrm{Lat}(T_i)$$

contains non-trivial spaces.

In Chapter 14, we used the Sz.-Nagy dilation theorem to prove von Neumann's inequality for contractions $T \in \mathcal{B}(H)$. Since the polynomials are uniformly dense in the ball algebra $A(\mathbb{B}) = \{f \in C(\overline{\mathbb{B}});\ f|\mathbb{B} \text{ is analytic}\}$, von Neumann's inequality on the unit ball, that is, the inequality

$$\|p(T)\| \leq |p|_{\overline{\mathbb{B}}} \quad (p \in \mathbb{C}[z_1, \ldots, z_n]),$$

is equivalent to the existence of a contractive ball algebra functional calculus for T. However, it is well known (see Drury 1978; or Arveson 1998) that, for each $n > 1$, there are spherical contractions $T \in \mathcal{B}(H)^n$ for which the polynomial functional calculus is unbounded, that is, for which

$$\sup\{\|p(T)\|;\ p \in \mathbb{C}[z_1, \ldots, z_n] \text{ with } |p|_{\overline{\mathbb{B}}} \leq 1\} = \infty .$$

Our aim in the following is to indicate that some of the invariant-subspace results valid for contractions with rich spectrum in the unit disc possess multivariable analogues, at least for spherical contractions that admit normal boundary dilations.

Definition 20.1.1 *Let $n \in \mathbb{N}$, let H be a Hilbert space, and let $T \in \mathcal{B}(H)^n$ be a commuting tuple.*

(i) *A* spherical unitary *on H is a commuting tuple $U \in \mathcal{B}(H)^n$ of normal operators such that $\sum_{i=1}^n U_i^* U_i = 1$.*

(ii) *A* spherical dilation *of T is a spherical unitary $U \in \mathcal{B}(K)^n$ on a larger Hilbert space K containing H such that*

$$T^k = P\, U^k | H \quad (k \in (\mathbb{Z}^+)^n),$$

where P is the orthogonal projection from K onto H and $T^k = T_1^{k_1} \cdots T_n^{k_n}$.

(iii) *A spherical dilation $U \in \mathcal{B}(K)^n$ of T is* minimal *if the only reducing subspace for U containing H is the space K itself.*

A commuting tuple $T = (T_1, \ldots, T_n) \in \mathcal{B}(H)^n$ is called a *spherical isometry* if $\sum_{i=1}^n T_i^* T_i = 1$. Let $T \in \mathcal{B}(H)^n$ be a spherical contraction with a spherical dilation $U \in \mathcal{B}(K)^n$. By the spectral theorem for commuting normal operators, the polynomial functional calculus of U extends to a unique C^*-algebra homomorphism $\pi : C(\partial\mathbb{B}) \to \mathcal{B}(K)$. Thus T satisfies von Neumann's inequality and possesses a contractive ball algebra functional calculus $\Psi : A(\mathbb{B}) \to \mathcal{B}(H)$. We call Ψ, or also T, *absolutely continuous* if Ψ extends to a weak-$*$ continuous algebra homomorphism $\Phi : H^\infty(\mathbb{B}) \to \mathcal{B}(H)$. If one wants to show that T has non-trivial joint invariant subspaces, then one may suppose that T is *completely non-unitary*, that is, possesses no non-zero reducing subspace M such that $T|M$ is a spherical unitary. In this case, T is absolutely continuous. Indeed, the following decomposition theorem can be proved.

Theorem 20.1.2 *Let $\Psi : A(\mathbb{B}) \to \mathcal{B}(H)$ be a contractive algebra homomorphism. Then there are contractive algebra homomorphisms*

$$\Psi_a, \Psi_s : A(\mathbb{B}) \to \mathcal{B}(H)$$

with $\Psi = \Psi_a + \Psi_s$ such that Ψ_a is absolutely continuous, Ψ_s extends to a C^*-algebra homomorphism $\Phi_s : C(\partial\mathbb{B}) \to \mathcal{B}(H)$, and

$$\Psi_a(f)\Psi_s(g) = 0 = \Psi_s(g)\Psi_a(f) \quad (f, g \in A(\mathbb{B})). \qquad \square$$

This theorem can be proved in the same way as Theorem 14.1.12. It suffices to replace the Lebesgue decomposition theorem and the F. and M. Riesz theorem by suitable multivariable generalizations. To be more precise, let $M(\partial\mathbb{B})$ be the space of all regular, complex Borel measures on $\partial\mathbb{B}$. Denote by M_0 its weak-$*$ compact, convex subset consisting of all probability measures $\rho \in M(\partial\mathbb{B})$ with

$$f(0) = \int_{\partial\mathbb{B}} f \, d\rho \quad (f \in A(\mathbb{B})) .$$

Definition 20.1.3 *Let $\mu \in M(\partial\mathbb{B})$. Then μ is called* absolutely continuous *if $\mu \ll \rho$ for some measure $\rho \in M_0$. The measure μ is called* singular *if it is concentrated on an F_σ-set $N \subset \partial\mathbb{B}$ with $\rho(N) = 0$ for all $\rho \in M_0$.*

By the Glicksberg–König–Seever decomposition theorem (for example, see Rudin 1980, Theorem 9.4.4), every measure $\mu \in M(\partial\mathbb{B})$ has a unique decomposition $\mu = \mu_a + \mu_s$ into an absolutely continuous part μ_a and a singular part μ_s. By Henkin's theorem (Exercise 20.2.1) and the Cole–Range theorem (see Rudin 1980, Theorem 9.6.1), a measure $\mu \in M(\partial\mathbb{B})$ is absolutely continuous if and only if it is a Henkin measure in the sense that, for every *Montel sequence* (f_k) in $A(\mathbb{B})$, we have

$$\lim_{k\to\infty} \int f_k \, d\mu = 0 .$$

Since every measure $\mu \in A(\mathbb{B})^\perp$ is obviously a Henkin measure, we obtain in particular that the only singular measure $\mu \in A(\mathbb{B})^\perp$ is the trivial measure $\mu = 0$. Now, to prove Theorem 20.1.2, it suffices to apply the Glicksberg–König–Seever theorem to an arbitrary family of representing measures of Ψ, and then to argue exactly as in the one-variable case.

As a useful byproduct of the above proof of Theorem 20.1.2, one obtains:

Corollary 20.1.4 *Let $U \in \mathcal{B}(K)^n$ be a minimal spherical dilation of a spherical contraction $T \in \mathcal{B}(H)^n$. If T is absolutely continuous, then so is U.* □

Let $T \in \mathcal{B}(H)^n$ be a commuting tuple with a continuous ball algebra functional calculus $\Psi : A(\mathbb{B}) \to \mathcal{B}(H)$. We say that T, or also Ψ, is *of type $C_{0\cdot}$* if $\Psi(f_k) \xrightarrow{\text{SOT}} 0$ for each Montel sequence (f_k) in $A(\mathbb{B})$. We say that T, or Ψ, is *of type $C_{\cdot 0}$* if $\widetilde{\Psi} : A(\mathbb{B}) \to \mathcal{B}(H)$ defined by

$$\widetilde{\Psi}(f) = \Psi(\widetilde{f})^* \quad (\widetilde{f}(z) = \overline{f(\overline{z})})$$

is of type $C_{0\cdot}$. Exactly as in the one-variable case, it is possible to prove that T has always non-trivial invariant subspaces if T is neither $C_{0\cdot}$ nor $C_{\cdot 0}$. A proof can be based on the ball analogue of Theorem 15.1.3.

Theorem 20.1.5 *For each function $g \in C(\overline{\mathbb{B}})$, the operator*

$$S_g : A(\mathbb{B}) \to C(\overline{\mathbb{B}})/A(\mathbb{B}), \quad f \mapsto [gf],$$

is compact. □

Theorem 20.1.5 follows from well-known compactness properties of suitable Hankel-type operators (see Rudin 1980, Theorem 6.5.4). A proof valid in more general situations can be found in Didas (2002; see also Cole and Gamelin 1982). A commuting tuple $S \in \mathcal{B}(H)^n$ is called *subnormal* if there is a commuting tuple $N \in \mathcal{B}(K)^n$ of normal operators on a Hilbert space $K \supset H$ such that $H \in \text{Lat}(N)$ and $S = N|H$. Exactly as in the one-variable case (Chapter 15), one can use Theorem 20.1.5 to prove the following corollaries.

Corollary 20.1.6 *Let $\Psi : A(\mathbb{B}) \to \mathcal{B}(H)$ be a continuous algebra homomorphism. Then Ψ is not of type $C_{\cdot 0}$ if and only if there is a non-zero bounded linear map $j : C(\partial\mathbb{B}) \to H$ with $\Psi(f)j = jM_{(f|\partial\mathbb{B})}$ for all $f \in A(\mathbb{B})$.* □

Corollary 20.1.7 *Let $\Psi : A(\mathbb{B}) \to \mathcal{B}(H)$ be a continuous algebra homomorphism, and let $T = (\Psi(z_1), \ldots, \Psi(z_n))$. If T is neither of type $C_{0\cdot}$ nor of type $C_{\cdot 0}$, then either all components of T are scalar multiples of the identity operator, or the hyperinvariant subspace lattice* $\text{Hyp}(T) = \text{Lat}((T)')$ *of T is non-trivial.* □

Corollary 20.1.8 *Let $S \in \mathcal{B}(H)^n$ be a subnormal tuple that possesses a continuous $A(\mathbb{B})$-functional calculus. Then S is completely non-unitary if and only if S is of type $C_{\cdot 0}$.* □

Using the above results, it is possible to prove an invariant-subspace result for spherical contractions that contains the corresponding one-variable result of Brown, Chevreau, and Pearcy (Theorem 16.1.2) as a special case. To formulate it, we need a suitable notion of joint spectrum. Let $T \in \mathcal{B}(H)^n$ be a commuting tuple on H. The *Harte spectrum* $\sigma^{\mathcal{H}}(T)$ of T consists of those points $\lambda \in \mathbb{C}^n$ for which the map

$$H \to H^n, \quad x \mapsto ((\lambda_i - T_i)x)_{i=1}^n,$$

is not bounded below or for which the map

$$H^n \to H, \quad (x_i)_{i=1}^n \mapsto \sum_{i=1}^n (\lambda_i - T_i)x_i,$$

is not surjective. The Harte spectrum is a compact subset of $\mathbb{C}^n$ which, for $n = 1$, reduces to the usual spectrum of an operator $T \in \mathcal{B}(H)$. For this and more results on joint spectra, see Eschmeier and Putinar (1996).

Theorem 20.1.9 *Let $T \in \mathcal{B}(H)^n$ be a commuting tuple that possesses a spherical dilation. Suppose that the Harte spectrum $\sigma^{\mathcal{H}}(T)$ of T is dominating in $\mathbb{B}$. Then T has non-trivial joint invariant subspaces.* □

Here a set $\sigma \subset \mathbb{C}^n$ is, of course, called *dominating* in $\mathbb{B}$ if $|f|_B = |f|_{\mathbb{B}\cap\sigma}$ for all $f \in H^\infty(\mathbb{B})$. The results preceding Theorem 20.1.9 can be used to reduce the assertion to the case where T possesses a weak-$*$ continuous $H^\infty(\mathbb{B})$-functional calculus of type $C_{\cdot 0}$. Then an application of the Scott Brown factorization technique leads to a proof of Theorem 20.1.9. An additional complication arises from the fact that one is not allowed to assume that the Harte spectrum of T coincides with the left essential spectrum of T, as we did in the one-dimensional case.

To overcome this difficulty, it can be shown that, in the above situation, the tuple T has an absolutely continuous co-isometric extension, that is, there is an absolutely continuous spherical isometry $V \in \mathcal{B}(K)^n$ on a Hilbert space K containing H such that $H \in \mathrm{Lat}(V^*)$ and $T = V^*|H$. The tuple V has a Wold-type decomposition $V = S \oplus R^*$ into a completely non-unitary subnormal tuple S and a spherical unitary R^*. By Corollary 20.1.8, the adjoint S^* is of type $C_{0\cdot}$, and T is the restriction of $S^* \oplus R$. Now a standard application of the Scott Brown factorization technique, using the $C_{\cdot 0}$-property of T and the $C_{0\cdot}$-property of S^*, allows one to complete the proof of Theorem 20.1.9. For details, we refer the reader to Eschmeier (1997).

An extension of the methods explained in Chapter 19 was used in Eschmeier (1999) to prove a reflexivity result for algebras of subnormal operators on the unit ball which extends Theorem 19.1.6.

Theorem 20.1.10 *Let $T \in \mathcal{B}(H)^n$ be subnormal spherical contraction with a weak-$*$ continuous isometric functional calculus $\Phi : H^\infty(\mathbb{B}) \to \mathcal{B}(H)$. Then T is reflexive and* $\mathrm{Alg\,Lat}(T) = \mathfrak{A}_T$. □

It is an open question whether the invariant-subspace result formulated in Theorem 20.1.9 remains true when the Harte spectrum $\sigma^{\mathcal{H}}(T)$ is replaced by the Taylor spectrum of T. The Taylor spectrum always contains the Harte spectrum, and it is, in many respects, the more natural joint spectrum (see Eschmeier and Putinar 1996). The answer to this question is positive in dimension $n = 2$ (see Didas 2002).

20.2 Exercises

1. Use Theorem 20.1.5 and the ball analogue of Corollary 15.1.4 to prove that, if $\mu \in M(\partial\mathbb{B})$ is a Henkin measure, then every measure $\nu \in M(\partial\mathbb{B})$ with $\nu \ll \mu$ is a Henkin measure. This result is usually referred to as Henkin's theorem.
2. Complete the proof of Theorem 20.1.2 as outlined in this chapter.
3. Let $\Psi : A(\mathbb{B}) \to \mathcal{B}(H)$ be a contractive algebra homomorphism. Show that there is at most one pair (Ψ_1, Ψ_2) consisting of contractive algebra homomorphisms $\Psi_i : A(\mathbb{B}) \to \mathcal{B}(H)$ with $\Psi = \Psi_1 + \Psi_2$ such that Ψ_1 has a representing family of measures that are absolutely continuous, and Ψ_2 has a representing family of measures that are singular on $\partial\mathbb{B}$.
4. Let $\Psi : A(\mathbb{B}) \to \mathcal{B}(H)$ be a continuous algebra homomorphism. Show that there is an extension of Ψ to a weak-$*$ continuous algebra homomorphism $\Phi : H^\infty(\mathbb{B}) \to \mathcal{B}(H)$ if and only if Ψ has a representing family of measures on $\partial\mathbb{B}$ that are absolutely continuous, or equivalently, if each representing family of measures for Ψ on $\partial\mathbb{B}$ consists of absolutely continuous measures.
5. Use Rudin (1980, Theorem 6.5.4) to prove Theorem 20.1.5.
6. Let $T \in \mathcal{B}(H)^n$ be a spherical contraction. Show that there is a unique orthogonal decomposition $H = H_0 \oplus H_1$ with reducing subspaces H_0, H_1 for T such that $T|H_0$ is completely non-unitary and $T|H_1$ is a spherical unitary.

20.3 Additional notes

Henkin's theorem and the Cole–Range theorem imply that on $\partial\mathbb{B}$ the set of all Henkin measures coincides with the band of measures generated by the set M_0 of all measures that represent the point evaluation at 0 on $A(\mathbb{B})$. Although Henkin's theorem fails on the unit polydisc $\mathbb{D}^n$, one can still show (see Eschmeier 2001*a*) that the band of measures generated by the set $M_0(\mathbb{T}^n)$ of all measures on $\mathbb{T}^n$ that represent the point evaluation at 0 on $A(\mathbb{D}^n)$ is the largest band in $M(\mathbb{T}^n)$ consisting entirely of Henkin measures. This result is used in Eschmeier (2001*a*, 2001*b*) to show that Theorem 20.1.9 remains true if the open unit ball is replaced by the open unit polydisc, and to prove an analogue of Theorem 20.1.10 on $\mathbb{D}^n$. Since by Ando's dilation theorem every pair $T = (T_1, T_2)$ of commuting contractions has a dilation to a commuting pair $U = (U_1, U_2)$ of unitary operators, we obtain in particular that every pair $T = (T_1, T_2)$ of commuting contractions with dominating Harte spectrum in $\mathbb{D}^2$ has a non-trivial joint invariant subspace. It is an open question whether the Harte spectrum can be replaced by the Taylor spectrum in this result. This question has a positive answer, even in arbitrary dimension n, if the tuple T is supposed to consist of doubly commuting contractions (see Albrecht and Ptak 1998).

Theorem 20.1.10 remains true if the condition that T possesses an isometric weak-$*$ continuous $H^\infty(\mathbb{B})$-functional calculus is replaced by the hypothesis that the Taylor spectrum of T is dominating in $\mathbb{B}$ (see Eschmeier 1999). Reflexivity results for spherical contractions that extend the above mentioned results for subnormal spherical contractions can be found in Eschmeier (2001*c*). In Didas (2002), many of the above results are shown to hold for commuting tuples $T \in \mathcal{B}(H)^n$ which possess an isometric, weak-$*$ continuous H^∞-functional calculus $\Phi : H^\infty(D) \to \mathcal{B}(H)$ and a normal boundary dilation over a strictly pseudoconvex domain D in $\mathbb{C}^n$, or more generally, a strictly pseudoconvex open set D in a Stein submanifold X of $\mathbb{C}^n$. In particular, every subnormal n-tuple with dominating Taylor spectrum in D, or with isometric weak-$*$ continuous $H^\infty(D)$-functional calculus, is shown to be reflexive.

By a result of Bercovici (1994) at least each commuting family of isometries on a Hilbert space is reflexive. Azoff and Ptak (1995) extended this result to the case of jointly quasinormal tuples. The question whether each spherical isometry is reflexive seems still to be open. Some partial results have been obtained in Müller and Ptak (1999) and Eschmeier (2001*c*). The general question, whether every subnormal tuple $S \in \mathcal{B}(H)^n$ on a Hilbert space is reflexive, is one of the major open questions in this area at the time of this writing.

Appendix to Part III

It is the purpose of this appendix to indicate a possible proof of Theorem 19.1.2.

For each open set U in $\mathbb{C}$, we denote by $P^\infty(U)$ the weak-$*$ closure of the polynomials in $L^\infty(U) \cong L^1(U)'$, where $L^1(U)$ and $L^\infty(U)$ are formed with respect to the planar Lebesgue measure. Then $P^\infty(U)$ is a weak-$*$ closed subalgebra of $H^\infty(U)$.

Let σ be a compact set in $\mathbb{C}$. We say that σ is *dominating* for $P^\infty(U)$ if

$$|f|_U = |f|_{U\cap\sigma} \quad (f \in P^\infty(U)) .$$

Let us assume that σ is dominating for $P^\infty(U)$ for some non-empty, bounded open set U in $\mathbb{C}$. Then

$$V = \bigcup\{U;\ U \subset \mathbb{C} \text{ is bounded and open, and } \sigma \text{ is dominating for } P^\infty(U)\}$$

is the largest bounded open set such that σ is dominating for $P^\infty(V)$.

Define $K = \sigma(M_z, P^\infty(V))$. If $\lambda \notin K$, then $1/(\lambda - z) \in P^\infty(V)$. In particular, $V \subset \mathrm{int}(K)$ and $R(K)|V \subset P^\infty(V)$. A standard result in rational approximation theory (*cf.* the proof of Theorem 3 in Brown (1987)) shows that there is a family $(L_i)_{i\in I}$ of bounded components of $\mathbb{C} \setminus K$ such that the set $C = K \cup \bigcup\{L_i;\ i \in I\}$ has the following properties:

(i) K is dominating in $\mathrm{int}(C)$ (i.e., for $H^\infty(\mathrm{int}(C))$),
(ii) $R(C)$ is a Dirichlet algebra.

By Conway (1991, Theorem VI.4.8) the algebra $R(C)$ is pointwise boundedly dense in $H^\infty(\mathrm{int}(C))$. Hence $H^\infty(\mathrm{int}(C))|V \subset P^\infty(V)$. We claim that the restriction map

$$H^\infty(\mathrm{int}(C)) \to P^\infty(V), \quad f \mapsto f|V ,$$

is isometric. Otherwise, $|g|_{\text{int}(C)} > |g|_V$ for some $g \in H^\infty(\text{int}(C))$, and there would be a point $a \in \text{int}(C) \cap K$ with $|g(a)| > |g|_V$. But then $a \notin \overline{V}$ and because of the fact that

$$|(z-a)f|_V \geq \text{dist}(a, V)|f|_V \quad (f \in P^\infty(V))$$

the proper ideal $(z-a)P^\infty(V) \subset P^\infty(V)$ is weak-$*$ closed. By the Hahn–Banach theorem this ideal is annihilated by some weak-$*$ continuous linear functional $\rho : P^\infty(V) \to \mathbb{C}$ with $\rho(1) = 1$. Since on polynomials ρ acts as the point evaluation at a, the functional ρ is multiplicative (*cf.* Exercise 14.2.3). By the cited result in Conway (1991), there is a sequence (r_n) of rational functions with no poles on C which is uniformly bounded on C and converges to g pointwise on $\text{int}(C)$. We obtain the contradiction

$$|g(a)| = |\lim_n r_n(a)| = |\lim_n \rho(r_n|V)| = |\rho(g|V)| \leq |g|_V .$$

The observation that

$$|f|_{\text{int}(C)} = |f|_V = |f|_{V\cap\sigma} \leq |f|_{\text{int}(C)\cap\sigma}$$

for all $f \in H^\infty(\text{int}(C))$ shows that σ is dominating in $\text{int}(C)$. The maximality of V implies that $V = \text{int}(C)$. It follows that

$$K = C, \quad H^\infty(V) = P^\infty(V),$$

and that V coincides with the largest bounded open set in $\mathbb{C}$ such that σ is dominating in V for $H^\infty(V)$.

Using for a third time Conway (1991, Theorem VI.4.8), we see that $R(\overline{V})$ is pointwise boundedly dense in $H^\infty(V)$. Using the fact that $\partial\overline{V} \subset \partial V \subset \partial K$ and that $R(K)$ is a Dirichlet algebra, we can easily deduce that $R(\overline{V})$ is a Dirichlet algebra. By Gamelin (1969, Corollary II.9.3), we have $R(\overline{V}) = A(\overline{V}) = A(V)$. The last equality follows because $V = \text{int}(\overline{V})$, or equivalently, $\partial V = \partial\overline{V}$. Since $R(\overline{V})$ is a Dirichlet algebra, the connected components of $V = \text{int}(\overline{V})$ are simply connected (see Conway 1991, Theorem VI.5.2).

To see that V is contained in the polynomially convex hull $\widehat{\sigma}$ of σ (see Definition 4.1.1) it suffices to observe that, for each point $z \in V$ and each polynomial p, we have

$$|p(z)| \leq |p|_V = |p|_{V\cap\sigma} \leq |p|_\sigma .$$

Thus the proof of Theorem 19.1.2 is complete.

References

Albrecht, E. and Chevreau, B. (1987). Invariant subspaces for ℓ^p-operators having Bishop's property (β) on a large part of their spectrum, *J. Operator Theory*, **18**, 339–72.

Albrecht, E. and Eschmeier, J. (1997). Analytic functional models and local spectral theory, *Proc. London Math. Soc.* (3), **75**, 323–48.

Albrecht, E. and Ptak, M. (1998). Invariant subspaces of doubly commuting contractions with rich Taylor spectrum, *J. Operator Theory*, **40**, 373–84.

Ambrozie, C. and Müller, V. (2003). Polynomially bounded operators and invariant subspaces, preprint.

Apostol, C. (1981). The spectral flavour of Scott Brown's techniques, *J. Operator Theory*, **6**, 3–12.

Apostol, C. and Chevreau, B. (1982). On M–spectral sets and rationally invariant subspaces, *J. Operator Theory*, **7**, 247–66.

Arveson, W. B. (1998). Subalgebras of C^*-algebras III: Multivariable operator theory, *Acta Math.,* **181**, 159–228.

Azoff, E. A. and Ptak, M. (1995). Jointly quasinormal families are reflexive, *Acta Sci. Math. (Szeged)*, **61**, 545–7.

Beauzamy, B. (1985). Un opérateur sans sous-espace invariant: simplification de l'exemple d'Enflo, *Integral Equations and Operator Theory*, **8**, 314–84.

Bercovici, H. (1990). Notes on invariant subspaces, *Bull. American Math. Soc.*, **23**, 1–36.

Bercovici, H. (1994). A factorization theorem with applications to invariant subspaces and the reflexivity of isometries, *Math. Res. Letters*, **1**, 511–18.

Bercovici, H., Foiaş, C., and Pearcy, C. (1985). *Dual algebras with applications to invariant subspaces and dilation theory*, CBMS Regional Conference Series in Mathematics, **56**, Providence, Rhode Island, American Mathematical Society.

Bercovici, H., Foiaş, C., and Pearcy, C. (1988). Two Banach space methods and dual operator algebras, *J. Functional Analysis*, **78**, 306–45.

Bernstein, A. R. and Robinson, R. (1966). Solution of an invariant subspace problem of K. T. Smith and P. R. Halmos, *Pacific J. Math.*, **16**, 421–31.

Bonsall, F. F. and Duncan, J. (1973). *Numerical ranges II*, London Mathematical Society Lecture Note Series, **10**, Cambridge University Press.

Brown, S. (1978). Some invariant subspaces for subnormal operators, *Integral Equations and Operator Theory*, **1**, 310–33.

Brown, S. (1987). Hyponormal operators with thick spectrum have invariant subspaces, *Annals of Math.*, **125**, 93–103.

Brown, S. and Chevreau, B. (1988). Toute contraction à calcul fonctionel isométrique est réflexive, *Comptes Rendus de l'academic des Sciences, Paris*, **307**, 185–8.

Brown, S., Chevreau, B., and Pearcy, C. (1979). Contractions with rich spectrum have invariant subspaces, *J. Operator Theory*, **1**, 123–36.

Brown, S., Chevreau, B., and Pearcy, C. (1988). On the structure of contraction operators, II, *J. Functional Analysis*, **76**, 30–55.

Clancey, K. F. and Putnam, C. R. (1972). Normal parts of certain operators, *J. Math. Soc. Japan*, **24**, 198–203.

Cole, B. J. and Gamelin, T. W. (1982). Tight uniform algebras and algebras of analytic functions, *J. Functional Analysis*, **46**, 158–220.

Conway, J. B. (1991). *The theory of subnormal operators*. Mathematical Surveys and Monographs, **38**, Providence, Rhode Island, American Mathematical Society.

Conway, J. B. and Dudziak, J. J. (1990). Von Neumann operators are reflexive, *J. reine angew. Math.*, **408**, 34–56.

Didas, M. (2002). On the structure of von Neumann n-tuples over strictly pseudoconvex sets, Dissertation, Universität des Saarlandes.

Drury, S. W. (1978). A generalization of von Neumann's inequality to the complex ball, *Proc. American Math. Soc.*, **68**, 300–4.

Enflo, P. (1987). On the invariant subspace problem in Banach spaces, *Acta Mathematica*, **158**, 213–313.

Eschmeier, J. (1997). Invariant subspaces for spherical contractions, *Proc. London Math. Soc.*, (3), **75**, 157–76.

Eschmeier, J. (1999). Algebras of subnormal operators on the unit ball, *J. Operator Theory*, **42**, 37–76.

Eschmeier, J. (2001*a*). Invariant subspaces for commuting contractions, *J. Operator Theory*, **45**, 413–43.

Eschmeier, J. (2001*b*). Algebras of subnormal operators on the unit polydisc, In K. D. Bierstedt, J. Bonet, M. Maestre, and H. Schmets (eds.), *Recent* *progress in functional analysis*, North-Holland Mathematics Studies **189**, Amsterdam, North-Holland, pp. 159–71.

Eschmeier, J. (2001*c*). On the structure of spherical contractions, In L. Kerchy, C. Foiaş, I. Gohberg, and M. Langer (eds.), *Operator theory: Advances and applications*, **127**, Basel, Birkhäuser, pp. 211–42.

Eschmeier, J. and Prunaru, B. (1990). Invariant subspaces for operators with Bishop's property (β) and thick spectrum, *J. Functional Analysis*, **94**, 196–222.

Eschmeier, J. and Prunaru, B. (2002). Invariant subspaces and localizable spectrum, *Integral Equations and Operator Theory*, **42**, 461–71.

Eschmeier, J. and Putinar, M. (1996). *Spectral decompositions and analytic sheaves*, London Mathematical Society Monographs, **10**, Oxford, Clarendon Press.

Finch, J. K. (1975). The single valued extension property on a Banach space, *Pacific J. Math.*, **58**, 61–9.

Gamelin, T. W. (1969). *Uniform algebras*, Englewood Cliffs, New Jersey, Prentice Hall.

Halmos, P. R. (1966). Invariant subspaces of polynomially compact operators, *Pacific J. Math.*, **16**, 433–7.

Hewitt, E. and Stromberg, K. (1965). *Real and abstract analysis*, Springer-Verlag, New York.

Kim, H. W., Pearcy, C., and Shields, A. L. (1975). Rank-one commutators and hyperinvariant subspaces, *Michigan Math. J.*, **22**, 193–4.

Lang, S. (1969). *Real analysis*, Reading, Massachusetts, Adison–Wesley.

Lomonosov, V. I. (1973). Invariant subspaces for operators commuting with compact operators, *Funct. Anal. Appl.*, **7**, 213–14.

Martin, M. and Putinar, M. (1989). *Lectures on hyponormal operators*, Basel, Birkhäuser.

Mlak, W. (1969). Decompositions and extensions of operator valued representations of function algebras, *Acta Sci. Math. (Szeged)*, **3**, 181–93.

Müller, V. and Ptak, M. (1999). Spherical isometries are hyporeflexive, *Rocky Mountain J. Math.*, **29**, 677–83.

Olin, R. and Thomson, J. E. (1980). Algebras of subnormal operators, *J. Functional Analysis*, **37**, 271–301.

Pearcy, C. and Shields, A. L. (1974). A survey of the Lomonosov technique in the theory of invariant subspaces, *Topics in operator theory*, Mathematical Surveys, **13**, pp. 219–29, Providence, Rhode Island, American Mathematical Society.

Prunaru, B. (1996). K-spectral sets and invariant subspaces, *Integral Equations and Operator Theory*, **26**, 367–70.

Read, C. J. (1984). A solution to the invariant subspace problem, *Bull. London Math. Soc.*, **16**, 337–401.

Rosenblum, M. (1956). On the operator equation $BX - XA = Q$, *Duke J. Math.*, **23**, 263–9.

Rudin, W. (1980). *Function theory in the unit ball of* $\mathbb{C}^n$, Heidelberg, Springer-Verlag.

Schaefer, H. (1966). *Topological vector spaces*, New York, Macmillan.

Singer, I. (1981). *Bases in Banach spaces, II*, Berlin, Springer-Verlag.

Stampfli, J. G. (1980). An extension of Scott Brown's invariant subspace theorem: K-spectral sets, *J. Operator Theory,* **3**, 3–21.

Stratila, S. and Zsido, L. (1979). *Lectures on von Neumann algebras*, Bucharest, Editura Academiei.

Sz.-Nagy, B. and Foiaş, C. (1970). *Harmonic analysis of operators on Hilbert spaces*, Amsterdam, North-Holland.

Thomson, J. E. (1986). Invariant subspaces for algebras of subnormal operators, *Proc. American Math. Soc.*, **96**, 462–4.

Thomson, J. E. (1991). Approximation in the mean by polynomials, *Annals of Math.*, **133**, 477–507.

Yan, K. (1988). Invariant subspaces for joint subnormal systems, *Chinese Ann. Math. Ser. A*, **9**, 561–6.

Part IV

Local spectral theory

KJELD BAGGER LAURSEN
University of Copenhagen, Denmark

21

Basic notions from operator theory

21.1 Introduction

This part of the book is intended as an invitation to the subject of local spectral theory. It contains the basics and some indications of the way the subject has developed. I would like to thank Garth Dales and Michael Neumann for their numerous good comments and suggestions. The entire story of the fascinating subject that Chapters 21–25 deal with may be found in Laursen and Neumann (2000), and I hope that after having been through these chapters you will want to go for more in that book, which also contains a full bibliography.

The phrase local spectral theory carries many connotations. Among the ones that are appropriate here you should expect to find concepts such as spectral subspaces, that is, invariant subspaces on which the restricted operator has a spectrum consisting of a chunk of the original spectrum. The archetypal conceptual framework is provided by the spectral theorem for normal operators on a Hilbert space, which specifies how this decomposition of the underlying space and of the spectrum is supposed to look. Another similar example is provided by the spectral theorem for compact operators on a Banach space.

Both of these examples may be traced back to what is often a high point of a first course in linear algebra, namely a result on diagonalizing symmetric matrices such as the following. (A *symmetric* matrix $[a_{ij}]$ satisfies the relations $a_{ij} = a_{ji}$ for all i, j, while for a symmetric operator T on a finite-dimensional, real inner-product space V with inner product $[\cdot, \cdot]$, it is true that we have $[Tx, y] = [x, Ty]$ for all $x, y \in V$.)

H. G. Dales, P. Aiena, J. Eschmeier, K. B. Laursen, and G. A. Willis, *Introduction to Banach Algebras, Operators, and Harmonic Analysis*. Published by Cambridge University Press.

Theorem 21.1.1 *A linear operator $T : V \to V$, where V is a finite-dimensional real inner-product space, is symmetric if and only if V has an orthonormal basis consisting of eigenvectors for T.* □

For a Hilbert space this may be turned into the spectral theorem for normal operators. To appreciate this from our chosen vantage point we need to know what a *spectral measure* is. An operator $N \in \mathcal{B}(H)$ is *normal* when it satisfies $NN^* = N^*N$ and *self-adjoint* when $N = N^*$. Here T^* is the adjoint of the operator $T \in \mathcal{B}(H)$ defined by the formula

$$[x, T^*y] = [Tx, y] \quad \text{for all } x, y \in H\,,$$

as in §3.5.

Definition 21.1.2 *Let $\mathfrak{B}$ denote the collection of Borel subsets of $\mathbb{C}$, and let H be a Hilbert space. Then a* spectral measure *E is a map from $\mathfrak{B}$ to the set of projections on H (i.e. all self-adjoint idempotents in $\mathcal{B}(H)$) which has these properties:*

- *$E(\emptyset) = 0$, $E(\mathbb{C}) = I_H$, the identity operator on H;*
- *$E(B_1 \cap B_2) = E(B_1)E(B_2)$ for all Borel sets B_1, B_2;*
- *if $\{B_j\}_{j=1}^{\infty}$ is a sequence of pairwise disjoint Borel sets, then*

$$E\left(\bigcup_{j=1}^{\infty} B_j\right) h = \sum_{j=1}^{\infty} E(B_j)h\,,$$

for every element $h \in H$, the sum converging in the Hilbert space norm (the terms are pairwise orthogonal).

Of course, in the statement to follow, the proper interpretation of the integral needs attention; we shall not dwell on it, because this point will not be needed elsewhere. The spectrum of an operator is defined in Part I, §2.1.

Theorem 21.1.3 *Let N be a normal operator on the Hilbert space H. Then there is a unique spectral measure E on the Borel subsets of the spectrum $\sigma(N)$ of N for which*

$$N = \int z\, \mathrm{d}E(z)\,.$$

Moreover, for every Borel set $\Delta \subseteq \mathbb{C}$ the subspace $E(\Delta)H$ is N-invariant and

$$\sigma(N|E(\Delta)H) \subseteq \overline{\Delta} \cap \sigma(N).$$ □

It is in this light that the modern definition of a *decomposable* operator should be viewed.

Throughout these Chapters 21–25, we shall be considering complex Banach spaces (as well as Banach algebras), and it will be assumed that, unless something else is explicitly stipulated, the term X refers to a complex Banach space, and the term A refers to a commutative complex Banach algebra.

Definition 21.1.4 *A bounded linear operator $T : X \to X$ is* decomposable *if, for every open cover $\{U, V\}$ of $\mathbb{C}$, there are closed T-invariant subspaces Y, Z of X for which:*

- $X = Y + Z$;
- $\sigma(T|Y) \subseteq U$ *and* $\sigma(T|Z) \subseteq V$.

Chapter 22 will look at examples in more detail (but it is an easy consequence of Theorem 21.1.3 that normal operators are decomposable – see Exercise 21.3.1). Here we shall concentrate on introducing the basic concepts.

If $U \subseteq \mathbb{C}$ is open, then $H(U)$ denotes the algebra (pointwise operations) of analytic functions defined on U. We give this algebra the topology of *locally uniform convergence*, that is, of uniform convergence on all compact subsets of U. It may be shown that this topology is locally convex and that it is induced by a complete and translation-invariant metric; thus $H(U)$ is a Fréchet algebra (see Part I, Examples 1.2 (x)). Among the elements of $H(U)$ we find the constant function 1 and the identity function Z.

The *analytic functional calculus* is defined in Chapter 4. We recall it this way: if $T \in \mathcal{B}(X)$ has spectrum $\sigma(T)$ and U is an open neighbourhood of $\sigma(T)$ on which the analytic function f is defined, then the map

$$\Phi : f \to f(T) := \frac{1}{2\pi \mathrm{i}} \int_{\Gamma} f(\lambda)(\lambda - T)^{-1}\, \mathrm{d}\lambda$$

(where Γ is a contour in U surrounding $\sigma(T)$, as explained in Chapter 4) is an algebra homomorphism from the Fréchet algebra $H(U)$ into $\mathcal{B}(X)$ for which $\Phi(1) = I_X$ and $\Phi(Z) = T$.

There are connections between functional calculi and decomposability, although in the case of the analytic functional calculus this connection is limited to the situations in which the spectrum is disconnected. To display an extreme case: this calculus may be used to show that, if $\sigma(T)$ is *totally disconnected*, then T is decomposable (Exercise 21.3.2). (Recall from general topology that a topological space is said to be totally disconnected if none of its connected subsets contains more than one point.) The reason the argument goes through is that a totally disconnected compact Hausdorff space has a base for its topology consisting of clopen sets, as shown, for instance, in Appendix A.7 of Rudin (1991). For more on totally disconnected spaces, see Engelking (1977).

A class that displays a better connection is that of the generalized scalar operators. To explain: look at the algebra $C^\infty(\mathbb{C})$ of complex-valued functions that have continuous partial derivatives of all orders with respect to the variables x and y (as usual for a complex number z, we write $z = x + \mathrm{i}y$, where x and y are real). This algebra may be equipped with the topology determined by the seminorms

$$\|f\|_{k,Q} := \sum_{|\alpha|\le k} \frac{1}{\alpha!} \sup\{|D^\alpha f(\lambda)| : \lambda \in Q\} \qquad \text{for all } f \in C^\infty(\mathbb{C}),$$

where Q ranges over all compact subsets of $\mathbb{C}$, k over $\mathbb{Z}^+$, and $\alpha = (\alpha_1, \alpha_2)$ is an arbitrary double index of integers which specifies the order of differentiation:

$$D^\alpha f(\lambda) = \partial_x^{\alpha_1} \partial_y^{\alpha_2} f(\lambda),$$

$|\alpha| := \alpha_1 + \alpha_2$, and $\alpha! := \alpha_1!\alpha_2!$; the factors $1/\alpha!$ are there to ensure submultiplicativity of the seminorm $\|\cdot\|_{k,Q}$. With the topology of this countable family of seminorms, $C^\infty(\mathbb{C})$ is a Fréchet algebra. Obviously the set $H(\mathbb{C})$ of entire functions is a subalgebra of $C^\infty(\mathbb{C})$.

Definition 21.1.5 *Let $T \in \mathcal{B}(X)$ be given. If there is a continuous algebra homomorphism Φ from $C^\infty(\mathbb{C})$ into $\mathcal{B}(X)$ for which $\Phi(1) = I$ and $\Phi(Z) = T$, then T is a* generalized scalar operator.

Note that the homomorphism of this definition will necessarily be an extension from $H(\mathbb{C})$ of the homomorphism that appears in the analytic functional calculus.

A generalized scalar operator is decomposable. This is a consequence of the fact that in $C^\infty(\mathbb{C})$ we have partitions of unity (cf. Exercise 21.3.3).

21.2 The properties (β) and (δ)

The main goal for the rest of this chapter is to introduce two seemingly merely technical conditions, and to relate them to decomposability. Our aim is Theorem 21.2.8, which shows that these two conditions together are equivalent to decomposability. It will become clear in the course of the next few chapters that both conditions have some truly astounding relations, which go well beyond this first result. Both properties (which are called properties (β) and (δ)) go back to Errett Bishop's PhD thesis (1959).

Definition 21.2.1 *An operator $T \in \mathcal{B}(X)$* has property (β) *(or just: T* has (β)*) if, for every open subset U of $\mathbb{C}$ and every sequence (f_n) of analytic functions $f_n : U \to X$ with the property that $(T - \lambda)f_n(\lambda) \to 0$ as $n \to \infty$, locally uniformly on U, it follows that $f_n(\lambda) \to 0$ as $n \to \infty$, also locally uniformly on U.*

This property may be conveniently rephrased in terms of an operator T_U, defined on the Fréchet space $H(U, X)$ of X-valued analytic functions, topologized by locally uniform convergence on U. This operator is specified by

$$(T_U f)(\lambda) := (T - \lambda)f(\lambda) \quad \text{for every } \lambda \in U .$$

It may then be shown that an operator T has (β) if and only if T_U is injective and has closed range, for every open set U in $\mathbb{C}$ (Exercise 21.3.4).

Incidentally, this last observation is also a possible lead-in to the notion of the *single-valued extension property* SVEP, which is as follows.

Definition 21.2.2 *An operator $T \in \mathcal{B}(X)$ has SVEP if, for every open subset U of $\mathbb{C}$, the equation $(T - \lambda)f(\lambda) = 0$ has only one analytic solution, namely $f = 0$.*

Thus T has SVEP if and only if T_U is injective for every open U. It is clear that (β) implies SVEP. The converse is not true, although you may have to think hard to come up with a counter-example (Exercise 21.3.5). We shall display some later. For more on SVEP, see Part V, Chapter 27.

To introduce property (δ) we need the *glocal subspaces.*

Definition 21.2.3 *Let $T \in \mathcal{B}(X)$ be given and suppose that $F \subseteq \mathbb{C}$ is closed. Then the* glocal spectral subspace *is*

$$\begin{aligned} \mathcal{X}_T(F) := \{x \in X : \quad & \text{there is an analytic function } f : \mathbb{C} \setminus F \to X \\ & \text{for which } (T - \lambda)f(\lambda) = x \text{ for all } \lambda \in \mathbb{C} \setminus F\}. \end{aligned}$$

In other, and more formal, terms,

$$\mathcal{X}_T(F) = \{x \in X : x \in T_{\mathbb{C}\setminus F} H(\mathbb{C} \setminus F, X)\}.$$

On the resolvent set $\rho(T)$ we obviously have that every $x \in X$ may be written as $x = (T - \lambda)(T - \lambda)^{-1}x$ for every $\lambda \in \rho(T)$, which shows immediately that $X = \mathcal{X}_T(\sigma(T))$. Also, it is an easy consequence of Liouville's theorem that $\mathcal{X}_T(\emptyset) = \{0\}$ (Exercise 21.3.6).

Definition 21.2.4 *An operator $T \in \mathcal{B}(X)$ has* property (δ) *(or just (δ)) if*

$$X = \mathcal{X}_T(\overline{U}) + \mathcal{X}_T(\overline{V})$$

for every open cover $\{U, V\}$ of $\mathbb{C}$.

If $T : X \to Y$ is a continuous linear operator between the Banach spaces X and Y, then T induces a linear map $T^{\natural} : H(U, X) \to H(U, Y)$, defined by the composition

$$(T^{\natural} f)(\lambda) := Tf(\lambda) \qquad \text{for all } \lambda \in U \text{ and } f \in H(U, X).$$

You may want to convince yourself that $T^{\natural}$ is continuous; this is not hard (Exercise 21.3.7).

Here we shall need this map only for open discs. We use $\overline{\mathbb{D}(a;b)}$ to denote the closed disc in $\mathbb{C}$ centred at a and of radius b, and $\mathbb{D}(a;b)$ to denote the corresponding open disc. The open unit disc is called just $\mathbb{D}$. If $U = \mathbb{D}(\lambda_0; r)$, then every $f \in H(U, X)$ is given by a power series

$$f(\lambda) = \sum_{n=0}^{\infty} a_n (\lambda - \lambda_0)^n \qquad \text{for all } \lambda \in U,$$

where the coefficients a_n are elements of X. Since the convergence of this series is locally uniform, the continuity of $T^{\natural}$ implies that

$$(T^{\natural} f)(\lambda) = \sum_{n=0}^{\infty} T(a_n)(\lambda - \lambda_0)^n \qquad \text{for all } \lambda \in U.$$

We shall need the fact that surjectivity is transferred from T to $T^{\natural}$.

Proposition 21.2.5 *Let $T : X \to Y$ be a continuous linear surjection from the Banach space X onto the Banach space Y. Then, for every open disc $U \subseteq \mathbb{C}$, the induced map $T^{\natural} : H(U, X) \to H(U, Y)$ is a continuous and open linear surjection.*

Proof Let $U = \mathbb{D}(\lambda_0; r)$, where $\lambda_0 \in \mathbb{C}$ and $r > 0$. Then, for every function g in $H(U, Y)$, we have a power series expansion $g(\lambda) = \sum_{n=0}^{\infty} b_n (\lambda - \lambda_0)^n$, valid for every $\lambda \in U$. The radius of convergence of this power series is at least r.

The open mapping theorem tells us there is a constant $c > 0$ for which, for every $y \in Y$, there is an $x \in X$ such that $Tx = y$ and $\|x\| \leq c\,\|y\|$. So, for every $n \in \mathbb{N}$, we may choose an element $a_n \in X$ so that $Ta_n = b_n$ and $\|a_n\| \leq c\,\|b_n\|$. The estimates

$$\limsup_{n\to\infty} \|a_n\|^{1/n} \leq \limsup_{n\to\infty} c^{1/n}\,\|b_n\|^{1/n} = \limsup_{n\to\infty} \|b_n\|^{1/n} \leq \frac{1}{r},$$

tell us that the power series $f(\lambda) := \sum_{n=0}^{\infty} a_n(\lambda - \lambda_0)^n$ converges for all λ in the disc U, hence defines a function $f \in H(U, X)$. Since

$$(T^{\natural} f)(\lambda) = \sum_{n=0}^{\infty} T(a_n)(\lambda - \lambda_0)^n = \sum_{n=0}^{\infty} b_n(\lambda - \lambda_0)^n = g(\lambda) \quad \text{for all } \lambda \in U,$$

$T^{\natural}$ is surjective, hence also, by the open mapping theorem, open. □

To obtain the proof of Theorem 21.2.8, we need first some technical observations.

Given $T \in \mathcal{B}(X)$, for a T-invariant closed linear subspace Y of X, let

$$T \mid Y \in \mathcal{B}(Y)$$

denote the operator given by the restriction of T to Y, and let

$$T/Y \in \mathcal{B}(X/Y)$$

denote the operator induced by T on the quotient space X/Y. The subscript $_{\mathrm{f}}$ (f as in *full*) refers to $\sigma(T)$ with all its holes filled in. This is the polynomially convex hull of $\sigma(T)$, as defined in Definition 4.1.4.

Proposition 21.2.6 *Let $T \in \mathcal{B}(X)$ be an operator, and suppose that Y and Z are T-invariant closed linear subspaces of X with the property that $X = Y + Z$. Then*

$$\sigma(T/Y) \subseteq \sigma(T) \cup \sigma(T \mid Y) \subseteq \sigma_{\mathrm{f}}(T) \quad \text{and} \quad \sigma(T/Z) \subseteq \sigma_{\mathrm{f}}(T \mid Y) \subseteq \sigma_{\mathrm{f}}(T).$$

Proof We know that $(T/Y)Q = QT$, where $Q : X \to X/Y$ is the natural quotient map. For arbitrary $\lambda \in \rho(T) \cap \rho(T \mid Y)$, $T/Y - \lambda$ is surjective because both Q and $T - \lambda$ are. Moreover, if $(T/Y - \lambda)Q_x = 0$ for some $x \in X$, then $(T - \lambda)x \in Y$, and hence $x \in Y$ because $\lambda \in \rho(T \mid Y)$; thus $T/Y - \lambda$ is invertible on X/Y. This shows that $\sigma(T/Y) \subseteq \sigma(T) \cup \sigma(T \mid Y)$.

Next, the classical identity

$$(T - \lambda)^{-1} = -\sum_{n=0}^{\infty} \lambda^{-n-1} T^n \qquad \text{for all } \lambda \in \mathbb{C} \text{ with } |\lambda| > \|T\|,$$

shows that $(T - \lambda)^{-1} Y \subseteq Y$ for all such λ and, since $\mathbb{C} \setminus \sigma_{\mathrm{f}}(T)$ is connected, the identity theorem for vector-valued functions tells us that $(T - \lambda)^{-1} Y \subseteq Y$ for all $\lambda \in \mathbb{C} \setminus \sigma_{\mathrm{f}}(T)$, hence that $\sigma(T \mid Y) \subseteq \sigma_{\mathrm{f}}(T)$.

Finally, the assumption that $X = Y + Z$ provides us with a canonical surjection from the space Y onto the quotient X/Z with kernel $Y \cap Z$. If the corresponding isomorphism is called $R : Y/(Y \cap Z) \to X/Z$, then

$(T/Z)R = R\,(T\mid Y)/(Y\cap Z)$, and so $\sigma(T/Z) = \sigma((T\mid Y)/(Y\cap Z))$. From the work already done, we know that

$$\sigma((T\mid Y)/(Y\cap Z)) \subseteq \sigma(T\mid Y)\cup\sigma(T\mid(Y\cap Z)) \subseteq \sigma_{\mathrm{f}}(T\mid Y) \subseteq \sigma_{\mathrm{f}}(T),$$

and hence $\sigma(T/Z) \subseteq \sigma_{\mathrm{f}}(T\mid Y) \subseteq \sigma_{\mathrm{f}}(T)$. □

Proposition 21.2.7 *Suppose that Y and Z are closed linear subspaces of X and that $U \subseteq \mathbb{C}$ is an open disc. Then the following properties hold.*

(i) *The quotient map from X onto X/Y induces a canonical topological isomorphism $H(U, X/Y) \cong H(U, X)/H(U, Y)$.*
(ii) *If $X = Y + Z$, then the operator $\Phi : H(U, Y) \times H(U, Z) \to H(U, X)$, given by $\Phi(f, g)(\lambda) := f(\lambda) + g(\lambda)$ for all $f \in H(U, Y)$, $g \in H(U, Z)$, and $\lambda \in U$, is a continuous and open linear surjection. In particular, if $X = Y \oplus Z$ (direct sum), then the map Φ yields a canonical identification $H(U, X) \cong H(U, Y) \oplus H(U, Z)$.*

Proof (i) This is immediate from Proposition 21.2.5: let $T : X \to X/Y$ be the canonical quotient map and observe that $\ker T^{\natural} = H(U, Y)$.

For (ii) first note that the definition $\Psi(f, g)(\lambda) := (f(\lambda), g(\lambda))$ for f in $H(U, Y)$, g in $H(U, Z)$, and λ in U yields a topological linear isomorphism Ψ from $H(U, Y) \times H(U, Z)$ onto $H(U, Y \times Z)$. Next, let $T : Y \times Z \to X$ be the canonical continuous linear surjection from the product space $Y \times Z$ onto X, given by $T(u, v) = u + v$ for all $u \in Y$ and $v \in Z$. Then Proposition 21.2.5 tells us that the corresponding map $T^{\natural} : H(U, Y \times Z) \to H(U, X)$ is continuous, surjective, open, and linear. This implies our claims for Φ, since $\Phi = T^{\natural} \circ \Psi$. □

And now the result itself.

Theorem 21.2.8 *Decomposability implies (β) and (δ) – and vice versa.*

Proof The first step is to show that decomposability implies (β). Let $U \subseteq \mathbb{C}$ be open, and consider a sequence of analytic functions $f_n : U \to X$ for which $(T - \lambda) f_n(\lambda) \to 0$ as $n \to \infty$, locally uniformly on U. It is enough to show that $f_n \to 0$ as $n \to \infty$ uniformly on any closed disc in in U. So take an arbitrary closed disc $D \subseteq U$ and choose an open disc E such that $D \subseteq E \subseteq \overline{E} \subseteq U$. Apply the definition of decomposability of T to the open cover $\{E, \mathbb{C}\setminus D\}$ of $\mathbb{C}$. This gives us T-invariant closed linear subspaces $Y, Z \subseteq X$ for which $\sigma(T\mid Y) \subseteq E$, $\sigma(T\mid Z)\cap D = \emptyset$, and $X = Y + Z$. Proposition 21.2.7 then yields functions $g_n \in H(U, Y)$ and $h_n \in H(U, Z)$ such that

$$f_n(\lambda) = g_n(\lambda) + h_n(\lambda) \qquad \text{for all } \lambda \in U \text{ and } n \in \mathbb{N}.$$

Additionally, by Proposition 21.2.6, $\sigma(T/Z) \subseteq \sigma_f(T \mid Y) \subseteq E$, so, for every λ in the boundary ∂E of E, the operator $T/Z - \lambda$ is invertible on the quotient space X/Z. By compactness and continuity, we obtain a $c > 0$ such that $\| (T/Z - \lambda)^{-1} \| \le c$ for all $\lambda \in \partial E$. Consequently, if $Q : X \to X/Z$ is the natural quotient map, then

$$Qg_n(\lambda) = Qf_n(\lambda) = (T/Z - \lambda)^{-1} Q(T - \lambda) f_n(\lambda) \qquad \text{for all } \lambda \in \partial E,$$

and therefore $\| Qg_n(\lambda) \| \le c \, \| (T - \lambda) f_n(\lambda) \|$ for all $\lambda \in \partial E$. This estimate, and our assumption on the functions f_n, then imply that the analytic functions $Q \circ g_n$ in $H(U, X/Z)$ converge to zero uniformly on ∂E and therefore, by the maximum modulus principle, uniformly on E. Now, by Proposition 21.2.7, $H(E, X/Z)$ may be identified with $H(E, X)/H(E, Z)$, and consequently we obtain functions $k_n \in H(E, Z)$ for which $g_n + k_n \to 0$ as $n \to \infty$, locally uniformly on E, hence uniformly on D.

Now, $f_n = g_n + h_n = (g_n + k_n) + (h_n - k_n)$ on D for all $n \in \mathbb{N}$, and so it remains to see that $h_n - k_n$ converges to 0 uniformly on D. Observe that, because $\sigma(T \mid Z) \cap D = \emptyset$, there is a constant $c_1 > 0$ such that $\| (T \mid Z - \lambda)^{-1} \| \le c_1$ for all $\lambda \in D$. Since both h_n and k_n map into the space Z and $h_n = f_n - g_n$, we conclude that

$$\begin{aligned} \| (h_n - k_n)(\lambda) \| &\le c_1 \; \| (T - \lambda)(h_n - k_n)(\lambda) \| \\ &\le c_1 \; \| (T - \lambda) f_n(\lambda) \| \; + \; c_1 \; \| (T - \lambda)(g_n + k_n)(\lambda) \| \end{aligned}$$

for all $\lambda \in D$ and $n \in \mathbb{N}$. We have now shown that $h_n - k_n \to 0$ uniformly on D.

That decomposability implies (δ) is immediate: if $\mathbb{C} = U \cup V$ and Y, Z are chosen in accordance with the definition of decomposability, then the inclusion $\sigma(T|Y) \subseteq U$ implies that $Y \subseteq \mathcal{X}_T(\overline{U})$. Similarly, $Z \subseteq \mathcal{X}_T(\overline{V})$. This establishes ($\delta$).

Finally, if T has both (β) and (δ) then decomposability will follow once we have seen that each $\mathcal{X}_T(F)$ is closed. And this follows from (β). Here is a sketch of the argument. Since $\mathcal{X}_T(F) := \{x \in X : x \in T_{\mathbb{C}\setminus F} H(\mathbb{C} \setminus F, X)\}$, we see that

$$\mathcal{X}_T(F) = T_{\mathbb{C}\setminus F} H(\mathbb{C} \setminus F, X) \cap X.$$

We leave it as an exercise (21.3.8) that the closedness of $\mathcal{X}_T(F)$ follows from this. Next, if $U \cup V = \mathbb{C}$ is an open cover, then choose open sets U_1 and V_1, still covering $\mathbb{C}$, so that $\overline{U_1} \subseteq U$ and $\overline{V_1} \subseteq V$. We have

$$X = \mathcal{X}_T(\overline{U_1}) + \mathcal{X}_T(\overline{V_1}).$$

It is then not too difficult, using the open mapping theorem, to see that

$$\sigma(T|\mathcal{X}_T(\overline{U_1})) \subseteq \overline{U_1} \subseteq U,$$

and also that

$$\sigma(T|\mathcal{X}_T(\overline{V_1})) \subseteq \overline{V_1} \subseteq V$$

which establishes the decomposability of T (Exercise 21.3.9). □

21.3 Exercises

1. Use Theorem 21.1.3 to show that normal operators are decomposable.
2. Show that, if $\sigma(T)$ is totally disconnected, then T is decomposable.
3. Show that a generalized scalar operator is decomposable *Hint:* in $C^\infty(\mathbb{C})$ we have partitions of unity.
4. Show that an operator T has (β) if and only if T_U is injective and has closed range, for every open set U in $\mathbb{C}$.
5. Find an example of a bounded linear operator which has SVEP, but not (β).
6. Use Liouville's theorem to show that $\mathcal{X}_T(\emptyset) = \{0\}$.
7. Prove that the map $T^\natural$ introduced after Definition 21.2.4 is continuous.
8. Show that, since $\mathcal{X}_T(F) = T_{\mathbb{C}\setminus F} H(\mathbb{C} \setminus F, X) \cap X$, $\mathcal{X}_T(F)$ is closed provided that T has (β). (cf. the proof of Theorem 21.2.8.)
9. Suppose that $\mathcal{X}_T(F)$ is closed. Show that $\sigma(T|\mathcal{X}_T(F)) \subseteq F$. *Hint:* use the open mapping theorem.

21.4 Additional notes

Decomposability may be described entirely in terms of spectral capacities (subspace-valued functions from the Borel sets of $\mathbb{C}$ which behave pretty much like the ranges of spectral projections, *cf.* Definition 21.1.2), which brings it conceptually even closer to the spectral theorem. The class of decomposable operators is but one of the natural extensions based in the spectral theorem. We shall define a couple of them in the next chapter, but there are others: spectral and scalar operators, going back to Dunford, as well as various strengthenings of the decomposability requirement. Chapter 1 of Laursen and Neumann (2000) tells more. Of course, Hilbert space operator theory has a history and dynamics of its own, displayed by the appearance of several classes containing the normal ones, such as sub- or hyponormal operators.

The history of decomposability is quite interesting. The subject has generated many classes of operators, as well as a conceptual evolution, for instance of the very definition of decomposability, which was quite complicated when Foiaş first proposed it, but now, thanks to Albrecht and Lange, reads very simply. Among the leading classical practitioners of this branch of operator theory we find people such as Dunford, Bishop, Foiaş, Vasilescu, Frunza and Albrecht, while the more recent developments have been strongly influenced by, inter alia, Albrecht, Lange, Eschmeier and Putinar. Chapter 1 of Laursen and Neumann (2000) contains an account of all this.

In Gleason (1962), it was shown that Proposition 21.2.5 actually holds for arbitrary open subsets U of $\mathbb{C}$. The proof of this general version is considerably more involved, and such a high level of generality is not needed in the present work.

22

Classes of decomposable operators

22.1 Local properties of operators

In this chapter we shall take a closer look at *super-decomposable* and at *generalized scalar operators*. Along the way we also obtain some examples of operators with (β), namely all isometries.

The companion concept of the glocal subspace is the *local analytic spectral subspace.* Here is first the definition of *local spectrum.*

Definition 22.1.1 *Let $T \in \mathcal{B}(X)$ and $x \in X$. The* local resolvent $\rho_T(x)$ *of T at x is the set of $\lambda \in \mathbb{C}$ for which there is an open neighbourhood $N(\lambda)$ on which the equation $(T - \mu)f(\mu) = x$ has an analytic solution $f : N(\lambda) \to X$. The* local spectrum *is*

$$\sigma_T(x) := \mathbb{C} \setminus \rho_T(x),$$

and, given a closed subset F of the complex plane, the local analytic subspace is

$$X_T(F) := \{x \in X : \sigma_T(x) \subseteq F\}.$$

It follows directly from the definition that $X_T(\bigcap F_\alpha) = \bigcap X_T(F_\alpha)$ for arbitrary collections $\{F_\alpha\}$ of closed subsets of $\mathbb{C}$.

There are some immediate connections between the glocal and the local spectral subspaces. In fact, it is clear that, if $x \in \mathcal{X}_T(F)$, then $\mathbb{C} \setminus F \subseteq \rho_T(x)$, and hence $\mathcal{X}_T(F) \subseteq X_T(F)$ for every closed set $F \subseteq \mathbb{C}$. As a matter of fact it is not difficult to tell when these two classes of subspaces coincide.

Proposition 22.1.2 *For $T \in \mathcal{B}(X)$, we have $\mathcal{X}_T(F) = X_T(F)$ for every closed set $F \subseteq \mathbb{C}$ if and only if $X_T(\emptyset) = \{0\}$, and this happens precisely when T has SVEP.*

Proof Exercise 22.4.1. □

The *surjectivity spectrum* is defined as

$$\sigma_{\mathrm{su}}(T) := \{\lambda \in \mathbb{C} : T - \lambda \text{ is not surjective}\} ,$$

while the *approximate point spectrum* is

$$\sigma_{\mathrm{ap}}(T) := \{\lambda \in \mathbb{C} : \text{ there is a sequence } \{x_n\} \text{ of unit vectors for which } (T - \lambda)x_n \to 0 \text{ as } n \to \infty\} .$$

Among the basic properties of the surjectivity spectrum, we note the following (the proofs are left as exercises – think of the open mapping theorem when you solve them). For every $\lambda \in \mathbb{C} \setminus \sigma_{\mathrm{su}}(T)$, there is a $c > 0$ for which $X = \mathcal{X}_T(\mathbb{C} \setminus \mathbb{D}(\lambda; c))$ (*cf.* Exercise 24.3.3). In particular, $\sigma(T) \setminus \sigma_{\mathrm{su}}(T) \subseteq \sigma_{\mathrm{p}}(T)$, and, if T has $SVEP$, then $\sigma(T) = \sigma_{\mathrm{su}}(T)$. The most immediate connection to local spectral theory is that

$$\sigma_{\mathrm{su}}(T) = \bigcup \{\sigma_T(x) : x \in X\} .$$

Similarly, for the approximate point spectrum it is a basic fact that it is a closed set which contains the topological boundary of the spectrum.

Proposition 22.1.3 *Let $T \in \mathcal{B}(X)$ be decomposable. Then*

$$\sigma(T) = \sigma_{\mathrm{ap}}(T) = \sigma_{\mathrm{su}}(T) = \bigcup \{\sigma_T(x) : x \in X\} .$$

Proof We know from Theorem 21.2.8 that T has (β), hence SVEP, so the remarks just made show us that $\sigma(T) = \sigma_{\mathrm{su}}(T) = \bigcup \{\sigma_T(x) : x \in X\}$. To see that $\sigma_{\mathrm{ap}}(T) = \sigma(T)$, we may argue this way: given any $\lambda \in \sigma(T)$ and $\varepsilon > 0$, let $U \subseteq \mathbb{C}$ be an open set for which $\lambda \notin U$ and $U \cup \mathbb{D}(\lambda; \varepsilon) = \mathbb{C}$. By decomposability, there are T-invariant closed linear subspaces $Y, Z \subseteq X$ for which $X = Y + Z$, $\sigma(T \mid Y) \subseteq U$ and $\sigma(T \mid Z) \subseteq \mathbb{D}(\lambda; \varepsilon)$. The subspace Z must be non-trivial, since otherwise $X = Y$, and hence $\sigma(T) = \sigma(T \mid Y) \subseteq U$, which would contradict that $\lambda \notin U$. Since Z is non-trivial and since

$$\partial\sigma(T \mid Z) \subseteq \sigma_{\mathrm{ap}}(T \mid Z) \subseteq \sigma_{\mathrm{ap}}(T) ,$$

it follows from the inclusion $\sigma(T \mid Z) \subseteq \mathbb{D}(\lambda; \varepsilon)$ that $\mathbb{D}(\lambda; \varepsilon) \cap \sigma_{\mathrm{ap}}(T) \neq \emptyset$. Thus we have $\lambda \in \sigma_{\mathrm{ap}}(T)$ because this set is closed. □

Here is another useful observation:

Lemma 22.1.4 *Let U be an open subset of $\mathbb{C}$. If $T \in \mathcal{B}(X)$, $x \in X$, and $f \in H(U, X)$ are chosen so that $T_U f \equiv x$ on U, then $f(\lambda) \in T_U H(U, X)$ for all $\lambda \in U$. Moreover, $\sigma_T(x) = \sigma_T(f(\lambda))$ for all $\lambda \in U$.*

Proof Given an arbitrary $\lambda \in U$, we define $g : U \to X$ by

$$g(\lambda) := f'(\lambda) \quad \text{and} \quad g(\mu) := (f(\mu) - f(\lambda))/(\mu - \lambda) \quad \text{for all} \quad \mu \in U \setminus \{\lambda\}.$$

Then $g \in H(U, X)$. Since $(T - \mu)f(\mu) = x$ for all $\mu \in U$, a short calculation reveals that $(T - \mu)g(\mu) = f(\lambda)$ for all $\mu \in U \setminus \{\lambda\}$, and, by continuity, also $(T - \lambda)g(\lambda) = f(\lambda)$. That settles the first part of the lemma. It also shows that $U \subseteq \rho_T(f(\lambda))$.

On the other hand, for any $\omega \in \rho_T(x) \setminus U$, we may choose an open neighbourhood W of ω such that $\lambda \notin W$ and an analytic function $h : W \to X$ for which $(T - \mu)h(\mu) = x$, identically on W. If we then define

$$k(\mu) := (h(\mu) - f(\lambda))/(\mu - \lambda)$$

for all $\mu \in W$, we have obtained an analytic function $k : W \to X$ for which

$$(T - \mu)k(\mu) = f(\lambda) \quad \text{for all} \quad \mu \in W.$$

This shows that $\omega \in \rho_T(f(\lambda))$, and hence that $\rho_T(x) \subseteq \rho_T(f(\lambda))$.

For the opposite inclusion, take any $\omega \in \rho_T(f(\lambda))$, and consider an analytic function $h : W \to X$ on an open neighbourhood W of ω such that $(T - \mu)h(\mu) = f(\lambda)$ for all $\mu \in W$. Substitution of this leads to

$$(T - \mu)(T - \lambda)h(\mu) = (T - \lambda)(T - \mu)h(\mu) = (T - \lambda)f(\lambda) = x$$

for all $\mu \in W$, which shows that $\omega \in \rho_T(x)$, and hence that $\rho_T(f(\lambda)) \subseteq \rho_T(x)$. All in all, $\sigma_T(x) = \sigma_T(f(\lambda))$ for all $\lambda \in U$. □

22.2 Super-decomposable operators

Here is our first class of examples of decomposable operators. The super-decomposable operators are based on the idea of operator-valued partitions of unity. More specifically, we are considering a class of operators for which the condition of decomposability is implemented by certain operator ranges.

Definition 22.2.1 *An operator $T \in \mathcal{B}(X)$ is* super-decomposable *if, for every open cover of $\mathbb{C}$ by two subsets U and V, there is an operator $R \in \mathcal{B}(X)$ which* *commutes with T and for which*

$$\sigma(T \mid \overline{RX}) \subseteq U \quad \text{and} \quad \sigma(T \mid \overline{(I-R)X}) \subseteq V\,.$$

This definition makes sense because, by the condition $RT = TR$, the spaces $\overline{RX}$ and $\overline{(I-R)X}$ are T-invariant. It is obvious that super-decomposable operators are decomposable.

Proposition 22.2.2 *Every operator $T \in \mathcal{B}(X)$ with totally disconnected spectrum is super-decomposable.*

Proof For any open cover $\{U, V\}$ of $\mathbb{C}$, $\sigma(T) \cap U$ is a relatively open neighbourhood of the closed subset $\sigma(T) \setminus V$ of $\sigma(T)$. Hence a standard compactness argument provides us with a clopen subset G of $\sigma(T)$ for which $G \subseteq U$ and $\sigma(T) \setminus G \subseteq V$. We may then choose disjoint open sets $U_1, V_1 \subseteq \mathbb{C}$ so that $U_1 \subseteq U$, $V_1 \subseteq V$, $U_1 \cap \sigma(T) = G$, and $V_1 \cap \sigma(T) = \sigma(T) \setminus G$.

Subsequently we pick a contour Γ in $U_1 \cup V_1$ that surrounds $\sigma(T)$, and let

$$R := \frac{1}{2\pi i} \int_\Gamma f(\lambda)\,(\lambda - T)^{-1}\,\mathrm{d}\lambda\,,$$

where now $f \in H(U_1 \cup V_1)$ is the characteristic function of U_1. The operator $R \in \mathcal{B}(X)$ is a projection, it commutes with T, and it is not difficult to show that

$$\sigma(T \mid RX) = G \subseteq U_1 \subseteq U \quad \text{and} \quad \sigma(T \mid (I-R)X) = \sigma(T) \setminus G \subseteq V_1 \subseteq V\,;$$

(see, e.g., Theorem 7.3.20 of Dunford and Schwartz 1958; or Part I, Theorem 4.3.1). This shows that T is super-decomposable. □

This result applies to all operators with countable spectrum, and hence, in particular, to all quasi-nilpotent operators (cf. Part I, Chapter 2). It also covers all *algebraic* operators $T \in \mathcal{B}(X)$, those for which $p(T) = 0$ for some non-trivial polynomial p (since the spectra of such operators are finite). Further examples are: compact operators and Riesz operators (where an operator $T \in \mathcal{B}(X)$ is a *Riesz operator* if, for each $\lambda \in \mathbb{C} \setminus \{0\}$, the spaces $\ker(T - \lambda)$ and $X/(T - \lambda)X$ are both of finite dimension; in other words, T is Riesz if $T - \lambda$ is Fredholm, for every $\lambda \in \mathbb{C} \setminus \{0\}$). See also Part V.

We have touched on the *analytic* functional calculus (cf. the account in Part I, Chapter 4). We now show that *non-analytic* functional calculi tend to generate super-decomposable operators. To be specific, let $\mathcal{A}$ be an algebra of complex-valued functions, defined on an arbitrary non-empty subset Ω of $\mathbb{C}$, so that the following two conditions are fulfilled. First, we suppose that $\mathcal{A}$ is *richly endowed*, which is to mean that $\mathcal{A}$ contains the restrictions to Ω of all polynomials, and that, for every $f \in \mathcal{A}$ and every $\lambda \in \mathbb{C} \setminus \operatorname{supp} f$, there exists a function $g \in \mathcal{A}$ for which $(Z - \lambda)g = f$. Here it is natural to define the *support* of a function $f \in \mathcal{A}$ as the set

$$\operatorname{supp} f := \overline{\{\lambda \in \Omega : f(\lambda) \neq 0\}},$$

where the closure is taken in $\mathbb{C}$, not just in the relative topology of Ω. Secondly, we suppose that $\mathcal{A}$ *admits partitions of unity*, in the sense that, for an arbitrary finite open cover $\{U_1, \ldots, U_m\}$ of $\overline{\Omega}$, there are functions $f_1, \ldots, f_m \in \mathcal{A}$ for which $\operatorname{supp} f_k \subseteq U_k$ for $k = 1, \ldots, m$ and $f_1 + \cdots + f_m \equiv 1$ on Ω.

Obvious examples of such algebras are the algebras $C^m(\Omega)$ of all m times continuously differentiable functions defined on an open set $\Omega \subseteq \mathbb{C}$, for arbitrary $m \in \{0, 1, \ldots, \infty\}$. Moreover, if, additionally, Ω is bounded, then examples are provided by the algebras $C^m(\overline{\Omega})$ of all functions in $C^m(\Omega)$ which, together with all their partial derivatives up to order m, have continuous extensions to the closure $\overline{\Omega}$.

Definition 22.2.3 *Suppose that the algebra $\mathcal{A}$ is richly endowed and admits partitions of unity. An operator $T \in \mathcal{B}(X)$ is an $\mathcal{A}$-*scalar operator *if there exists an algebra homomorphism $\Phi : \mathcal{A} \to \mathcal{B}(X)$ for which $\Phi(1) = I_X$ and $\Phi(Z) = T$.*

Here I_X is the identity operator on X. There is no continuity assumption on Φ in this definition (although it may be shown that such homomorphisms have quite remarkable automatic continuity properties). The map Φ is called a *non-analytic functional calculus for T*. An important special case occurs when $\mathcal{A} = C^\infty(\mathbb{C})$, and we also require that Φ be continuous with respect to the natural Fréchet algebra topology on $C^\infty(\mathbb{C})$: in this case, we have already introduced the terminology that the operator T is *generalized scalar*. We shall give more details on these operators below.

It is immediate that the spectrum of any $\mathcal{A}$-scalar operator is contained in $\overline{\Omega}$, where Ω is the domain of the functions in the algebra $\mathcal{A}$. But our ambition now is to show that $\mathcal{A}$-scalar operators are super-decomposable and to give an explicit description of the local spectral subspaces.

Theorem 22.2.4 *Let $\mathcal{A}$ be an algebra of functions on an arbitrary non-empty set $\Omega \subseteq \mathbb{C}$, and suppose that $\mathcal{A}$ is richly endowed and admits partitions of unity. Also, let the $\mathcal{A}$-scalar operator $T \in \mathcal{B}(X)$ be given by an algebra* *homomorphism* $\Phi : \mathcal{A} \to \mathcal{B}(X)$ *for which* $\Phi(1) = I$ *and* $\Phi(Z) = T$. *Then* T *is super-decomposable, and the local spectral subspace* $X_T(F)$ *for a closed set* $F \subseteq \mathbb{C}$ *is given by*

$$X_T(F) = \bigcap \{\ker \Phi(f) : f \in \mathcal{A} \text{ and } \operatorname{supp} f \cap F = \emptyset\} .$$

Proof The idea of the proof is, for arbitrary closed $F \subseteq \mathbb{C}$, to establish enough properties of the right-hand set

$$E(F) := \bigcap \{\ker \Phi(f) : f \in \mathcal{A} \text{ and } \operatorname{supp} f \cap F = \emptyset\} .$$

to see that it is indeed $X_T(F)$. Evidently, $E(F)$ is a closed linear subspace of X and $E(F)$ is invariant under $\Phi(g)$ for all $g \in \mathcal{A}$, hence, in particular, under T. An argument involving both partitions of unity and the fact that $\mathcal{A}$ is richly endowed then shows that $\sigma(T \mid E(F)) \subseteq F$. It is now clear that $E(F) \subseteq X_T(F)$. The next step then establishes that T is super-decomposable: let $\{U, V\}$ be an open cover of $\mathbb{C}$, and choose $f \in \mathcal{A}$ for which $\operatorname{supp} f \subseteq U$ and $\operatorname{supp}(1 - f) \subseteq V$. The operator $R := \Phi(f)$ commutes with T, and

$$RX \subseteq E(\operatorname{supp} f) \quad \text{and} \quad (I - R)X \subseteq E(\operatorname{supp}(1 - f)) .$$

By the spectral inclusions from the preceding paragraph, it follows that T is super-decomposable.

Finally, the inclusion $E(F) \supseteq X_T(F)$ is a bit of a technical *tour de force*: we first establish that $E(\bigcap_{n=1}^{\infty} F_n) = \bigcap_{n=1}^{\infty} E(F_n)$ for every countable family $\{F_n\}$ of closed sets in $\mathbb{C}$. Here the inclusion $\subseteq$ is obvious. For the other one, let $x \in \bigcap_{n=1}^{\infty} E(F_n)$, and let $f \in \mathcal{A}$ be any function for which $\operatorname{supp} f \cap F = \emptyset$, where $F := \bigcap_{n=1}^{\infty} F_n$. By compactness, there exists $k \in \mathbb{N}$ such that

$$\sigma(T) \cap \operatorname{supp} f \subseteq (\mathbb{C} \setminus F_1) \cup \cdots \cup (\mathbb{C} \setminus F_k) .$$

If we supplement this union with the complement of $\sigma(T) \cap \operatorname{supp} f$, we have a finite open cover of $\mathbb{C}$. Consequently, we may choose functions $f_0, \ldots, f_k \in \mathcal{A}$ for which $f_0 + \cdots + f_k$ is 1 on Ω, $\sigma(T) \cap \operatorname{supp} f \cap \operatorname{supp} f_0 = \emptyset$, and $\operatorname{supp} f_j \cap F_j = \emptyset$ for $j = 1, \ldots, k$. It is clear that

$$\Phi(f f_0)X \subseteq E(\operatorname{supp}(f f_0)) \subseteq X_T(\operatorname{supp}(f f_0) \cap \sigma(T)) = \{0\},$$

and hence $\Phi(f f_0)x = 0$. Moreover, since $x \in E(F_j)$ and $\operatorname{supp} f_j \cap F_j = \emptyset$, we obtain $\Phi(f_j)x = 0$ for $j = 1, \ldots, k$. All in all, we conclude that

$$\Phi(f)x = \sum_{j=0}^{k} \Phi(f f_j)x = \sum_{j=1}^{k} \Phi(f)(f_j)x = 0 ,$$

and therefore $x \in E(F)$.

With this knowledge about $E(F)$ we are now in a position to show that $E(F) \supseteq X_T(F)$. To do this, we fix an arbitrary open neighbourhood U of F, and choose an open set $V \subseteq \mathbb{C}$ so that $U \cup V = \mathbb{C}$ and $\overline{V} \cap F = \emptyset$. From what we did before, we see that $X = E(\overline{U}) + E(\overline{V})$. Since

$$E(\overline{U} \cap \overline{V}) = E(\overline{U}) \cap E(\overline{V}),$$

the quotient spaces $X/E(\overline{U})$ and $E(\overline{V})/E(\overline{U} \cap \overline{V})$ are seen to be canonically isomorphic. This implies that

$$\sigma(T/E(\overline{U})) \subseteq \sigma(T \mid E(\overline{V})) \cup \sigma(T \mid E(\overline{U} \cap \overline{V})) \subseteq \overline{V} \cup (\overline{U} \cap \overline{V}) = \overline{V},$$

by Proposition 21.2.6. On the other hand, we know that $\sigma(T \mid X_T(F)) \subseteq F$. Now, if $Q : X \to X/E(\overline{U})$ denotes the canonical quotient map, then clearly

$$(T/E(\overline{U}))\, Q \mid X_T(F) = Q\,(T \mid X_T(F)).$$

By the analytic functional calculus, it follows that

$$f(T/E(\overline{U}))\, Q \mid X_T(F) = Q\, f(T \mid X_T(F))$$

for every function f analytic on some open neighbourhood of $\overline{V} \cup F$. Since $\overline{V} \cap F = \emptyset$, there is an analytic function f for which $f \equiv 1$ on an open neighbourhood of $\overline{V}$ and $f \equiv 0$ on an open neighbourhood of F. For this function, it follows that $f(T/E(\overline{U}))$ is the identity operator on $X/E(\overline{U})$, while $f(T \mid X_T(F))$ is the zero operator on $X_T(F)$. We conclude that $Q \mid X_T(F) = 0$, and hence that $X_T(F) \subseteq E(\overline{U})$. Since F may be written as an intersection of countably many open neighbourhoods, and E preserves countable intersections, we have finally shown that $X_T(F) \subseteq E(F)$. □

Example 22.2.5 This example is based on Laursen and Neumann (2000, Example 1.4.11). Consider a commutative Banach algebra A with totally disconnected character space Φ_A and let Θ denote an algebra homomorphism from A to some $\mathcal{B}(X)$. We claim that $\Theta(a) \in \mathcal{B}(X)$ is super-decomposable for every element $a \in A$. Once this is established, we see (taking Θ to be the left regular representation of A) that all multiplication operators on such an algebra are super-decomposable. This covers the case of the group algebra $L^1(G)$ of a compact abelian group G. (Since the character space of $L^1(G)$ may be identified with the dual group of G, and since the dual group of a compact abelian group is discrete, we have that, in this case, $\Theta_{L^1(G)}$ is discrete, and hence totally disconnected.) For a brief discussion of the algebras $L^1(G)$, see Part I, §3.4 and Part II.

The proof uses the Šilov idempotent theorem (see Part I, Additional note 4.5.4). If A is as stipulated, and we also suppose (to cover the more complicated case) that A has no unit element, let $A^\# := A \oplus \mathbb{C}$ be the unitization of A, and let $\Theta_1 : A^\# \to \mathcal{B}(X)$ denote the canonical extension of the given Θ, so that $\Theta_1(a + \lambda) := \Theta(a) + \lambda I$ for all $a \in A$ and $\lambda \in \mathbb{C}$. An operator $T \in B(X)$ is super-decomposable if and only if this holds for $T - \lambda$, for any fixed $\lambda \in \mathbb{C}$, so we may replace a by $a - \lambda$, if necessary, and hence may add the assumption that $a \in A^\#$ be invertible in $A^\#$. To establish that $\Theta_1(a) \in \mathcal{B}(X)$ is super-decomposable, let $\{U, V\}$ be an open cover of $\mathbb{C}$. If $\mu \in \mathbb{C}$ is the scalar for which $a - \mu \in A$, we may suppose that $\mu \in V$. The Gelfand transform of $a - \mu$ vanishes at infinity on Φ_A, so the set

$$K := \{\varphi \in \Phi_A : \varphi(a) \in \mathbb{C} \setminus V\}$$

is compact, and it is clear that K is contained in the open set

$$L := \{\varphi \in \Phi_A : \varphi(a) \in U\} .$$

By local compactness and total disconnectedness of Φ_A, its Gelfand topology has a base consisting of compact and open sets. Consequently, there is a compact and open set $C \subseteq \Phi_A$ for which $K \subseteq C \subseteq L$. Now we invoke the Šilov idempotent theorem: there is an idempotent $r \in A$ for which $\varphi(r) = 1$ for all $\varphi \in C$ and $\varphi(r) = 0$ for all $\varphi \in \Phi_A \setminus C$. Evidently, $R := \Theta(r)$ commutes with T. Moreover, for the spectrum $\sigma(ar)$ of $ar \in A$, elementary Gelfand theory tells us that

$$\sigma(ar) = \{\varphi(ar) : \varphi \in \Phi_A\} \cup \{0\} \subseteq U \cup \{0\} .$$

Hence $ar - \lambda$ is invertible in $A^\#$ if $0 \neq \lambda \in \mathbb{C} \setminus U$. This implies that, for every such λ, there is an element $a_\lambda \in A^\#$ for which $(ar - \lambda)a_\lambda = r$, and therefore $(a - \lambda)a_\lambda r = r$. This equality also holds for $\lambda = 0$ if we choose $a_0 := a^{-1}$. Now apply the homomorphism Θ_1: we see that $(T - \lambda)\Theta_1(a_\lambda)R = R$, and hence that $(T - \lambda)\Theta_1(a_\lambda)x = x$ for all $x \in \overline{RX}$ and $\lambda \in \mathbb{C} \setminus U$. Since $\Theta(a_\lambda)$ commutes with T and leaves $\overline{RX}$ invariant, this establishes that $\sigma(T \mid \overline{RX}) \subseteq U$. An analogous argument gives $\sigma(T \mid \overline{(I - R)X}) \subseteq V$. So T is super-decomposable.

□

It is interesting that for many super-decomposable operators it is possible to give a description of their local spectral subspaces in purely algebraic terms.

This is important in applications of local spectral theory to automatic continuity questions, where we want topological conclusions, but not topological assumptions. The way to do this is to relate the local spectral subspaces to spaces (dating back to Johnson and Sinclair (1969) which are called the *algebraic spectral subspaces*.

For a subset F of $\mathbb{C}$ and a linear map T on a vector space X, consider the collection of all linear subspaces Y of X with the property that $(T-\lambda)Y = Y$ for every $\lambda \in \mathbb{C} \setminus F$. The linear span $E_T(F)$ of all such subspaces Y will evidently itself have the property that $(T-\lambda)E_T(F) = E_T(F)$ for every $\lambda \in \mathbb{C} \setminus F$. This shows that $E_T(F)$ is the largest linear subspace with respect to this surjectivity condition.

A simple example: if the map T is injective then

$$E_T(\mathbb{C} \setminus \{0\}) = \bigcap_{n=1}^{\infty} T^n X.$$

This intersection is often called *the generalized range* and denoted by $T^{\infty}X$. Indeed, the inclusion $\subseteq$ is immediate, and the other one follows from the observation that, by injectivity, $\bigcap_{n=1}^{\infty} T^n X = T(\bigcap_{n=1}^{\infty} T^n X)$. The space $E_T(\mathbb{C} \setminus \{0\})$ is also studied in Part V, where it is called the algebraic core (Definition 26.1.5; the above formula for $E_T(\mathbb{C} \setminus \{0\})$, when T is injective, is a special case of Proposition 26.1.10).

The space $E_T(\emptyset)$ is of particular interest. It is the largest linear subspace Y of X for which $(T-\lambda)Y = Y$ for all $\lambda \in \mathbb{C}$, and is called the *largest divisible subspace* for T. In the theory of automatic continuity, it is often crucial to exclude the existence of non-trivial divisible subspaces, because their presence tends to preclude the desired continuity conclusions. For this sort of reason, we are particularly interested in classes of operators T for which $E_T(\emptyset) = \{0\}$.

Now getting back to the case of a bounded linear operator $T \in \mathcal{B}(X)$, it is easy to see that $E_T(F) = E_T(F \cap \sigma(T))$ for all sets $F \subseteq \mathbb{C}$. Furthermore, it is clear that $X_T(F) \subseteq E_T(F)$ for all sets $F \subseteq \mathbb{C}$. In general, these inclusions will be strict, even when $F = \emptyset$. Indeed, $X_T(\emptyset) = \{0\}$ precisely when T has SVEP, but even super-decomposable operators may well have non-trivial divisible subspaces.

For an example of this, let $T \in \mathcal{B}(X)$ be both quasi-nilpotent and injective. By Proposition 22.2.2, T is super-decomposable. Thus $X_T(\emptyset) = \{0\}$, while it follows from the observations we have just made that

$$E_T(\emptyset) = E_T\left((\mathbb{C} \setminus \{0\}) \cap \{0\}\right) = E_T(\mathbb{C} \setminus \{0\}) = T^{\infty}X\,.$$

The generalized range may be quite large, as illustrated, for instance, by the Volterra operator:

Example 22.2.6 *The Volterra operator* (*cf.* Part I, Exercise 1.5.6) We define $X := C([0, 1])$, and let $T \in \mathcal{B}(X)$ be given by

$$(Tf)(t) := \int_0^t f(s)\,\mathrm{d}s \quad \text{for all} \quad f \in X \quad \text{and} \quad t \in [0, 1].$$

Then T is injective, compact, and quasi-nilpotent. In particular, it follows that T is a super-decomposable operator with $X_T(\emptyset) = \{0\}$. On the other hand, from the identity $E_T(\emptyset) = T^{\infty}X$, we may conclude that $E_T(\emptyset)$ is the set of all infinitely differentiable functions on $[0, 1]$ which, along with all their derivatives, vanish at 0. Thus $E_T(\emptyset)$ is strictly larger than $X_T(\emptyset)$. □

As another typical example (where $E_T(\emptyset)$ is small) we mention the case of decomposable multiplication operators on semisimple commutative Banach algebras. They will be discussed in Chapter 25, but if T is any multiplication operator on a commutative Banach algebra A, then it is easily seen that the intersection $\bigcap\{(T-\lambda)A : \lambda \in \mathbb{C}\}$ is contained in the radical of A, and hence must be trivial when X is semisimple.

Here is a simple special case in which we can readily give an explicit, algebraic description of the local spectral subspaces. We shall mention a similar one a bit later, for generalized scalar operators (Theorem 22.3.2).

Proposition 22.2.7 *Let $T \in \mathcal{B}(X)$ be super-decomposable and suppose that*

$$\bigcap_{\lambda\in\mathbb{C}} (T-\lambda)^p X = \{0\} \quad \text{for some } p \in \mathbb{N}.$$

Then, for every closed set $F \subseteq \mathbb{C}$,

$$X_T(F) = E_T(F) = \bigcap_{\lambda\in\mathbb{C}\setminus F} (T-\lambda)^p Xx.$$

Proof The idea of proof is similar to that of Theorem 22.2.4: given a closed set $F \subseteq \mathbb{C}$, let $G_T(F)$ be the intersection of the spaces $(T-\lambda)^p X$ where λ ranges over $\mathbb{C}\setminus F$. The containments $X_T(F) \subseteq E_T(F) \subseteq G_T(F)$ are immediate and, to show the converse, it is enough to show that $G_T(F) \subseteq X_T(U)$ for an arbitrary open neighbourhood U of F.

By definition of super-decomposability and properties of the spectral subspaces, there is an operator $R \in \mathcal{B}(X)$ for which $TR = RT$, $RX \subseteq X_T(\mathbb{C}\setminus F)$, and such that $(I-R)X \subseteq X_T(U)$. Hence to establish that $G_T(F) \subseteq X_T(U)$, it remains to be seen that $RG_T(F) = \{0\}$. By the assumption on T, for this it suffices to show that $RG_T(F) \subseteq (T-\lambda)^p X$ for every $\lambda \in \mathbb{C}$. If $\lambda \in \mathbb{C}\setminus F$, this inclusion follows from the very definition of $G_T(F)$ and the fact that the operators R and T commute. On the other hand, if $\lambda \in F$, the inclusion $RG_T(F) \subseteq (T-\lambda)^p X$ follows from

$$RX \subseteq X_T(\mathbb{C} \setminus F) = (T-\lambda)X_T(\mathbb{C} \setminus F).$$

Thus $RG_T(F) = \{0\}$, and consequently $G_T(F) \subseteq X_T(U)$. □

22.3 Generalized scalar operators

We have mentioned already that an operator $T \in \mathcal{B}(X)$ is *generalized scalar* if $T = \Phi(Z)$ for some continuous algebra homomorphism $\Phi : C^\infty(\mathbb{C}) \to \mathcal{B}(X)$ with $\Phi(1) = I_X$. We have also observed that, as $\mathcal{A}$-scalar operators, generalized scalar operators are super-decomposable, hence also decomposable. Much more is true.

The continuity of Φ means that there exist a compact subset Q of $\mathbb{C}$, a constant $c \geq 0$, and $k \in \mathbb{Z}^+$ such that

$$\|\Phi(f)\| \leq c\,\|f\|_{k,Q} \qquad \text{for all } f \in C^\infty(\mathbb{C}),$$

where $\mathbb{C}$ is canonically identified with $\mathbb{R}^2$, and

$$\|f\|_{k,Q} := \sum_{|\alpha|\leq k} \frac{1}{\alpha!} \sup\{|D^\alpha f(\lambda)| : \lambda \in Q\} \qquad \text{for all } f \in C^\infty(\mathbb{C}).$$

To start off, here is a specific example of a generalized scalar operator.

Example 22.3.1 Every surjective isometry $T \in B(X)$ is generalized scalar: a functional calculus on $C^\infty(\mathbb{C})$ for such an operator is given by

$$\Phi(f) := \sum_{n=-\infty}^{\infty} \widehat{f}(n)\,T^n \qquad \text{for all } f \in C^\infty(\mathbb{C}),$$

where $\widehat{f}(n)$ is the nth Fourier coefficient of the restriction of the function f to the unit circle $\mathbb{T}$, that is,

$$\widehat{f}(n) := \frac{1}{2\pi} \int_{-\pi}^{\pi} f(e^{i\theta})\,e^{-in\theta}\,d\theta \qquad \text{for all } n \in \mathbb{Z}.$$

If, as in Part I, Exercise 1.5.4, we let $g(\theta) := f(e^{i\theta})$ for all $\theta \in [-\pi, \pi]$, then, for every $n \in \mathbb{Z} \setminus \{0\}$, repeated integrations by parts lead to the formula

$$\widehat{f}(n) = \frac{1}{(in)^p} \frac{1}{2\pi} \int_{-\pi}^{\pi} g^{(p)}(\theta)\, e^{-in\theta}\, d\theta \qquad \text{for all } p \in \mathbb{Z}^+.$$

This shows that $|\widehat{f}(n)|$ is $O(|n|^{-p})$ for any given $p \in \mathbb{Z}^+$. Since $\|T^n\| = 1$ for all $n \in \mathbb{Z}$, it follows that the infinite series

$$\sum_{n=-\infty}^{\infty} \widehat{f}(n)\, T^n$$

converges in the operator norm of $\mathcal{B}(X)$, and hence there is a well-defined continuous linear map $\Phi : C^\infty(\mathbb{C}) \to \mathcal{B}(X)$ with $\Phi(T) = \sum \widehat{f}(n) T^n$. (This argument works not just for invertible isometries, but also for any $T \in \mathcal{B}(X)$ that satisfies $\sigma(T) \subseteq \mathbb{T}$ and has polynomial growth in the sense that there is a bound $\|T^n\| = O(|n|^\alpha)$ for some $\alpha \geq 0$, as $|n| \to \infty$. In particular, every doubly power-bounded operator is generalized scalar.) □

Incidentally, knowledge of the *order* of a generalized scalar operator T in $\mathcal{B}(X)$ (which is defined to be the smallest $k \in \mathbb{Z}^+$ for which the estimates involved in the above specification of continuity of Φ hold, for appropriate choice of Φ, Q, and c) makes it possible to give a very precise description of the local spectral subspaces. The proof of that part of the next theorem is rather involved, and we omit it here. It may be found in Laursen and Neumann (2000, Theorem 1.5.4).

Theorem 22.3.2 *Let $T \in \mathcal{B}(X)$ be a generalized scalar operator of order $k \in \mathbb{Z}^+$. Then T is super-decomposable, and the representation*

$$X_T(F) = E_T(F) = \bigcap_{\lambda \in \mathbb{C} \setminus F} (T - \lambda)^p X$$

holds for every closed set $F \subseteq \mathbb{C}$ and every integer p for which $p \geq k + 3$. In particular, T has no non-trivial divisible subspace. □

Finally, we may combine the observation of Example 22.3.1 with the fact (noted in Chapter 21) that a restriction of a decomposable operator to a closed invariant subspace will have (β). The conclusion is that (β) holds for any isometry. A simple proof comes from the following extension result, due to Douglas (1969).

Proposition 22.3.3 *Let $T \in \mathcal{B}(X)$ be an isometry. Then there is a Banach space Y into which X may be isometrically embedded and an invertible isometry $S \in \mathcal{B}(Y)$ which (via the embedding) extends T.*

Proof (sketched) Define Y_0 to be the vector space (pointwise operations) of all sequences $u = \{u_1, u_2, \dots\}$ in X for which $u_{m+1} = Tu_m$ for all sufficiently large m. Since T is an isometry, the norms $\|u_m\|$ are, for each such sequence, eventually constant, so the definition $p(u) = \lim_{m\to\infty} \|u_m\|$ provides us with a seminorm on Y_0. If we quotient out $N = \{u \in Y_0 | p(u) = 0\}$ and complete the quotient space Y_0/N, we obtain a Banach space, say Y. It is easy to see that the assignment $x \to \{T^n x\}_{n=1}^{\infty}/N$ yields an isometric embedding of X into Y.

In Y_0 we may define the map S_0 by $S_0(\{u_m\}_{m=1}^{\infty}) = \{Tu_m\}_{m=1}^{\infty}$. Clearly, S_0 induces an isometry S on Y. This isometry is invertible and extends T (Exercise 22.4.5). □

Corollary 22.3.4 *Every isometry has property (β).* □

22.4 Exercises

In these exercises, let X be a Banach space, and consider an operator T in $\mathcal{B}(X)$.

1. Show that, $\mathcal{X}_T(F) = X_T(F)$ for every closed set $F \subseteq \mathbb{C}$ if and only if $X_T(\emptyset) = \{0\}$, and this happens precisely when T has SVEP.
2. Show that, for each $\lambda \in \mathbb{C} \setminus \sigma_{\text{su}}(T)$, there is a constant $c > 0$ for which $X = \mathcal{X}_T(\mathbb{C} \setminus \mathbb{D}(\lambda; c))$. In particular, show that $\sigma(T) \setminus \sigma_{\text{su}}(T) \subseteq \sigma_{\text{p}}(T)$, and $\sigma(T) = \sigma_{\text{su}}(T)$, if T has SVEP.
3. Show that $\sigma_{\text{su}}(T) = \bigcup \{\sigma_T(x) : x \in X\}$.
4. Show that $E_T(F) = E_T(F \cap \sigma(T))$ for all sets $F \subseteq \mathbb{C}$.
5. Fill in the details of the proof of Proposition 22.3.3.

22.5 Additional notes

As already mentioned, a general source for virtually everything that is touched on here is Laursen and Neumann (2000). A specific instance of this: as we saw in Example 22.3.1, if the operator T has powers (negative and positive) of polynomial growth, then T must be generalized scalar; in Laursen and Neumann (2000) there are more specific results along these lines: it turns out that, for operators with spectrum in $\mathbb{T}$, these two properties are essentially equivalent.

This result was originally due to Colojoara and Foiaş, but Laursen and Neumann (2000, Theorem 1.5.12) has a simpler proof. There it is also shown that these conditions are equivalent to the condition that $\sigma(T) \subseteq \mathbb{T}$ provided that the following norm estimate holds for the growth of the resolvent near the spectrum: there exist a constant $d > 0$ and an $m \in \mathbb{N}$ such that

$$\|(T - \lambda)^{-1}\| \leq d\,|1 - |\lambda||^{-m}$$

for all $\lambda \in \mathbb{C}$ with $|\lambda| \neq 1$.

23
Duality theory

In this chapter we shall explore the surprising relations between decomposability and properties (β) and (δ).

23.1 Duality between (β) and (δ)

We have already seen, in Theorem 21.2.8, that T is decomposable if and only if it has both (β) and (δ). But the main reason for emphasizing these two properties is not just that they together describe decomposability – surely (β), in particular, is too technical, and too non-intuitive, to gain fame just for that! Their main conceptual *raison d'être* is that they possess a remarkable dual relationship: an operator will have one of them (either one!) precisely when its adjoint operator has the other one. This is the significant conclusion that is provided by the duality theory. (You should note that the adjoint operator is called the *dual* operator in Part I.)

The duality theory for operators that we are talking about here goes back to Errett Bishop's PhD thesis, published in Bishop (1959), where he developed a spectral theory for an arbitrary bounded linear operator on a reflexive Banach space. Bishop called his development a duality theory, because the operator and its adjoint are involved. In Bishop (1959) we can see, in more or less fully developed form, much of what are now basic tools and concepts of the field, such as the glocal spectral subspaces, and conditions (β) and (δ). There was even a precursor of decomposability (which Bishop called duality theory of type 3).

Theorem 21.2.8 (that T is decomposable if and only if it has both (β) and (δ)) was of course not available to Bishop. But he did show that, if both T and T' satisfy (β), then T has a duality theory of type 3, that is, T is decomposable. This result was established without Bishop's reflexivity condition by Frunză (1971). Bishop also showed that, if T' has (β), then T has (δ). This was a result that rather glaringly used reflexivity, because it showed that $(T')'$ has (δ).

The full duality theory owes its spectacular completion to Albrecht and Eschmeier (1997). It contains the following results. They are not complicated to state, but a lot of ingenuity lies behind their proofs. You will find reasonably accessible proofs in Laursen and Neumann (2000, Chapter 2).

Theorem 23.1.1 *A continuous linear operator on a Banach space has* (δ) *if and only if it is a quotient of a decomposable operator with respect to a closed invariant subspace.* □

Theorem 23.1.2 *A continuous linear operator on a Banach space has* (β) *if and only if it is a restriction of a decomposable operator to a closed invariant subspace.* □

Incidentally, the latter theorem explains why operators with (β) are often called *sub-decomposable* operators.

In both theorems, the *if* parts are easy enough. The approach to proving, say, Theorem 23.1.2 may be explained in fairly non-technical terms as follows. At its root is a pair of *3-space lemmas* both of which may be combined into the following statement. The proof for the (β) case is easy, the (δ) case is not.

Lemma 23.1.3 *Consider a commutative diagram of continuous linear operators between Banach spaces*

$$\begin{array}{ccccccccc} 0 & \longrightarrow & X & \stackrel{J}{\longrightarrow} & Y & \stackrel{Q}{\longrightarrow} & Z & \longrightarrow & 0 \\ & & A\downarrow & & B\downarrow & & C\downarrow & & \\ 0 & \longrightarrow & X & \underset{J}{\longrightarrow} & Y & \underset{Q}{\longrightarrow} & Z & \longrightarrow & 0 \end{array}$$

with exact rows. If both A and C have (β) *(or* (δ)*) then so does B.* □

Now, through considerable technical effort, it may be established that for an arbitrary T there is a particular commutative diagram

$$\begin{array}{ccccccccc} 0 & \longrightarrow & X & \stackrel{J}{\longrightarrow} & Y & \stackrel{Q}{\longrightarrow} & Z & \longrightarrow & 0 \\ & & T\downarrow & & B\downarrow & & C\downarrow & & \\ 0 & \longrightarrow & X & \underset{J}{\longrightarrow} & Y & \underset{Q}{\longrightarrow} & Z & \longrightarrow & 0 \end{array}$$

with exact rows, in which the operator C is generalized scalar, in particular decomposable, and B, as a quotient of a generalized scalar operator (not displayed in this diagram), has (δ). Now, if we suppose that T has (β), then the 3-space lemma implies that B will also have (β), and hence B is decomposable. This establishes that T is the restriction of a decomposable operator.

Although the next result would be an easy consequence of Theorem 23.1.5, we state it separately, because in the actual development of the duality theory it is used in the proof of Theorem 23.1.5 – and doing so makes it possible for us to sketch quite a bit of the arguments.

Theorem 23.1.4 *An operator $T \in \mathcal{B}(X)$ is decomposable if and only if the adjoint $T' \in \mathcal{B}(X')$ is decomposable.* □

The proof may be found in Laursen and Neumann (2000, Section 2.5). Not too surprisingly it makes extensive use of annihilators and preannihilators of local spectral subspaces. The *only if* part begins by showing that, if T has the property that all its local spectral subspaces $X_T(F)$ are closed, for every closed $F \subseteq \mathbb{C}$ (this is called property (C) for T), then $X' = X_T(\mathbb{C}\backslash U)^\perp + X_T(\mathbb{C}\backslash V)^\perp$ for every open cover $\{U, V\}$ of $\mathbb{C}$. This is followed by establishing that, if T is decomposable, then $X'_{T'}(F) = X_T(\mathbb{C}\backslash F)^\perp$. Finally, this then implies that T' is decomposable. To establish that decomposability of T' implies the same property for T is somewhat more delicate, but the main step consists in showing that, on the assumption that T' have SVEP, $X'_{T'}(F)$ is norm-closed if and only if it is weak-$*$ closed.

Theorem 23.1.5 *A operator $T \in \mathcal{B}(X)$ will have one of the properties (β) or (δ) if and only if its adjoint $T' \in \mathcal{B}(X')$ has the other one.*

Proof We sketch how to show that, if T has (β), then T' has (δ). If T has (β), then, by Theorem 23.1.2, T has a decomposable extension, say S. By Theorem 23.1.4, the adjoint operator S' is also decomposable. Moreover, T' may be realized as a quotient of S'. Hence, by Theorem 23.1.1, T' has (δ). □

The hardest part of all this, by far, is the proof that, if T' has (β), then T has (δ). You may wish to consult Laursen and Neumann (2000, Chapter 2).

23.2 Exercises

1. Fill in the details of the sketch of the part of the proof of Theorem 23.1.5 that is included above.
2. Use Theorems 23.1.2, 23.1.1, and 23.1.4 to show that, if T has (δ), then T' has (β). Then prove that, if T' has (δ), then T has (β).

23.3 Additional notes

The actual functional model is explained in some detail in the Appendix to Part IV, 'The functional model'. A fairly elementary account, from which this background material is culled, is in Laursen and Neumann (2000, Chapter 2). The complete duality theory was first described in Eschmeier's Habilitationsschrift and in Albrecht and Eschmeier (1997).

The properties (β) and (δ) have been used in some of the most general solutions of the invariant-subspace problem. Here is a sample, from Eschmeier and Prunaru (1990): if $T \in B(X)$ has (β) or (δ) and if $V \subseteq \mathbb{C}$ is open and bounded, then if $\sigma(T)$ is dominating in V (a technical condition that is satisfied, for instance, when $\sigma(T)$ has non-empty interior), then T has a non-trivial closed invariant subspace. Moreover, if the smaller set consisting of the essential spectrum of T is dominating in V (the essential spectrum is defined at the beginning of the next chapter), then T has many closed invariant subspaces. These invariant-subspace results, and much more, are described in Part III of this book.

24

Preservation of spectra and index

It is well known that, if two linear operators on a finite-dimensional space are similar, then their spectra are equal, and that this equality extends to distinguished parts of the spectra, such as the point spectra, etc. Here we consider natural extensions of this situation, where similarity is replaced by weaker conditions, and where concepts and techniques from local spectral theory are used.

24.1 An intertwining result

We will concentrate on proving the following result.

Theorem 24.1.1 *Suppose that $S \in \mathcal{B}(Y)$ and $T \in \mathcal{B}(X)$ both have property (β) and that $A \in \mathcal{B}(X, Y)$ and $B \in \mathcal{B}(Y, X)$ are operators with dense ranges for which $SA = AT$ and $TB = BS$. Then $\sigma(T) = \sigma(S)$, $\sigma_e(T) = \sigma_e(S)$ and* $\operatorname{ind}(T - \lambda) = \operatorname{ind}(S - \lambda)$ *for all $\lambda \in \rho_e(T)$.*

First we shall explain the terms. Recall that an operator $T \in \mathcal{B}(X)$ is a *Fredholm operator* (or just Fredholm) if its null space $\ker T$ is of finite dimension and its range TX is of finite codimension in X; the range is then automatically closed (Exercise 24.3.1). The *essential spectrum* $\sigma_e(T)$ of T is the set of complex numbers λ for which $T - \lambda$ is not Fredholm (often expressed this way: λ is not a Fredholm point for T.) Fredholm theory is developed more fully in Part V.

A relation such as $SA = AT$ is described by saying that A *intertwines* the pair (S, T). If A is a bijection, then S and T are *similar,* and A is a *similarity*. As we mentioned, a similarity preserves spectra, and even the finer structure of the spectra (point spectrum, approximate point spectrum, Fredholm points, etc.). Here we shall see what happens if the conditions on the intertwining operator are weakened, for example to that of being a *quasi-affinity*, that is, an injective map with dense range, or even if both (S, T) and (T, S) are intertwined by quasi-affinities; in this latter case the operators S and T are said to be *quasi-similar.* Note, incidentally, that the answer provided by Theorem 24.1.1 is made on intertwining assumptions that are slightly weaker than quasi-similarity.

The *index* of a Fredholm operator $T \in \mathcal{B}(X)$ is

$$\operatorname{ind}(T) := \dim \ker T - \dim(X/TX).$$

Here are a few basic results from classical Fredholm theory: an operator $T \in \mathcal{B}(X)$ is Fredholm if and only if its adjoint operator $T' \in \mathcal{B}(X')$ is Fredholm; in fact, since the dimension of the null space of one equals the codimension of the range of the other, we see that

$$\operatorname{ind}(T') = -\operatorname{ind}(T).$$

The product of two Fredholm operators is again Fredholm. In particular, every power of a Fredholm operator is Fredholm. The *index theorem* says that the index is a homomorphism, that is,

$$\operatorname{ind}(TS) = \operatorname{ind}(T) + \operatorname{ind}(S)$$

for all Fredholm operators $T, S \in \mathcal{B}(X)$. Since invertible operators are Fredholm operators of index zero, it follows that the Fredholm property as well as the index are preserved under similarity.

Does this hold for quasi-similarities also?

Example 24.1.2 (Fialkow (1977), here reproduced from Laursen and Neumann 2000, Example 3.7.12). Let T be the unilateral right shift on $\ell^2(\mathbb{N})$, and let S be the unilateral weighted right shift on $\ell^2(\mathbb{N})$ with weight sequence $(1/(n+1))_{n\in\mathbb{N}}$. Then T has (β), $\sigma(T) = \overline{\mathbb{D}}$, and $\sigma_e(T) = \mathbb{T}$, while S is compact with $\sigma(S) = \sigma_e(S) = \{0\}$. Thus $\sigma_e(T)$ and $\sigma_e(S)$ are disjoint. However, the pair (S, T) is intertwined by a quasi-affinity, namely the multiplication operator A on $\ell^2(\mathbb{N})$ given by

$$Ax := (x_n/n!)_{n\in\mathbb{N}} \qquad \text{for all } x = (x_n)_{n\in\mathbb{N}} \in \ell^2(\mathbb{N}).$$

It is easy to see that A is injective, has dense range, and that $SA = AT$. □

Even when (β) or (δ) is assumed, very little is true about inclusions for the essential spectra of one-sidedly intertwined operators,

Example 24.1.3 Let $H^p(\mathbb{D})$ be the classical Hardy space of all complex-valued analytic functions f defined on the open unit disc $\mathbb{D}$ for which

$$\sup\left\{\int_{-\pi}^{\pi} |f(re^{i\theta})|^p \, d\theta : 0 \le r < 1\right\} < \infty.$$

Observe that, for any choice of p, q such that $1 < p < q < \infty$, the canonical embedding A of $H^q(\mathbb{D})$ into $H^p(\mathbb{D})$ is injective and it has dense range (the polynomials on $\mathbb{D}$ are dense in $H^p(\mathbb{D})$). For arbitrary $1 < p < \infty$, the Cesàro operator C_p on the Hardy space $H^p(\mathbb{D})$ is defined by

$$(C_p f)(\lambda) := \frac{1}{\lambda}\int_0^\lambda \frac{f(\zeta)}{1-\zeta}\, d\zeta \quad \text{for all } f \in H^p(\mathbb{D}) \quad \text{and } \lambda \in \mathbb{D}\,.$$

In Miller, Miller and Smith (1998) it is shown that C_p has (β), that

$$\sigma(C_p) = \overline{\mathbb{D}(p/2; p/2)},$$

and that $\sigma_e(C_p) = \partial\mathbb{D}(p/2; p/2)$. The intertwining $C_p\, A = A\, C_q$ is clear, so here we have one-sided intertwining, but no nice containment for the essential spectra; in fact, we have $\sigma_e(C_p) \cap \sigma_e(C_q) = \{0\}$. □

The proof of Theorem 24.1.1 is relatively elementary, but lengthy. It begins by using an observation about restrictions of Fredholm operators to certain invariant subspaces.

Proposition 24.1.4 *Suppose that $T \in \mathcal{B}(X)$, and that $Z \subseteq X$ is a T-invariant closed linear subspace of finite codimension in X. Then T is a Fredholm operator on X if and only if $T \mid Z$ is a Fredholm operator on Z. Moreover, if the two operators are both Fredholm, they have the same index.*

Proof Pick a projection $P \in \mathcal{B}(X)$ with range Z, and observe that the space $(I - P)X = \ker P$ is finite-dimensional. Since $T(I - P)$ is of finite-rank, and T equals $TP + T(I - P)$, it follows that T and TP are Fredholm simultaneously, and that, in this case, $\operatorname{ind}(T) = \operatorname{ind}(TP)$. Moreover, since an easy calculation will reveal that we have $\ker(TP) = \ker(T \mid Z) \oplus \ker P$ and $(TP)(X) = (T \mid Z)(Z)$, we conclude that TP and $T \mid Z$ also are Fredholm simultaneously and that, when this happens, they will have the same index. □

Another fundamental fact of Fredholm theory, the *punctured disc theorem,* says that, if T is Fredholm, then there is a positive number c with the property that, for each $\lambda \in \mathbb{D}(0; c)$, the operator $T - \lambda$ is Fredholm, and the quantities $\dim\ker(T - \lambda)$ and $\dim(X/(T - \lambda)X)$ are constant on the punctured disc $\mathbb{D}(0; c) \setminus \{0\}$. Moreover,

$$\dim\ker(T - \lambda) \le \dim\ker T \quad \text{and} \quad \dim(X/(T - \lambda)X) \le \dim(X/TX)$$

for all $\lambda \in \mathbb{D}(0; c)$. The *jump* of T is then the non-negative integer

$$\dim \ker T - \lim_{\lambda \to 0} \dim \ker(T - \lambda) = \dim(X/TX) - \lim_{\lambda \to 0} \dim(X/(T - \lambda)X),$$

and the *zero-jump theorem* says that, for a Fredholm operator T with positive jump, there exists an invariant closed linear subspace of finite codimension on which the restriction of T is Fredholm, with jump zero. Versions more general than we need here may be found in Kato (1958) and West (1990). See also Part V.

The zero-jump theorem that we will use, Proposition 24.1.6, has a simple local-spectral-theory-based proof. We make one observation first.

Lemma 24.1.5 *Every surjective operator with complemented kernel is right invertible.*

Proof If $T \in \mathcal{B}(X)$ is surjective and if there is a closed subspace Y for which $X = \ker T \oplus Y$, then, for every $x \in X$, there is a unique element $y \in Y$ such that $Ty = x$. It is clear that $T|Y : Y \to X$ is a bijection, so it has an inverse, say $W \in \mathcal{B}(X, Y)$. We may regard W as a bounded linear operator on X. Obviously $TW = I_X$. □

The converse of the above lemma happens to be true as well, but we shall not need this.

Proposition 24.1.6 *Given a Fredholm operator $T \in \mathcal{B}(X)$, there are numbers $n \in \mathbb{Z}^+$ and $c > 0$ for which the dimension of* $\ker(T - \lambda) \mid T^n X$ *is constant for all $\lambda \in \mathbb{D}(0; c)$. This constant value is* $\dim \ker(T - \lambda)$ *for all $\lambda \in \mathbb{D}(0; c) \setminus \{0\}$. If T also has SVEP, then this constant value is* 0.

Proof If we add the assumption that T be surjective, this is quite easy: by Lemma 24.1.5 there is an operator $S \in \mathcal{B}(X)$ for which $TS = I$. The invertibles form an open subset of $\mathcal{B}(X)$, so the operator $(T - \lambda)S$ is invertible for all λ in some open disc $\mathbb{D}(0; c)$; in fact, by standard arguments (see Part I, Theorem 1.4.2 (iii)), this holds for $c = 1/\|S\|$. By shrinking the radius $c > 0$, if necessary, we may suppose that $\mathbb{D}(0; c) \subseteq \rho_e(T)$. Thus $T - \lambda$ is a surjective Fredholm operator for all $\lambda \in \mathbb{D}(0; c)$. Because the index is locally constant, it follows that $\dim \ker(T - \lambda) = \dim \ker T$ for every $\lambda \in \mathbb{D}(0; c)$.

Now we drop the surjectivity assumption on T. Recall the generalized range

$$T^\infty X := \bigcap \{T^n X : n \in \mathbb{N}\}.$$

It can be proved that $R := T \mid T^\infty X$ is surjective and Fredholm (Exercise 24.3.2). So the first part shows that there is a $c > 0$ such that $\dim \ker(R - \lambda)$ is constant for all $\lambda \in \mathbb{D}(0; c)$. If we choose an $n \in \mathbb{Z}^+$ so that

$$\ker T \cap T^n X = \ker T \cap T^\infty X,$$

then we see that the constant $\dim \ker(R - \lambda)$ is $\dim \ker T \mid T^n X$. However, for every non-zero $\lambda \in \mathbb{C}$, $\ker(T - \lambda) \subseteq T^\infty X$, and therefore

$$\ker(T - \lambda) = \ker(T \mid T^n X - \lambda) = \ker(R - \lambda).$$

This shows that the constant value is $\dim \ker(T - \lambda)$ for all $\lambda \in \mathbb{D}(0; c) \setminus \{0\}$.

Finally, if T also has SVEP, then its restriction R is a surjective operator with SVEP. Thus $\ker R = \{0\}$ (Exercise 24.3.3. Exercise before Proposition 22.1.3: if $\ker R$ is non-trivial then surjectivity of R allows the construction of a power series with sum f, say, for which $(R - \mu)f(\mu) = 0$ in a neighbourhood of 0) and the last assertion follows. □

The principal tool in the rest of the proof is the next lemma. For an open set $U \subseteq \mathbb{C}$ and a $W \in \mathcal{B}(X, Y)$, recall from Chapter 21 the notation

$$W^\natural : H(U, X) \to H(U, Y)$$

for the operator given by $(W^\natural f)(\lambda) = W(f(\lambda))$ for all $f \in H(U, X)$ and $\lambda \in U$.

Lemma 24.1.7 *Suppose that $T \in \mathcal{B}(X)$ is Fredholm, and that the operator $C \in \mathcal{B}(X)$ commutes with T and has dense range. Then there is an open disc $V := \mathbb{D}(0; r)$ for which C induces a surjection*

$$C^\diamond : H(V, X)/T_V H(V, X) \to H(V, X)/T_V H(V, X)$$

by the assignment

$$C^\diamond(f + T_V H(V, X)) := C^\natural f + T_V H(V, X) \quad \text{for all} \quad f \in H(V, X).$$

Additionally, if T is also injective, then $r > 0$ may be chosen so that $C^\diamond$ is bijective.

Proof Evidently, since $CT = TC$, the definition of $C^\diamond$ makes sense. The range TX is of finte codimension, so there is a finite-dimensional subspace F of X for which $X = TX \oplus F$. Using this splitting, we observe that

$$CX = CTX + CF = TCX + CF \subseteq TX + CF.$$

The space $TX + CF$ is closed and C has dense range, so $X = TX + CF$. In particular, the dimension of CF is at least as big as that of F, that is, $C \mid F$ is injective and $X = TX \oplus CF$. Similarly, $X = TX \oplus C^n F$ for every $n \in \mathbb{N}$.

If we now define an analytic operator function η on $\mathbb{C}$ taking its values in $\mathcal{B}(X \times F, X)$ by

$$\eta(\lambda)(x, u) := (T - \lambda)x + Cu \quad \text{for all } \lambda \in \mathbb{C} \quad \text{and} \quad (x, u) \in X \times F,$$

then we have shown that $\eta(0)$ is surjective. Moreover, since $TX \cap CF = \{0\}$, the space $\ker \eta(0) = \ker T \times \{0\}$ is finite-dimensional. It follows, by Lemma 24.1.5, that $\eta(0)$ has a continuous linear right inverse, so that $\eta(0)S = I_X$ for some $S \in \mathcal{B}(X, X \times F)$. Continuity of η then implies that invertibility holds for $\eta(\lambda)S$ for all λ near 0, that is, there is an $r > 0$ such that $\eta(\lambda)S$ is invertible for every $\lambda \in V := \mathbb{D}(0; r)$. Thus, given any $f \in H(V, X)$, the assignment

$$(f_1(\lambda), f_2(\lambda)) := S(\eta(\lambda)S)^{-1} f(\lambda) \qquad \text{for all } \lambda \in V$$

yields analytic functions $f_1 \in H(V, X)$ and $f_2 \in H(V, F)$ such that we have $f = T_V f_1 + C^\natural f_2$. The surjectivity of $C^\diamond$ is now obvious.

Next we add the assumption that T is also injective, and we define

$$\xi(\lambda)(x, u) := (T - \lambda)x + u \quad \text{for all } \lambda \in \mathbb{C} \quad \text{and} \quad (x, u) \in X \times F.$$

The function $\xi : \mathbb{C} \to \mathcal{B}(X \times F, X)$ is analytic, and $\xi(0)$ is bijective. Hence, shrinking the radius of $V = \mathbb{D}(0; r)$, if necessary, we may suppose that, for every λ in V, the operator $\xi(\lambda)$ is both right and left invertible, that is bijective. This implies that the operator $T - \lambda$ is injective and Fredholm, and that $(T - \lambda)X + F = X$ for all $\lambda \in V$. Since, by continuity of the index,

$$\dim(X/(T - \lambda)X) = -\operatorname{ind}(T - \lambda) = -\operatorname{ind}(T) = \dim F \quad \text{for all } \lambda \in V,$$

we conclude that, in fact, $X = (T - \lambda)X \oplus F$ for all $\lambda \in V$, and the same argument as before then shows that $(T - \lambda)X \cap C^n F = \{0\}$ for every $n \in \mathbb{N}$. Now, if $f, g \in H(V, X)$ are functions for which $C^\natural f = T_V g$, then, writing, as above, $f = T_V f_1 + C^\natural f_2$ for suitable $f_1 \in H(V, X)$ and $f_2 \in H(V, F)$, we see that $(C^2)^\natural f_2 \in T_V H(V, X)$, and consequently $f_2 = 0$, so that $f \in T_V H(V, X)$. This shows that $C^\diamond$ is injective, and therefore bijective. □

Proof of Theorem 24.1.1 We begin by establishing the equality of the two spectra. For this we need two useful observations:

1. Let $T \in \mathcal{B}(X)$ be arbitrary. For each closed F for which $F \cap \sigma(T) = \emptyset$, we have $\mathcal{X}_T(F) = \{0\}$. This follows from Liouville's classic theorem: in fact, if $x = (T - \lambda)f(\lambda)$ for all $\lambda \in \mathbb{C} \setminus F$, where f is analytic, then in a neighbourhood of F we must have $f(\lambda) = (T - \lambda)^{-1}x$. This implies that f may be extended analytically (by means of the expression $(T - \lambda)^{-1}x$) to all of $\mathbb{C}$, and since this yields a bounded entire function, f must be constant. The only possible constant is 0.

2. If $T \in \mathcal{B}(X)$, $S \in \mathcal{B}(Y)$, and if $A \in \mathcal{B}(X, Y)$ intertwines (S, T), then $\sigma_S(Ax) \subseteq \sigma_T(x)$ for every $x \in X$. This is immediate from the fact that, if the equation $x = (T - \lambda)f(\lambda)$ holds, with f analytic, in some open subset of $\mathbb{C}$, then $Ax = (S - \lambda)Af(\lambda)$ in that same subset, and Af is analytic. By the same argument we conclude that, if $F \subseteq \mathbb{C}$ is an arbitrary closed set, then $A\mathcal{X}_T(F) \subseteq \mathcal{Y}_S(F)$.

Now suppose that T and S are intertwined by A, as in Observation 2, and suppose that A is injective (this part of the argument will soon be applied to the adjoint operators, thus the injectivity assumption, which is dual to the density assumption of Theorem 24.1.1). If $\lambda \in \sigma_p(T)$, then

$$\{0\} \neq \ker(T - \lambda) \subseteq \mathcal{X}_T(\{\lambda\})$$

(the last inclusion is Exercise 24.3.4), and hence $\{0\} \neq A\mathcal{X}_T(\{\lambda\}) \subseteq \mathcal{Y}_S(\{\lambda\})$. But then, by Observation 1, $\lambda \in \sigma(S)$. We have shown that $\sigma_p(T) \subseteq \sigma(S)$. Since we have $\sigma(T) = \sigma_p(T) \cup \sigma_{su}(T)$, we next consider points of the surjectivity spectrum $\sigma_{su}(T)$.

Continuing for a little while in our dual mode, we impose the assumption that T has (δ). Let U be an open neighbourhood of $\sigma(S)$ and pick another open set V with closure $\overline{V}$ disjoint from $\sigma(S)$ such that $\mathbb{C} = U \cup V$. By Observation 1, $\mathcal{Y}_S(\overline{V}) = \{0\}$, and consequently, by the injectivity of A, also $\mathcal{X}_T(\overline{V}) = \{0\}$. By using (δ), we see that

$$X = \mathcal{X}_T(\overline{U}) + \mathcal{X}_T(\overline{V}),$$

and hence $X = \mathcal{X}_T(\overline{U})$. An immediate consequence of this is that $\sigma_{su}(T) \subseteq \overline{U}$. Since the set U is an arbitrary neighbourhood of $\sigma(S)$, we may conclude that $\sigma_{su}(T) \subseteq \sigma(S)$. All in all $\sigma(T) \subseteq \sigma(S)$, if T has (δ) and A is injective.

Now we position ourselves in the setting of Theorem 24.1.1, and suppose that S and T have (β) and that the intertwining maps A and B have dense ranges. Then S' and T' have (δ), they are intertwined by B' and A', and these two maps are both injective. It follows that $\sigma(T') = \sigma(S')$, and consequently $\sigma(T) = \sigma(S)$.

To establish the claim about the essential spectra and the index, we must show that if T is Fredholm, then so is S, and $\operatorname{ind}(T) \leq \operatorname{ind}(S)$.

If we replace X by T^nX, where n is specified in Proposition 24.1.6, and T by $T \mid T^nX$, then this restriction also has (β). Moreover, by Proposition 24.1.4, $T \mid T^nX$ is a Fredholm operator with the same index as T.

What will change in the passage from S to $S \mid \overline{S^nY}$? First of all, $S \mid \overline{S^nY}$ inherits (β) from S. Moreover, obviously $A \mid T^nX$ maps into $\overline{S^nY}$, densely, and $B \mid \overline{S^nY}$ maps into T^nX, also densely. The originally assumed intertwining relationships hold for the four restrictions. Since T'^n is Fredholm and $A' \ker S'^n \subseteq \ker T'^n$, it follows from the injectivity of the adjoint A' that $\ker S^{*n}$ is finite-dimensional. The space $\overline{S^n Y}$ is the pre-annihilator of $\ker S'^n$, hence it is of finite codimension (this is a consequence of the annihilator theorem; see, for example, Laursen and Neumann (2000, Theorem A.1.8)). Thus, by Proposition 24.1.4, S is Fredholm precisely when $S \mid \overline{S^n Y}$ is Fredholm. Moreover, when they are, S and $S \mid \overline{S^n Y}$ have the same index.

Consequently, we may continue the argument with $T \mid T^n X$. Because T has SVEP, this means that, by Proposition 24.1.6, we may invoke the additional assumption that T is injective. We shall do this, but continue using the shorter symbols T, S, A, B, rather than those of their restrictions.

The operator BA commutes with T, and has dense range. Consequently, since T is injective, Lemma 24.1.7 tells us that there is an open disc $V := \mathbb{D}(0; r)$ for which

$$(BA)^\diamond : H(V, X)/T_V H(V, X) \to H(V, X)/T_V H(V, X),$$

defined, as before, by $(BA)^\diamond(f + T_V H(V, X)) := (BA)^\# f + T_V H(V, X)$ for all $f \in H(V, X)$, is a bijection.

Since T and S have (β), we know that the ranges $T_V H(V, X)$ and $S_V H(V, Y)$ are closed (cf. the remark immediately after Definition 21.2.1). Hence

$$H(V, X)/T_V H(V, X) \quad \text{and} \quad H(V, Y)/S_V H(V, Y)$$

are Fréchet spaces in the natural quotient space topology. By the open mapping theorem, the continuous bijection $(BA)^\diamond$ is a topological linear isomorphism.

The canonically defined map

$$A^\diamond : H(V, X)/T_V H(V, X) \to H(V, Y)/S_V H(V, Y)$$

is obviously continuous, and we now show that $A^\diamond$ is a bijection. Since $(BA)^\diamond = B^\diamond A^\diamond$, and $(BA)^\diamond$ is bijective, $A^\diamond$ is certainly injective. Moreover, as a map from the range of $A^\diamond$ to $H(V, X)/T_V H(V, X)$, $B^\diamond$ is a continuous bijection. Since the linear span of functions of the form $\lambda \mapsto f(\lambda)y$, $V \to Y$ (as f ranges through $H(V)$ and y through Y) is dense in $H(V, Y)$, and A has dense range, so does $A^\diamond$. But this range is closed because $(BA)^\diamond$ is a topological isomorphism (so that the range of $A^\diamond$ is the pre-image with respect to $B^\diamond$ of $H(V, X)/T_V H(V, X)$), and therefore $A^\diamond$ is surjective. It then follows that also

$$B^\diamond : H(V, Y)/S_V H(V, Y) \to H(V, X)/T_V H(V, X)$$

is bijective.

Since $A^{\diamond}$ is surjective, there is, for every $g \in H(V, Y)$, a function f in $H(V, X)$ for which

$$g - A^{\#} f \in S_V H(V, Y).$$

In particular, if we evaluate this expression at 0, and use the fact that g is arbitrary, we see that, for every $y \in Y$, there are elements $x \in X$ and $z \in Y$ so that $y = Sz + Ax$. Thus $Y = SY + AX$. If F, as before, is a finite-dimensional subspace of X for which $X = TX \oplus F$, then

$$AX \subseteq ATX + AF = SAX + AF \subseteq SY + AF,$$

and therefore $Y = SY + AF$. This shows that SY is of finite codimension in Y. In fact, $\dim(Y/SY) \leq \dim(AF) \leq \dim F$.

Moreover, if $y \in \ker S$ and we view y as an analytic function on V, then its class $y + S_V H(V, Y)$ satisfies $B^{\diamond}(y + S_V H(V, Y)) = 0$; this holds because $TBy = BSy = 0$, and T is assumed to be injective. We have already established that $B^{\diamond}$ is injective, so it follows that $y \in S_V H(V, Y)$, and therefore

$$y \in Y_S(\{0\}) \cap Y_S(\mathbb{C} \setminus V) = Y_S(\emptyset) = \{0\},$$

because S has SVEP. This shows that S is injective, and hence establishes that S is Fredholm. Moreover, it follows that

$$\operatorname{ind}(S) = -\dim(Y/SY) \geq -\dim F = \operatorname{ind}(T),$$

and therefore $\operatorname{ind}(T) \leq \operatorname{ind}(S)$. By symmetry, we are done. □

There is an analogous result for operators with (δ). It says that if $S \in \mathcal{B}(Y)$ and $T \in \mathcal{B}(X)$ both have (δ), and if $A \in \mathcal{B}(X, Y)$ and $B \in \mathcal{B}(Y, X)$ are injective operators for which $SA = AT$ and $TB = BS$, then $\sigma(T) = \sigma(S)$, $\sigma_e(T) = \sigma_e(S)$ and

$$\operatorname{ind}(T - \lambda) = \operatorname{ind}(S - \lambda) \quad \text{for all } \lambda \in \rho_e(T).$$

This is proved by relying heavily on the natural dualities; but you should be aware that, although the adjoint of an injective operator always has weak-$*$ dense range, this range is not necessarily norm-dense. Hence the (δ) result is *not* an immediate consequence of Theorem 23.1.5 via the duality between the properties (β) and (δ). The details may be found in Laursen and Neumann (2000, Section 3.7).

24.2 Automatic continuity

Automatic continuity theory for homomorphisms and derivations is discussed in Part I, Chapter 5. Just to give an impression of the kind of automatic continuity results that may be established on the basis of local spectral theory, I quote one theorem. There is not sufficient space here to dwell on how it is proved, but it and others like it may be found in Laursen and Neumann (2000, Chapter 5). However, significant in the arguments is the simple observation that the algebraic spectral subspaces, discussed immediately after Example 22.2.5, have the property (analogous to what is true for both the local and the glocal spectral subspaces and *continuous* Θ, in the notation of this next theorem) that $\Theta E_T(F) \subseteq E_S(F)$, when $S\Theta = \Theta T$.

Theorem 24.2.1 *Suppose that $T \in \mathcal{B}(X)$ is decomposable, while $S \in \mathcal{B}(Y)$ is generalized scalar. Suppose also that $\Theta : X \to Y$ is a linear transformation for which $S\Theta = \Theta T$. Then there is a finite set $\Lambda \subset \mathbb{C}$ such that $\Theta|X_T(F)$ is continuous for every closed $F \subseteq \mathbb{C}$ which is disjoint from Λ. Moreover, there is a non-zero polynomial p, all of whose roots are eigenvalues of S, such that the composition $p(S)\Theta : X \to Y$ is continuous. In particular, Θ is continuous if S has no eigenvalues.* □

The finite set Λ is called the *singularity set*. The condition on T may be relaxed to that of assuming (δ), while that on S could be replaced by super-decomposable and $E_S(\emptyset) = \{0\}$, and even more general conditions. For other results on automatic continuity, see Part I, Chapter 5.

24.3 Exercises

1. Show that, if the range TX of the operator T is of finite codimension in X, then the range is automatically closed.
2. Referring to the proof of Theorem 24.1.6, assume that T is Fredholm. Prove that $R := T \mid T^{\infty}X$ is surjective and Fredholm.
3. (cf. the exercise before Proposition 22.1.3) Show that, if R is a surjective, non-injective operator, then there is a power series with sum f, say, for which $(R - \mu)f(\mu) = 0$ in a neighbourhood of 0.
4. Prove the inclusion $\ker(T - \lambda) \subseteq \mathcal{X}_T(\{\lambda\})$ for an arbitrary operator T.

24.4 Additional notes

The issue of spectral overlaps has received a lot of attention. For general Banach spaces, the classical Lumer and Rosenblum (1959) result says that two operators

(S, T) intertwined by a non-zero intertwiner must have overlapping spectra, and this result was refined by Davis and Rosenthal (1974) to $\sigma_{su}(T) \cap \sigma_{ap}(S) \neq \emptyset$. Both of these results are provable, in even greater generality, by local spectral methods. Laursen and Neumann (2000, Chapter 3) has details, and work on this may also be found in Laursen and Neumann (1994): it is not even necessary to suppose that $SA = AT$. There are ways of saying how close parts of the two spectra must be in terms of the commutator $SA - AT$.

Fialkow (1977) also showed that, in contrast to Example 24.1.2, the essential spectra of arbitrary quasi-similar operators on a separable Hilbert space always have non-empty intersection. Moreover, Herrero (1988) showed that, for every pair of quasi-similar operators $T \in B(X)$ and $S \in B(Y)$ on the Banach spaces X and Y, each connected component of $\sigma_e(T)$ touches $\sigma_e(S)$, and vice versa. This result now has an attractive sheaf-theoretic proof (see Eschmeier 2000).

In this chapter we presented a proof of the (β) part of Theorem 24.1.1. This proof is quite elementary compared with previous ones. Theorem 24.1.1 has a background that may be traced to a problem from Clary (1975): having established the equality of the spectra for quasi-similar hyponormal operators (hyponormal T means that the commutator $T^*T - TT^* \geq 0$), Clary asked whether this extends to the essential spectra. This problem received considerable attention. The final solution, due to Putinar (1992) and, independently, to Yang (1993), revealed that the assumption of hyponormal is not really needed: all that matters is that hyponormal operators have (β). But generally, of course, Hilbert space operators have their own theory.

25

Multipliers on commutative Banach algebras

25.1 Multipliers

In this chapter I want to give an impression of what sort of information becomes available via local spectral theory when it is applied to a particular class of operators, namely the *multipliers* on a commutative Banach algebra. Throughout this chapter, the letter A will denote a commutative, complex Banach algebra.

Definition 25.1.1 *A linear map $T : A \to A$ is a* multiplier *if $aT(b) = T(a)b$ for all $a, b \in A$. The set of multipliers on A is denoted by $M(A)$.*

The most obvious example, given A, is the *multiplication operator* L_a induced by a fixed element $a \in A$, that is, the operator $L_a(b) := ab$ for all $b \in A$. If A has a unit e (A is *unital*) then every multiplier T is a multiplication operator. In this case $T = L_{T(e)}$ (Exercise 25.3.2).

If the map $a \to L_a : A \to \mathcal{B}(A)$ is injective (faithful), then A is said to be a *faithful* algebra. Every unital algebra is faithful, as is every semisimple, and also every *semiprime* algebra – the latter term means, in our commutative case, that the algebra contains no non-zero nilpotent elements.

Example 25.1.2 Let $A := C_0(\Omega)$ be the Banach algebra of all continuous complex-valued functions vanishing at ∞ on the locally compact Hausdorff space Ω, and let $f \in C^{\mathrm{b}}(\Omega)$ be a bounded continuous function on Ω. Then $T := L_f$ (notation self-explanatory by now) is a multiplier. Conversely, it can be shown that a multiplier gives rise to a bounded continuous function on Ω with respect to which the multiplier acts by pointwise multiplication (Exercise 25.3.4). □

The group algebras are perhaps the most important class of examples.

Example 25.1.3 Let G be a locally compact abelian group, with Haar measure m, and let $L^1(G)$ denote the group algebra of G, as described in Parts I and II. Also let $M(G)$ denote the measure algebra, that is, the algebra of all bounded, regular, complex-valued measures on G. Equipped with the total variation norm, given by $\|\mu\| := |\mu|(G)$, $M(G)$ is a commutative unital Banach algebra (point mass at the unit $e \in G$ is the identity) and since the convolution of an element $f \in L^1(G)$ and a measure $\mu \in M(G)$,

$$(f \star \mu)(s) = \int_G f(st^{-1})\,\mathrm{d}\mu(t),$$

is an element in $L^1(G)$, we see that every $\mu \in M(G)$ defines a multiplier on $L^1(G)$. It is a classical result of Wendel (1952) and of Helson (1953) (see Theorem 9.1.11) that all multipliers on $L^1(G)$ arise this way. This shows that $M(L^1(G)) = M(G)$. □

The next result (the proof of which is left to the reader, Exercise 25.3.5) claims, *inter alia*, that $M(A)$ is a *full* subalgebra of $\mathcal{B}(A)$. This means that, if $T \in M(A)$ is invertible as an operator on A, that is, invertible in $\mathcal{B}(A)$, then $T^{-1} \in M(A)$. You should also note that the definition of a multiplier does not require the map to be a priori continuous, so the proposition contains an automatic continuity result.

Proposition 25.1.4 *Let A be faithful. Then the multiplier algebra $M(A)$ is a closed, commutative, unital, full subalgebra of $\mathcal{B}(A)$ which contains A as an ideal.* □

Here is our first explicit allusion to local spectral theory.

Proposition 25.1.5 *Let A be semiprime. Then every multiplier $T \in M(A)$ has SVEP.*

Proof We observe, to begin with, that $\ker T \cap TA = \{0\}$: if $u \in \ker T \cap TA$, then $Tu = 0$ and $u = Tv$ for some $v \in A$; hence $(Tv)^2 = (T^2v)v = 0v = 0$, and therefore $u = Tv = 0$, since A contains no non-zero nilpotent elements.

Now, on an open subset U of the complex plane we consider an analytic function $f : U \to A$ such that $(T - \lambda)f(\lambda) = 0$ for all $\lambda \in U$. For each $\lambda \in U$, it is clear that $f(\lambda) \in \ker(T - \lambda)$; moreover $f(\lambda) \in (T - \lambda)A$, by Lemma 22.1, and so

$$f(\lambda) \in \ker(T - \lambda) \cap (T - \lambda)A = \{0\},$$

since $T - \lambda$ is also a multiplier. Thus $f(\lambda) = 0$ for all $\lambda \in U$. This proves that T has SVEP. □

We next proceed to study the extent to which decomposability is transferred by algebra homomorphisms. This development has some of the same flavour as Theorem 22.2.4. We suppose that A is faithful and that $\varphi : A \to \mathcal{B}(X)$ is an algebra homomorphism. This turns X into a left A-module, with module action defined by $a \cdot x := \varphi(a)x$ for all $a \in A$ and $x \in X$. Although the following facts may be established without continuity assumptions about φ, the simpler proofs presented here cover only this case.

As our first step we extend the homomorphism φ from A to its multiplier algebra $M(A)$. For this to go through, we shall impose two conditions. The first condition is one of non-degeneracy.

Definition 25.1.6 *The left A-module X is* non-degenerate *if, for each non-zero $x \in X$, there exists an element $a \in A$ for which $a \cdot x$ is non-zero.*

We also say that the homomorphism φ is *non-degenerate*. Note that when the algebra A is viewed as a module over itself (with module action given by the algebra multiplication), non-degeneracy is the same as faithfulness.

In the non-degenerate case, there is at most one algebra homomorphism

$$\Phi : M(A) \to \mathcal{B}(X)$$

that extends φ, in the sense that $\Phi(L_a) = \varphi(a)$ for all $a \in A$. Indeed, if Φ_1 and Φ_2 are two such extensions, then

$$\varphi(a)\ \ (\Phi_1(S)x - \Phi_2(S)x) = \varphi(Sa)x - \varphi(Sa)x = 0$$
$$\text{for all } S \in M(A),\ x \in X, \quad \text{and} \quad a \in A,$$

and therefore $\Phi_1 = \Phi_2$ if φ is non-degenerate. Non-degenerateness will also, with much the same argument, establish that the extension Φ of φ maps the identity operator on A to the identity operator on X.

We shall also assume about the module action that $AX = X$. Here, for arbitrary subsets $B \subseteq A$ and $Y \subseteq X$, we define

$$BY = \text{lin}\{b \cdot y;\ b \in B \text{ and } y \in Y\},$$

where lin denotes the linear span. Evidently, the condition that $AX = X$ is fulfilled whenever $\bigcup \{\text{ran}\, \varphi(a) : a \in A\} = X$.

If A has an identity element e, and if $\varphi(e) = I_X$, then X is clearly non-degenerate and satisfies $AX = X$. Actually, these two conditions are fulfilled in lots of situations: for instance, if the homomorphism φ is continuous, and if A has a bounded approximate identity $(e_\lambda)_{\lambda \in \Lambda}$ for which $\varphi(e_\lambda)x \to x$ for each

$x \in X$, then the left A-module X is obviously non-degenerate, and the Cohen factorization theorem implies that $AX = X$.

Lemma 25.1.7 *Let A be faithful, and consider an algebra homomorphism $\varphi : A \to \mathcal{B}(X)$ which provides us with the module action $a \cdot x := \varphi(a)x$ for all $a \in A$ and $x \in X$. Suppose that X is non-degenerate and that $AX = X$. If $x \in X$ is represented as $x = \sum_{k=1}^{n} \varphi(a_k)x_k$ with $a_k \in A$ and $x_k \in X$ for each $k = 1, \dots, n$, and if we define*

$$\Phi(S)x := \sum_{k=1}^{n} \varphi(Sa_k)x_k$$

for all $S \in M(A)$, then $\Phi : M(A) \to \mathcal{B}(X)$ is a unital algebra homomorphism that extends φ, in the sense that $\Phi(L_a) = \varphi(a)$ for every $a \in A$. The extension Φ is continuous on $M(A)$ if and only if φ is continuous on A.

Proof Let $S \in M(A)$. First we must make sure that the above definition of $\Phi(S) : X \to X$ makes sense: since $AX = X$, every element $x \in X$ has a representation $x = \sum_{k=1}^{n} \varphi(a_k)x_k$, so if $\sum_{k=1}^{n} \varphi(a_k)x_k = 0$, then, for every $a \in A$, we see that

$$\varphi(a)\left(\sum_{k=1}^{n} \varphi(Sa_k)x_k\right) = \varphi(Sa)\left(\sum_{k=1}^{n} \varphi(a_k)x_k\right) = 0,$$

because φ is a homomorphism and S is a multiplier. The module X is non-degenerate, so this implies that $\sum_{k=1}^{n} \varphi(Sa_k)x_k = 0$.

The linearity of the operator $\Phi(S)$ is immediate, and, by the closed graph theorem, $\Phi(S)$ is continuous: if $u_n \to 0$ in A and $\Phi(S)u_n \to v$ in X, then clearly, for each $a \in A$, $\varphi(Sa)u_n \to 0$ and $\varphi(a)\Phi(S)u_n \to \varphi(a)v$ as $n \to \infty$. But a computation similar to the one above shows that

$$\varphi(a)\Phi(S)u_n = \varphi(Sa)u_n \quad \text{for all } n \in \mathbb{N}.$$

Thus $\varphi(a)v = 0$ for each $a \in A$, and therefore $v = 0$ by non-degeneracy.

It is straightforward to check that Φ is a unital algebra homomorphism that extends φ, and it is clear that φ is continuous whenever Φ is. On the other hand, if φ is continuous on A, then, for every $x \in X$, the defining expression for $\Phi(S)x$ shows that the map $S \mapsto \Phi(S)x$ is continuous on $M(A)$. This, by the principle of uniform boundedness, is enough to establish continuity of Φ. □

The following theorem is a key result for the preservation of decomposability under homomorphisms. I have chosen to show only a special case; there is a rather more elaborate version of it, avoiding continuity assumptions on φ, in Laursen and Neumann (2000, Lemma 4.2.4).

Theorem 25.1.8 *Let A have a bounded approximate identity $(e_\lambda)_{\lambda\in\Lambda}$ and $\varphi : A \to \mathcal{B}(X)$ be a continuous algebra homomorphism for which $\varphi(e_\lambda)x \to x$ for each $x \in X$. If $S \in M(A)$ is a multiplier with (δ) and SVEP, then the operator $T := \Phi(S) \in \mathcal{B}(X)$ is decomposable, and*

$$X_T(F) = \{x \in X : \varphi(a)x = 0 \text{ for all } a \in A \text{ for which } \sigma_S(a) \cap F = \emptyset\} ,$$

for all closed sets $F \subseteq \mathbb{C}$.

Proof We shall establish the decomposability of T by checking the very definition, and show that the spaces $Z_T(F) := \{x \in X : \varphi(a)x = 0$ for all $a \in A$ for which $\sigma_S(a) \cap F = \emptyset\}$ give us the claimed spectral subspaces.

Right away we note that, for every closed set $F \subseteq \mathbb{C}$, $Z_T(F)$ is a closed linear subspace of X with the property that $\Phi(R)\, Z_T(F) \subseteq Z_T(F)$ for each $R \in M(A)$. In particular, $Z_T(F)$ is T-invariant.

We first show that $X = Z_T(\overline{U}) + Z_T(\overline{V})$ for every open cover $\{U, V\}$ of $\mathbb{C}$. By Cohen's factorization theorem, $AX = X$, and we may represent an arbitrary $x \in X$ as $x = \varphi(a_0)x_0$ with $a_0 \in A$ and $x_0 \in X$. Since S has (δ), there exist elements $u_0, v_0 \in A$ for which $\sigma_S(u_0) \subseteq \overline{U}$, $\sigma_S(v_0) \subseteq \overline{V}$, and $a_0 = u_0 + v_0$. Thus $x = \varphi(u_0)x_0 + \varphi(v_0)x_0$. We claim that $\varphi(u_0)x_0$ belongs to $Z_T(\overline{U})$: consider an arbitrary $a \in A$ with the property that $\sigma_S(a) \cap \overline{U} = \emptyset$. Since

$$\sigma_S(au_0) \subseteq \sigma_S(a) \cap \sigma_S(u_0) \subseteq \sigma_S(a) \cap \overline{U} = \emptyset,$$

$au_0 = 0$, because S has SVEP. This implies that $\varphi(a)\varphi(u_0)x_0 = 0$, and hence we have $u \in Z_T(\overline{U})$. The same argument shows that $v \in Z_T(\overline{V})$. We have proved that

$$X = Z_T(\overline{U}) + Z_T(\overline{V}).$$

We now claim that $\sigma(T \mid Z_T(F)) \subseteq F$ for every closed set $F \subseteq \mathbb{C}$. To establish this, fix an arbitrary open neighbourhood U of F, and observe that $AZ_T(F) = Z_T(F)$, again by Cohen factorization. Hence, given any $x \in Z_T(F)$, we have a representation of the form $x = \varphi(a)z$ with $a \in A$ and $z \in Z_T(F)$. Since S has (δ) and $\mathbb{C} = U \cup (\mathbb{C} \setminus F)$, we may write $a = u + v$ with $u, v \in A$ chosen so that $\sigma_S(u) \subseteq U$ and $\sigma_S(v) \cap F = \emptyset$. Because z is in $Z_T(F)$, we know that $\varphi(v)z = 0$, and hence $x = \varphi(u)z$. Since S has SVEP, and $\sigma_S(u) \subseteq U$, there is an analytic function $f : \mathbb{C} \setminus \overline{U} \to A$ for which $(S - \lambda)f(\lambda) = u$ for every

$\lambda \in \mathbb{C} \setminus \overline{U}$. Now apply the homomorphism Φ; this gives

$$(T - \lambda)\varphi(f(\lambda)) = \varphi(u),$$

and therefore $(T - \lambda)y_\lambda = x$, where we set $y_\lambda := \varphi(f(\lambda))z \in Z_T(F)$ for all $\lambda \in \mathbb{C} \setminus \overline{U}$. Consequently, for each $\lambda \in \mathbb{C} \setminus \overline{U}$, the operator $T - \lambda$ maps $Z_T(F)$ onto itself.

Additionally, it does it injectively. To see this, let

$$\lambda \in \mathbb{C} \setminus \overline{U} \quad \text{and} \quad z \in Z_T(F) \cap \ker(T - \lambda).$$

Since X is non-degenerate, it suffices to show that $\varphi(a)z = 0$ for each $a \in A$. Clearly $x := \varphi(a)z$ belongs to $Z_T(F)$. Hence the preceding argument supplies an analytic function $f : \mathbb{C} \setminus \overline{U} \to A$ for which $(T - \lambda)\varphi(f(\lambda))z = x$. Thus

$$x = \varphi(f(\lambda))(T - \lambda)z = 0,$$

proving that $(T - \lambda) \mid Z_T(F)$ is injective. It follows that $\sigma(T \mid Z_T(F)) \subseteq \overline{U}$ for every open neighbourhood U of F, and therefore that $\sigma(T \mid Z_T(F)) \subseteq F$.

We conclude that T is decomposable, and that $Z_T(F) \subseteq X_T(F)$ for every closed set $F \subseteq \mathbb{C}$.

The opposite inclusion $Z_T(F) \supseteq X_T(F)$ follows easily from the continuity of φ. In fact, given any $x \in X_T(F)$ and $a \in A$ for which $\sigma_S(a) \cap F = \emptyset$, we obtain from the continuity of φ that $\rho_S(a) \subseteq \rho_T(\varphi(a)x)$, and hence that

$$\sigma_T(\varphi(a)x) \subseteq \sigma_S(a) \cap \sigma_T(x) \subseteq \sigma_S(a) \cap F = \emptyset.$$

Since T has SVEP, this implies that $\varphi(a)x = 0$, and consequently $x \in Z_T(F)$. □

Here is a pleasing observation which is immediate from what we have said so far.

Corollary 25.1.9 *Let A have a bounded approximate identity. Then a multiplier on A is decomposable if and only if it has (δ) and SVEP. In particular, if A is also semiprime, a multiplier on A is decomposable precisely when it has (δ).*

Proof Simply let $\varphi : A \to \mathcal{B}(A)$ be the left regular representation given by $\varphi(a) := L_a$ for all $a \in A$. Then φ is a continuous algebra homomorphism, and the extension of φ to the multiplier algebra $M(A)$, as in Lemma 25.1.7, is the identity map on $M(A)$. By Theorem 25.1.8, every multiplier on A with (δ) and SVEP is decomposable. The last statement is immediate from this in conjunction with Proposition 25.1.5. □

25.2 The hull-kernel topology

The second main result of this chapter is Theorem 25.2.2. To appreciate all of it, it is necessary to realize that, in addition to the Gelfand topology, there is another topology on the character space Φ_A of A. And it is also worth knowing that the condition of continuity of the Gelfand transform of a given algebra element with respect to this other topology, which is the main assumption of Theorem 25.2.2, is really a decomposability condition in disguise: in Proposition 25.2.5 it is established that, if A is faithful and if $T \in M(A)$ is a multiplier with (δ), then the restriction $\widehat{T} \mid \Phi_A$ is $\mathfrak{hk}$-continuous. This of course applies in particular to a multiplication operator.

The topology that we are referring to is introduced in the next proposition, stated here without proof; it is called the *hull-kernel topology*, or $\mathfrak{hk}$*-topology*, of Φ_A. This topology is defined via specification of the closure of any given subset of the character space. This involves the following steps, already given as Exercise 3.6.1 in Part I. For a set $S \subseteq \Phi_A$, the *kernel* of S is the ideal of A defined as

$$\mathfrak{k}(S) := \{a \in A : \varphi(a) = 0 \text{ for all } \varphi \in S\} ,$$

and, symmetrically, for an ideal $J \subseteq A$, the *hull* of J in Φ_A is

$$\mathfrak{h}(J) := \{\varphi \in \Phi_A : \varphi(a) = 0 \text{ for all } a \in J\} .$$

For an arbitrary subset S of Φ_A, we then define the $\mathfrak{hk}$*-closure of* S by

$$\mathfrak{hk}(S) := \mathfrak{h}(\mathfrak{k}(S)) = \{\varphi \in \Phi_A : \widehat{a}(\varphi) = 0 \text{ for all } a \in A \text{ with } \widehat{a} \equiv 0 \text{ on } S\} .$$

Proposition 25.2.1 *Let A be a commutative complex algebra. The sets of the form $\mathfrak{hk}(S)$, where S ranges over the subsets of the character space Φ_A, are the closed sets of a topology, the $\mathfrak{hk}$-topology of Φ_A. This topology has the following properties:*

(i) *for any $S \subseteq \Phi_A$, the closure of S in the $\mathfrak{hk}$-topology is precisely $\mathfrak{hk}(S)$;*
(ii) *the $\mathfrak{hk}$-topology is coarser than (or equal to) the Gelfand topology.* □

As noted in Part I, these two topologies coincide exactly when A is regular (regular: whenever F is a Gelfand closed subset of Φ_A and $\varphi \in \Phi_A \backslash F$ then there is $a \in A$ for which $\widehat{a}$ vanishes on F, but not at φ).

With these basic facts listed, we can state the theorem.

Theorem 25.2.2 *Suppose that A is semisimple and that $a \in A$ is an element for which $\widehat{a}$ is $\mathfrak{hk}$-continuous on Φ_A. Then, for an arbitrary homomorphism*

$\varphi : A \to \mathcal{B}(X)$, the operator $\varphi(a) \in \mathcal{B}(X)$ is super-decomposable on X. If, additionally, φ is non-degenerate, then

$$X_{\varphi(a)}(F) = \{x \in X : \varphi(u)x = 0 \text{ for every } u \in A \text{ for which } \overline{\widehat{a}\,(\operatorname{supp} u)} \cap F = \emptyset\}.$$

Note that one consequence of this is that, if $a \in A$ is an element for which $\widehat{a}$ is $\mathfrak{hk}$-continuous on Φ_A, then a defines a super-decomposable multiplication operator on A. The converse is also true, as we shall see in Proposition 25.2.5.

The proof of Theorem 25.2.2 comes after the next two lemmas. Both of these are classical; see Laursen and Neumann (2000, Section 4.3).

Lemma 25.2.3 *The following statements are true for every commutative Banach algebra A.*

(i) *If S is a compact hull in Φ_A, and if $S \subseteq U_1 \cup \cdots \cup U_n$ for finitely many $\mathfrak{hk}$-open sets $U_1, \ldots, U_n \subseteq \Phi_A$, then there are elements $u_1, \ldots, u_n \in A$ such that $\widehat{u}_1 + \cdots + \widehat{u}_n \equiv 1$ on S and $\widehat{u}_k \equiv 0$ on $\Phi_A \setminus U_k$ for $k = 1, \ldots, n$.*

(ii) *If S_1 and S_2 are two disjoint hulls in Φ_A, and if S_1 is compact in the Gelfand topology, then there is an element $u \in A$ for which $\widehat{u} \equiv 1$ on S_1 and $\widehat{u} \equiv 0$ on S_2.*

(iii) *If $S \subseteq \Phi_A$ is a hull such that there exist $v \in A$ and $\delta > 0$ for which $|\widehat{v}| \geq \delta$ on S, then there is an element $w \in A$ for which $\widehat{vw} \equiv 1$ on S.* □

Lemma 25.2.4 *Suppose that A is semisimple and has an identity element e, and suppose that $\varphi : A \to \mathcal{B}(X)$ is a homomorphism such that $\varphi(e) = I$. Moreover, let $a \in A$ be an element for which $\widehat{a}$ is $\mathfrak{hk}$-continuous on Φ_A. Then, for every open cover $\{U, V\}$ of $\mathbb{C}$, there exists an element $r \in A$ for which $\widehat{r} \equiv 0$ on $\widehat{a}^{-1}(\mathbb{C} \setminus U)$, and for which $\widehat{r} \equiv 1$ on $\widehat{a}^{-1}(\mathbb{C} \setminus V)$, and is such that $\sigma(T \mid \overline{RX}) \subseteq U$ and $\sigma(T \mid \overline{(I - R)X}) \subseteq V$, where $R, T \in \mathcal{B}(X)$ denote the commuting operators given by $R := \varphi(r)$ and $T := \varphi(a)$.*

Proof Start with the open cover $\{U, V\}$ of $\mathbb{C}$. Choose open sets $G, H \subseteq \mathbb{C}$ for which $\overline{G} \subseteq U$, $\overline{H} \subseteq V$, and $G \cup H = \mathbb{C}$. Since the complements $\mathbb{C} \setminus G$ and $\mathbb{C} \setminus H$ are closed and disjoint, $\mathfrak{hk}$-continuity of $\widehat{a}$ implies that their preimages $\widehat{a}^{-1}(\mathbb{C} \setminus G)$ and $\widehat{a}^{-1}(\mathbb{C} \setminus H)$ are hulls in Φ_A, of course also disjoint. Moreover, since A has an identity, both sets are compact in the Gelfand topology of the compact space Φ_A. Consequently, by Lemma 25.2.3 (ii), there is $r \in A$ for which $\widehat{r} \equiv 0$ on $\widehat{a}^{-1}(\mathbb{C} \setminus G)$ and $\widehat{r} \equiv 1$ on $\widehat{a}^{-1}(\mathbb{C} \setminus H)$. Evidently, r satisfies the first two conditions of this lemma, and the operator $R := \varphi(r)$ commutes with $T := \varphi(a)$.

To show that $\sigma(T \mid \overline{RX}) \subseteq U$, look at an arbitrary $\lambda \in \mathbb{C} \setminus U$. The distance δ from λ to $\overline{G}$ is strictly positive. Furthermore, $|\widehat{a} - \lambda\widehat{e}| \geq \delta$ on the hull $\widehat{a}^{-1}(\overline{G})$. Hence, by Lemma 25.2.3 (iii), there is a $u_\lambda \in A$ for which

$$(\widehat{a} - \lambda)\widehat{u_\lambda} \equiv 1 \quad \text{on} \quad \widehat{a}^{-1}(\overline{G}).$$

Since $\widehat{r} \equiv 0$ on $\widehat{a}^{-1}(\mathbb{C} \setminus G)$, the identity $(\widehat{a} - \lambda)\widehat{u_\lambda}\widehat{r} \equiv \widehat{r}$ holds everywhere on the character space Φ_A. Thus $(a - \lambda)u_\lambda r = r$, by semisimplicity of A. Now apply the homomorphism φ to the last identity. If $U_\lambda := \varphi(u_\lambda) \in \mathcal{B}(X)$, then

$$(T - \lambda)U_\lambda(Rx) = U_\lambda(T - \lambda)(Rx) = Rx \qquad \text{for all } x \in X,$$

and therefore $(T - \lambda)U_\lambda = U_\lambda(T - \lambda) = I$ on $\overline{RX}$. This space is U_λ-invariant, so $\lambda \in \rho(T \mid \overline{RX})$. This proves that $\sigma(T \mid \overline{RX}) \subseteq U$. A similar argument will show that $\sigma(T \mid \overline{(I - R)X}) \subseteq V$. □

Proof of Theorem 25.2.2 First extend the homomorphism φ to the unitization $A^\# = A \oplus \mathbb{C}$ by defining

$$\Phi(u + \lambda e) := \varphi(u) + \lambda I \qquad \text{for all } u \in A \text{ and } \lambda \in \mathbb{C}.$$

Clearly, $\Phi : A^\# \to \mathcal{B}(X)$ is a homomorphism with $\Phi(e) = I$ and $\Phi(u) = \varphi(u)$ for all $u \in A$. It may be shown that $\widehat{a}$ remains $\mathfrak{hk}$-continuous on $\Phi_{A^\#}$, so Lemma 25.2.4 establishes the super-decomposability of T.

To prove the formula for the local spectral subspaces of T when φ is assumed non-degenerate, we consider, for each closed set $F \subseteq \mathbb{C}$, the space

$$Z_T(F) := \{x \in X : \varphi(u)x = 0 \text{ for all } u \in A \text{ for which } \overline{\widehat{a}\,(\operatorname{supp}\widehat{u})} \cap F = \emptyset\}.$$

This is the space which is claimed to equal $X_T(F)$. Note that $Z_T(F)$ consists of all $x \in X$ which satisfy $\varphi(u)x = 0$ for all $u \in A$ for which $\widehat{u} \equiv 0$ on $\widehat{a}^{-1}(U)$ for some open neighbourhood U of F, depending on u.

The equality $X_T(F) = Z_T(F)$ is established in several steps. The first step is to show that $Z_T(F) \cap Z_T(G) = \{0\}$ for arbitrary closed and disjoint F, G in $\mathbb{C}$. No number can belong to both F and G, so we may as well suppose that $0 \notin F$. Then we may select open sets $U, V \subseteq \mathbb{C}$ such that $F \subseteq U$, $G \subseteq V$, $0 \notin \overline{U}$, and $\overline{U} \cap \overline{V} = \emptyset$. By the $\mathfrak{hk}$-continuity of $\widehat{a}$ on Φ_A, the sets $\widehat{a}^{-1}(\overline{U})$ and

$\widehat{a}^{-1}(\overline{V})$ are disjoint hulls in Φ_A. Moreover, if δ denotes the positive distance from 0 to $\overline{U}$, then $|\widehat{a}| \geq \delta$ on $\widehat{a}^{-1}(\overline{U})$. This shows that $\widehat{a}^{-1}(\overline{U})$ is Gelfand compact in Φ_A. By Lemma 25.2.3 (ii), there is an element $u \in A$ for which

$$\widehat{u} \equiv 1 \text{ on } \widehat{a}^{-1}(\overline{U}) \quad \text{and} \quad \widehat{u} \equiv 0 \text{ on } \widehat{a}^{-1}(\overline{V}).$$

Hence, for arbitrary $c \in A$, we have that $\widehat{c} - \widehat{cu} \equiv 0$ on $\widehat{a}^{-1}(\overline{U})$ and $\widehat{cu} \equiv 0$ on $\widehat{a}^{-1}(\overline{V})$. For every $x \in Z_T(F) \cap Z_T(G)$, we get both that $\varphi(c - cu)x = 0$, and $\varphi(cu)x = 0$, and consequently that $\varphi(c)x = 0$ for all $c \in A$. The homomorphism φ is non-degenerate, so it follows that $x = 0$. Thus we have as our first step that the spaces $Z_T(F)$ and $Z_T(G)$ have trivial intersection.

The next step consists in showing that $X_T(F) \subseteq Z_T(F)$ for every closed set $F \subseteq \mathbb{C}$. To this end, consider an arbitrary $u \in A$ for which $\widehat{u} \equiv 0$ on $\widehat{a}^{-1}(U)$ for some open neighbourhood U of F, and choose an open set $V \subseteq \mathbb{C}$ so that $F \cap \overline{V} = \emptyset$ and $U \cup V = \mathbb{C}$. We now apply Lemma 25.2.4 to the open cover $\{U, V\}$ of $\mathbb{C}$ and to the unitization of A and φ. Let $r \in A^\#$ denote the corresponding element provided by Lemma 25.2.4. Since $\widehat{r} \equiv 0$ on $\widehat{a}^{-1}(\mathbb{C} \setminus U)$, we have that $\widehat{ur} \equiv 0$ on Φ_A, and hence, by semisimplicity, $ur = 0$. Furthermore, with $R := \varphi(r) \in \mathcal{B}(X)$, $\sigma(T \mid \overline{(I - R)X}) \subseteq V$, and so $(I - R)X \subseteq X_T(\overline{V})$. Because $RT = TR$, this implies that

$$(I - R)X_T(F) \subseteq X_T(F) \cap X_T(\overline{V}) = X_T(F \cap \overline{V}) = X_T(\emptyset) = \{0\}.$$

For each $x \in X_T(F)$, we conclude that $(I - R)x = 0$, hence $x = Rx$, and therefore $\varphi(u)x = \varphi(u)Rx = \varphi(ur)x = 0$. Thus $x \in Z_T(F)$, and this shows that $X_T(F) \subseteq Z_T(F)$.

It remains to establish that $Z_T(F) \subseteq X_T(F)$, for every closed set $F \subseteq \mathbb{C}$. For this it is enough to prove that $Z_T(F) \subseteq X_T(\overline{U})$, for every open neighbourhood U of F. As in the preceding paragraph, given U, we may choose an open $V \subseteq \mathbb{C}$ for which $F \cap \overline{V} = \emptyset$ and $U \cup V = \mathbb{C}$, and put $R := \Phi(r) \in \mathcal{B}(X)$, where $r \in A^\#$ denotes the element provided by Lemma 25.2.4, again applied to the open cover $\{U, V\}$ of $\mathbb{C}$ and to the unitization of A and φ. The two spectral inclusions from Lemma 25.2.4 imply that $R\,X \subseteq X_T(\overline{U})$ and $(I - R)\,X \subseteq X_T(\overline{V})$. Moreover, it is clear that the space $Z_T(F)$ is invariant under R. Gathering the results of the preceding two paragraphs, we obtain

$$(I - R)\,Z_T(F) \subseteq Z_T(F) \cap X_T(\overline{V}) \subseteq Z_T(F) \cap Z_T(\overline{V}) = \{0\},$$

and consequently $Z_T(F) = R\,Z_T(F) \subseteq X_T(\overline{U})$. This completes the proof of the inclusion $Z_T(F) \subseteq X_T(F)$. □

I want to conclude this chapter by stating a converse to Theorem 25.2.2. As a consequence of this, and because (as pointed out in connection with Proposition 25.2.1) there is a close connection between equality of the two topologies on Φ_A and regularity of A, we obtain a nice result that expresses this connection in terms of decomposability.

Proposition 25.2.5 *Suppose that A is faithful and that $T \in M(A)$ is a multiplier with (δ). Then the restriction $\widehat{T} \mid \Phi_A$ is $\mathfrak{hk}$-continuous.*

Proof Given any closed $F \subseteq \mathbb{C}$. Our task is to show that

$$E := \{\varphi \in \Phi_A : \widehat{T}(\varphi) \in F\}$$

is $\mathfrak{hk}$-closed.

So consider any $\psi \in \Phi_A \setminus E$, and choose an $x \in A$ so that $\psi(x) = 1$. Since $\lambda := \widehat{T}(\psi)$ does not belong to F, there exists an open cover $\{U, V\}$ of $\mathbb{C}$ for which $\overline{U} \subseteq \mathbb{C} \setminus \{\lambda\}$ and $\overline{V} \subseteq \mathbb{C} \setminus F$. By (δ), we may decompose $x = y + z$, with $y \in A_T(\overline{U})$ and $z \in A_T(\overline{V})$. Since $\lambda \notin \overline{U}$ and $y \in A_T(\overline{U})$, we know that $y = (T - \lambda)u$ for some $u \in A$, and therefore

$$\psi(y) = (\widehat{T}(\psi) - \lambda)\, \psi(u) = 0\,.$$

Thus $\psi(z) = \psi(x) = 1$. On the other hand, for any $\varphi \in E$, clearly $\mu := \widehat{T}(\varphi)$ belongs to F, and hence not to $\overline{V}$. Since $z \in A_T(\overline{V})$, $z = (T - \mu)v$ for suitable $v \in A$, and therefore

$$\varphi(z) = (\widehat{T}(\varphi) - \mu)\, \varphi(v) = 0\,.$$

Since $\varphi(z) = 0$ for all $\varphi \in E$ and $\psi(z) = 1$, we see that $\psi \notin \mathfrak{hk}(E)$. Thus the set $E = \mathfrak{hk}(E)$ is closed, and we have established the $\mathfrak{hk}$-continuity of $T \mid \Phi_A$. □

Corollary 25.2.6 *Suppose that A is semisimple. Then A is regular if and only if, for every $a \in A$, the multiplication operator L_a is decomposable.*

Proof If A is regular then, by the remark immediately after Proposition 25.2.1, $\widehat{a}$ is $\mathfrak{hk}$-continuous for every $a \in A$.

Conversely, if every L_a is decomposable then Proposition 25.2.5 implies that $\widehat{a}$ is $\mathfrak{hk}$-continuous for every $a \in A$. But the Gelfand topology is, by definition, the coarsest topology on Φ_A for which all these functions are continuous. Consequently, the $\mathfrak{hk}$-topology coincides with the Gelfand topology on Φ_A. By the remark after Proposition 25.2.1, this establishes the regularity of A. □

25.3 Exercises

1. Show that a multiplier on a faithful commutative Banach algebra is continuous.
2. Show that if A is unital then every multiplier T is a multiplication operator, by showing that in this case $T = L_{T(e)}$.
3. Show that, if A has approximate units, that is, for every $a \in A$ and every $\varepsilon > 0$ there is $u \in A$ such that $\|a - ua\| < \varepsilon$, then A is faithful.
4. Let f be a bounded continuous function on Ω. Show that $T := L_f$ is a multiplier on $C_0(\Omega)$. Establish also that the converse is true: every multiplier is given by pointwise multiplication by a bounded continuous function on Ω.
5. Let A be faithful. Prove that the multiplier algebra $M(A)$ is a closed, commutative, unital, full subalgebra of $\mathcal{B}(A)$ and contains A as an ideal.
6. Show that $\mathfrak{hk}_{M(A)}(\Phi_A) = \mathfrak{h}(\{0\}) = \Phi_{M(A)}$, hence that Φ_A is $\mathfrak{hk}$-dense in $\Phi_{M(A)}$.
7. Show that if C is a commutative Banach algebra, if $c \in C$, if $\widehat{c}$ is $\mathfrak{hk}$-continuous on Φ_C, and if B is any commutative algebra that contains C as a subalgebra, then $\widehat{c}$ is $\mathfrak{hk}$-continuous on Φ_B.

25.4 Additional notes

Concerning examples of non-degenerate modules, it is easily seen that X is non-degenerate and satisfies $AX = X$ whenever the left A-module X is irreducible; see Proposition 24.4 of Bonsall and Duncan (1973).

The best results about decomposability of multipliers are obtained when certain vanishing-at-infinity conditions are imposed.

Definition 25.4.1 *Suppose that A is semisimple and has multiplier algebra $M(A)$, and character space Φ_A (respectively, $\Phi_{M(A)}$). Then*

$$M_0(A) := \{T \in M(A) : \widehat{T}|\Phi_A \text{ vanishes at infinity on } \Phi_A \subseteq \Phi_{M(A)}\}$$

and

$$M_{00}(A) := \{T \in M(A) : \widehat{T} \text{ vanishes on } \Phi_{M(A)} \backslash \Phi_A\}.$$

In the simple case of $A = C(\Omega)$, where Ω is a locally compact Hausdorff space, it is easy to picture these two ideals. Evidently, we always have

$$A \subseteq M_{00}(A) \subseteq M_0(A) \subseteq M(A),$$

and in general all these containments are strict; notably, this holds when $A = L^1(G)$ is the group algebra for a locally compact, non-discrete, abelian group.

The next result, and its proof, may be found in Laursen and Neumann (2000, Theorem 4.5.4).

Theorem 25.4.2 *Let A be semisimple. For $T \in M_0(A)$, the following are equivalent:*

(a) *T is super-decomposable on A;*
(b) *T is decomposable on A;*
(c) *T has (δ);*
(d) *$T \in M_{00}(A)$ and $\widehat{T}|\Phi_A$ is hk-continuous on Φ_A;*
(e) *$\widehat{T}$ is hk-continuous on $\Phi_{M(A)}$;*
(f) *multiplication by T is decomposable on $M(A)$.*

In the case where A is also regular, $T \in M_0(A)$ satisfies all these conditions exactly when $T \in M_{00}(A)$. □

Appendix to Part IV: The functional model

The following material forms a key background to our subject; a fuller account of it is given in Laursen and Neumann (2000, Chapter 2).

We first review some notions from the theory of scalar-valued functions and distributions in two variables. Fuller expositions of this may also be found in Rudin (1991) and Trèves (1967).

As before, we identify $\mathbb{C}$ and $\mathbb{R}^2$, write $z = x + \mathrm{i}\, y = (x, y)$ for $z \in \mathbb{C}$ and $x, y \in \mathbb{R}$, and let ∂_x and ∂_y denote the usual operators of partial differentiation with respect to x and y, acting on functions of the real variables x and y. In Chapter 21 we touched on $C^\infty(\mathbb{C})$, in particular the notation

$$D^\alpha := \partial_x^{\alpha_1} \partial_y^{\alpha_2}, \; |\alpha| := \alpha_1 + \alpha_2 \quad \text{and} \quad \alpha! := \alpha_1!\, \alpha_2!$$

for every pair $\alpha = (\alpha_1, \alpha_2) \in (\mathbb{Z}^+)^2$. We then consider, for a compact subset K of $\mathbb{C}$, the space of *smooth* functions

$$\mathcal{D}(K) := \{ f \in C^\infty(\mathbb{C}) : \operatorname{supp} f \subseteq K \},$$

topologized by the submultiplicative seminorms given by

$$\| f \|_{m,K} := \sum_{|\alpha| \le m} \frac{1}{\alpha!} \sup \{ |D^\alpha f(z)| \, : \, z \in K \}$$

for all $f \in C^\infty(\mathbb{C})$ and $m \in \mathbb{Z}^+$. Thus the convergence of a sequence $(f_n)_{n\in\mathbb{N}}$ in $\mathcal{D}(K)$ with respect to this topology is that of uniform convergence of all sequences of derivatives $(D^\alpha f_n)_{n\in\mathbb{N}}$, for every $\alpha \in (\mathbb{Z}^+)^2$. It is a standard fact that $\mathcal{D}(K)$ is a Fréchet space.

For an open subset U of $\mathbb{C}$, we choose a sequence of compact sets $(K_n)_{n\in\mathbb{N}}$ such that $K_n \subset \operatorname{int} K_{n+1}$ for all $n \in \mathbb{N}$ and $U = \bigcup \{ K_n : n \in \mathbb{N} \}$, and let $\mathcal{D}(U) := \bigcup_{n=1}^\infty \mathcal{D}(K_n)$. This space we topologize as the locally convex inductive limit of the spaces $\mathcal{D}(K_n)$, which means that a set in $\mathcal{D}(U)$ is open if and only if its intersection with every $\mathcal{D}(K_n)$ is open (this definition does not depend on the choice of the sequence $(K_n)_{n\in\mathbb{N}}$). Moreover, a linear map from $\mathcal{D}(U)$ into an arbitrary locally convex topological vector space is continuous if and only if its restriction to $\mathcal{D}(K)$ is continuous for every compact subset K of U. Topologized this way, the space $\mathcal{D}(U)$ is the space of *test functions* on U.

The space of continuous linear functionals on $\mathcal{D}(U)$, customarily denoted by $\mathcal{D}'(U)$, is the *space of distributions* on U.

A distribution $\varphi \in \mathcal{D}'(U)$ is said to *vanish on an open subset* V of U if $\varphi(f) = 0$ for all $f \in \mathcal{D}(V)$. There is a largest open set $V \subseteq U$ on which φ vanishes, and the complement of this set with respect to U is called the *support* of φ, and will be denoted by $\operatorname{supp}\varphi$. Evidently, $\operatorname{supp}\varphi$ is a relatively closed subset of U.

If $g : U \to \mathbb{C}$ is locally integrable with respect to the restriction of two-dimensional Lebesgue measure μ, then g induces a distribution $\varphi \in \mathcal{D}'(U)$ via the formula

$$\varphi(f) := \int_U g f \, \mathrm{d}\mu \quad \text{for all } f \in \mathcal{D}(U).$$

Since two locally integrable functions induce the same distribution exactly when they coincide almost everywhere with respect to μ, this formula leads to a canonical embedding of the space $L^1_{\text{loc}}(U)$ into $\mathcal{D}'(U)$, where $L^1_{\text{loc}}(U)$ consists of all equivalence classes modulo μ of the locally integrable functions on U. In particular, each element of $L^1_{\text{loc}}(U)$ has a support in the sense of distributions.

It also follows that each of the spaces $C(U)$ and $L^p(U)$ $(1 \le p < \infty)$ may be regarded as a linear subspace of $\mathcal{D}'(U)$. Here of course $L^p(U)$ is the classical Banach space of all pth power integrable complex-valued functions on U (with two functions identified when they coincide almost everywhere with respect to Lebesgue measure). The fact that all pth power integrable functions on U are locally integrable is a consequence of Hölder's inequality. For the space $L^\infty(U)$ of all μ-essentially bounded measurable complex-valued functions on U, it is immediate that $L^\infty(U)$ is contained in $L^1_{\text{loc}}(U)$, and hence in $\mathcal{D}'(U)$.

We equip $\mathcal{D}'(U)$ with the strong topology, that is, the topology of uniform convergence on the bounded subsets of $\mathcal{D}(U)$.

It is immediate that, for every $\alpha \in (\mathbb{Z}^+)^2$, the differential operator D^α defines a continuous linear map on $\mathcal{D}(K)$ for each compact subset K of U, and hence on $\mathcal{D}(U)$. To extend this differential operator from $\mathcal{D}(U)$ to the larger space $\mathcal{D}'(U)$, one uses the notion of a signed adjoint: for each $\varphi \in \mathcal{D}'(U)$, let

$$(D^\alpha \varphi)(f) := (-1)^{|\alpha|}\, \varphi(D^\alpha f) \qquad \text{for all } f \in \mathcal{D}(U).$$

It is obvious that $D^\alpha \varphi \in \mathcal{D}'(U)$.

Given a function $g \in C^m(U)$ for some $m \in N$, integration by parts leads to

$$\int_U (D^\alpha g)\, f \,\mathrm{d}\mu = (-1)^{|\alpha|} \int_U g\, D^\alpha f \,\mathrm{d}\mu$$

for all $f \in \mathcal{D}(U)$ and $\alpha \in (\mathbb{Z}^+)^2$ with $|\alpha| \leq m$. It follows that, for $|\alpha| \leq m$, the distribution induced by the function $D^\alpha g \in C(U)$ coincides with the distributional derivative $D^\alpha \varphi$ of the distribution φ corresponding to g. This justifies the factor $(-1)^{|\alpha|}$ in the definition of the distributional derivative, and shows that differentiation in the sense of distributions extends the usual notion of differentiation for continuously differentiable functions.

An important tool in the theory of partial differential equations is the class of Sobolev spaces. Following Trèves (1967), the classical Sobolev space $H^m(U)$ of order $m \in \mathbb{Z}^+$ consists of all functions $f \in L^2(U)$ for which the distributional derivative $D^\alpha f$ belongs to $L^2(U)$ for every $\alpha \in (\mathbb{Z}^+)^2$ with $|\alpha| \leq m$. Of course, here the space $L^2(U)$ is considered as a linear subspace of $\mathcal{D}'(U)$.

We shall need a variant of this, more convenient for the study of the $\overline{\partial}$-operator. As usual, let

$$\overline{\partial} := (\partial_x + i\,\partial_y)/2 \quad \text{and} \quad \partial := (\partial_x - i\,\partial_y)/2.$$

As we have just explained, these differential operators (and their powers) are canonically defined on the space $\mathcal{D}'(U)$, and hence act, in the sense of distributions, on functions in $L^2(U)$. Be aware, however, that, for arbitrary $f \in L^2(U)$, the distributions $\overline{\partial} f$ and ∂f need not be implemented by functions in $L^2(U)$.

For every $m \in \mathbb{Z}^+$, our Sobolev-like space of order m is defined as

$$W^m(U) := \{f \in L^2(U) : \overline{\partial}^k f \in L^2(U) \text{ for } k = 0, \ldots, m\}.$$

The proof of the following result may be found in Laursen and Neumann (2000).

Proposition A.1 *For each open set $U \subseteq \mathbb{C}$ and each $m \in \mathbb{Z}^+$, the space $W^m(U)$ is a Hilbert space when equipped with the inner product given by*

$$[f, g] := \sum_{k=0}^{m} \int_U \overline{\partial}^k f\, \overline{\overline{\partial}^k g} \,\mathrm{d}\mu$$

for all $f, g \in W^m(U)$. The corresponding norm on $W^m(U)$ satisfies

$$\|f\| = \left(\sum_{k=0}^{m} \left\| \overline{\partial}^k f \right\|_2^2 \right)^{1/2}$$

for all $f \in W^m(U)$, *where* $\|\cdot\|_2$ *denotes the standard norm of the Hilbert space* $L^2(U)$. □

The connection between distributional and old-fashioned differentiation is described very well by the Sobolev embedding theorem.

Proposition A.2 *The inclusion* $W^{m+2}(U) \subseteq C^m(U)$ *holds for every open set* $U \subseteq \mathbb{C}$ *and every* $m \in \mathbb{Z}^+$. □

We proceed now to the vector-valued situation, and let $L^2(U, X)$ denote the space of all X-valued measurable functions on U (here X is a given complex Banach space and U an open subset of $\mathbb{C}$) which are Bochner square-integrable with respect to two-dimensional Lebesgue measure μ on U. With respect to the norm given by

$$\|f\|_2 := \left(\int_U \|f(\zeta)\|^2 \, d\mu(\zeta)\right)^{1/2} \quad \text{for all } f \in L^2(U, X),$$

$L^2(U, X)$ is a Banach space. Concerning integration of functions with values in a Banach space, it may be sufficient here to indicate what is involved by referring to a classical result due to Pettis (Dunford and Schwartz 1958, Theorem 3.6.11), which shows that an X-valued function f on U is measurable (with respect to μ) if and only if it is μ-essentially separably valued in X and has μ-measurable composition $\varphi \circ f$ with every continuous linear functional φ on X.

For every $m \in \mathbb{Z}^+$, we define the vector-valued Sobolev-like space

$$W^m(U, X) := \{f \in L^2(U, X) : \overline{\partial}^k f \in L^2(U, X) \quad \text{for } k = 0, \ldots, m\},$$

where the differential operator $\overline{\partial} := (\partial_x + \partial_y)/2$ is of course taken distributionally. A result like Proposition A.1 will then show that $W^m(U, X)$ is a Banach space when endowed with the obvious norm.

If X is a Hilbert space, then so are the two spaces $L^2(U, X)$ and $W^m(U, X)$ with respect to the canonical inner product.

Next we mention a vector analogue of the Sobolev embedding theorem, sufficient for our purposes. This one relates the Sobolev-like spaces to the space $C^\infty(U, X)$ of all infinitely differentiable X-valued functions on U. This space is a Fréchet space with respect to the topology of locally uniform convergence for the functions and all their derivatives.

We let $\mathcal{D}(U, X)$ denote the space of all functions in $C^\infty(U, X)$ with compact support in U. One convenient thing about the space $C^\infty(U, X)$ is that a function $f : U \to X$ belongs to $C^\infty(U, X)$ if and only if $\varphi \circ f \in C^\infty(U)$ for every

$\varphi \in X'$ (this follows from the nuclearity of the Fréchet space $C^\infty(U)$ and is included in Théorème 2.13 of Grothendieck (1955)).

Proposition A.3 *For every open set $U \subseteq \mathbb{C}$ and every Banach space X, we have*

$$\bigcap_{m=0}^{\infty} W^m(U, X) \subseteq C^\infty(U, X).$$

Sketch of proof We refer the interested reader to Laursen and Neumann (2000, Proposition 2.3.6). The argument is a beautiful piece of functional analysis: it makes use of the Pettis result mentioned before. It also relies on the Krein–Smulian theorem: a linear functional ψ on X' is weak-$*$ continuous on X' exactly when the restriction $\psi\,|(X')_{[r]}$ is weak-$*$ continuous for each $r > 0$. Additional classic tools employed are Lebesgue's dominated convergence theorem, the result of Grothendieck (1955) mentioned above, and the Hahn–Banach theorem. □

Proposition A.3 may be used to show that the vector-valued Bergman space

$$A^2(U, X) := H(U, X) \cap L^2(U, X)$$

consists of all $f \in L^2(U, X)$ for which $\overline{\partial} f = 0$ in the distributional sense (exercise). Consequently, $A^2(U, X)$ is a closed linear subspace of $W^m(U, X)$ for every $m \in \mathbb{Z}^+$. Moreover, it is easily seen that the Banach space $A^2(U, X)$ is continuously embedded in the Fréchet space $H(U, X)$. As in the scalar-valued setting, it is clear that the $\overline{\partial}$-operator acts as a continuous linear map from $W^m(U, X)$ into $W^{m-1}(U, X)$.

From now on, let $T \in \mathcal{B}(X)$ be given, and suppose that U is a bounded open subset of $\mathbb{C}$ containing the spectrum $\sigma(T)$. We need the definition of the local analytic functional calculus

$$\Phi_\ell(f) = \frac{1}{2\pi i} \int_\Gamma (\lambda - T)^{-1} f(\lambda)\, \mathrm{d}\lambda \quad \text{for all } f \in H(U, X),$$

where Γ is a contour in U surrounding $\sigma(T)$. The map $\Phi_\ell : H(U, X) \to X$ is continuous, and the Bergman space $A^2(U, X)$ is continuously embedded in $H(U, X)$, so that

$$B^2(U, X) := \{f \in A^2(U, X) : \Phi_\ell(f) = 0\}$$

is a closed linear subspace of $A^2(U, X)$. The null space of the natural surjection from

$$W^m(U, X)/B^2(U, X)$$

onto $W^m(U,X)/A^2(U,X)$ may be identified with $A^2(U,X)/B^2(U,X)$. Thus we obtain a short exact sequence

$$0 \to A^2(U,X)/B^2(U,X) \to W^m(U,X)/B^2(U,X)$$
$$\to W^m(U,X)/A^2(U,X) \to 0$$

of continuous linear operators between Banach spaces.

Here the surjectivity of the local analytic functional calculus allows us to identify the quotient $A^2(U,X)/B^2(U,X)$ with the given Banach space X. Moreover, the $\overline{\partial}$-operator induces an identification of the spaces $W^m(U,X)/A^2(U,X)$ and $W^{m-1}(U,X)$. Hence the preceding exact sequence may be rephrased as

$$0 \to X \to W^m(U,X)/B^2(U,X) \to W^{m-1}(U,X) \to 0.$$

This explains the meaning of the spaces appearing in this next theorem.

Theorem A.4 *Let $T \in \mathcal{B}(X)$ be an arbitrary operator, let U be a bounded open subset of $\mathbb{C}$ such that U contains the spectrum $\sigma(T)$, and let $m \in \mathbb{N}$. Then the assignments*

- $J(x) = 1 \otimes x + B^2(U,X)$ *for all* $x \in X$,
- $Q(f + B^2(U,X)) = \overline{\partial} f$ *for all* $f \in W^m(U,X)$,
- $S(f + B^2(U,X)) = Zf + B^2(U,X)$ *for all* $f \in W^m(U,X)$,
- $R(g) = Zg$ *for all* $g \in W^{m-1}(U,X)$

yield a commutative diagram with exact rows of continuous linear operators between Banach spaces:

$$\begin{array}{ccccccccc} 0 & \longrightarrow & X & \xrightarrow{J} & W^m(U,X)/B^2(U,X) & \xrightarrow{Q} & W^{m-1}(U,X) & \longrightarrow & 0 \\ & & T\downarrow & & S\downarrow & & R\downarrow & & \\ 0 & \longrightarrow & X & \underset{J}{\longrightarrow} & W^m(U,X)/B^2(U,X) & \underset{Q}{\longrightarrow} & W^{m-1}(U,X) & \longrightarrow & 0. \end{array}$$

Moreover, R is generalized scalar with $\sigma(R) = \overline{U}$, while S is a quotient of a generalized scalar operator with $\sigma(S) = \overline{U}$. □

Since the statement of Theorem A.4 includes the specific forms of the maps involved, carrying out the proof of this result is more a matter of tenacity than of ingenuity, as may be seen by looking at Laursen and Neumann (2000, Theorem 2.4.3). I have elected to leave out the details, and instead concentrate on our main uses for this result.

It is interesting that the above theorem establishes a certain kind of universality of generalized scalar operators: every bounded linear operator on a Banach space is a restriction of a quotient of a generalized scalar operator. However, our main concern is expressed in the next result, which is now simple. Its proof follows by applying the 3-space lemma, Lemma 23.1.3, for (β).

Theorem A.5 *A continuous linear operator on a Banach space has (β) if and only if it is similar to the restriction of a decomposable operator to one of its closed invariant subspaces. Moreover, if T is an operator with (β), then the operator S, in the functional model of Theorem A.4 for T, is a decomposable extension of T.* □

To get a handle on how to establish the remaining duality property we have to mention a second functional model.

Theorem A.6 *Let $T \in \mathcal{B}(X)$ be an arbitrary operator on a Banach space X, let $m \in \mathbb{N}$ be given, and suppose that U is a bounded open subset of $\mathbb{C}$ such that U contains $\sigma(T)$. Let*

$$Y := \{V \in \mathcal{B}(W^m(U), X) : V(f) = f(T)V(1) \quad \text{for all} \quad f \in A^2(U)\},$$

where the operator $f(T) \in \mathcal{B}(X)$ is given by the analytic functional calculus for T. Then the assignments:

- $J(C) := C \circ \overline{\partial}$;
- $R(C) := C \circ M_Z$ *for all* $C \in \mathcal{B}(W^{m-1}(U), X)$;
- $Q(V) := V(1)$;
- $S(V) := V \circ M_Z$ *for all* $V \in Y$

yield a commutative diagram with exact rows of continuous linear operators between Banach spaces:

$$\begin{array}{ccccccc} 0 \to & \mathcal{B}(W^{m-1}(U), X) & \xrightarrow{J} & Y & \xrightarrow{Q} & X & \to 0 \\ & R\downarrow & & S\downarrow & & T\downarrow & \\ 0 \to & \mathcal{B}(W^{m-1}(U), X) & \xrightarrow[J]{} & Y & \xrightarrow[Q]{} & X & \to 0. \end{array}$$

Moreover, R is generalized scalar, and S is subscalar. □

The 3-space lemma for (δ), also contained in Lemma 23.1.3, is used in proving the 'only if' part.

Finally, here are the detailed duality statements. Evidently, it is an immediate consequence of the fact (Theorem 21.2.8) that an operator is decomposable if and only if it has both (β) and (δ), combined with Theorem A.8, that an operator and its adjoint are simultaneously decomposable. However, the proof of A.8 proceeds by showing this latter fact first, so we state it separately. That will also serve to emphasize its importance.

Theorem A.7 *An operator $T \in \mathcal{B}(X)$ is decomposable if and only if the adjoint $T' \in B(X')$ is decomposable.* □

Theorem A.8 *For every operator $T \in \mathcal{B}(X)$, the following assertions hold.*

(i) *If T has (β), then T' has (δ).*
(ii) *If T has (δ), then T' has (β).*
(iii) *If T' has (δ), then T has (β).*
(iv) *If T' has (β), then T has (δ).*

Proof We sketch the proof of (i). If T has (β), then T has a decomposable extension, say S. By Theorem 23.1.4, the adjoint operator S' is also decomposable. Moreover, T' may be realized as a quotient of S'. Hence, by Theorem 23.1.1, T' has (δ). Exercise: fill in the details.

(ii) and (iii): exercises.

We leave out entirely the details of how to establish part (iv) because this is so much harder than the rest. This result is one of J. Eschmeier's great achievements. It can be done via a tensor version of the functional model of Theorem A.4. The details are omitted here, except for the diagram itself. For full details, see Laursen and Neumann (2000). □

For an arbitrary operator $T \in \mathcal{B}(X)$ on a Banach space X, the following diagram is commutative and has exact rows:

$$\begin{array}{ccccccccc}
0 \to & W^{m-1}(U)'\widehat{\otimes}_\varepsilon X & \xrightarrow{\overline{\partial}'\widehat{\otimes}_\varepsilon I} & {}^\perp\ker \breve{\Psi}_\ell & \xrightarrow{Q} & X & \to 0 \\
& M_Z'\widehat{\otimes}_\varepsilon I \big\downarrow & & S \big\downarrow & & T \big\downarrow & \\
0 \to & W^{m-1}(U)'\widehat{\otimes}_\varepsilon X & \underset{\overline{\partial}'\widehat{\otimes}_\varepsilon I}{\to} & {}^\perp\ker \breve{\Psi}_\ell & \underset{Q}{\to} & X & \to 0,
\end{array}$$

where $S := (M_Z' \widehat{\otimes}_\varepsilon I)|{}^\perp\ker \breve{\Psi}_\ell$ is a restriction of a generalized scalar operator, and the operator $M_Z' \widehat{\otimes}_\varepsilon I$ on the left-hand side is generalized scalar.

Theorem A.9 *The dual of the above diagram may be identified with the following commutative diagram with exact rows:*

$$\begin{array}{ccccccccc}
0 \longrightarrow & X' & \overset{\check{J}}{\longrightarrow} & \left(W^m(U)\widehat{\otimes}_\pi X'\right) / \ker \check{\Psi}_\ell & \overset{[\overline{\partial}\widehat{\otimes}_\pi I']}{\longrightarrow} & W^{m-1}(U)\widehat{\otimes}_\pi X' & \longrightarrow 0 \\
& T' \downarrow & & (M_Z \widehat{\otimes}_\pi I') / \ker \check{\Psi}_\ell \downarrow & & M_Z \widehat{\otimes}_\pi I' \downarrow & \\
0 \longrightarrow & X' & \underset{\check{J}}{\longrightarrow} & \left(W^m(U)\widehat{\otimes}_\pi X'\right) / \ker \check{\Psi}_\ell & \underset{[\overline{\partial}\widehat{\otimes}_\pi I']}{\longrightarrow} & W^{m-1}(U)\widehat{\otimes}_\pi X' & \longrightarrow 0.
\end{array}$$

Finally, for every $\lambda \in \mathbb{C}$ and every sequence of vectors $(x_n)_{n\in\mathbb{N}}$ in X for which $(T-\lambda)x_n$ converges to 0 as $n \to \infty$, there exist $y_n \in {}^{\perp}\ker \check{\Psi}_\ell$ such that $Q y_n = x_n$ for all $n \in \mathbb{N}$ and $(S-\lambda) y_n \to 0$ as $n \to \infty$. In particular, we have $\sigma_{\mathrm{ap}}(T) \subseteq \sigma_{\mathrm{ap}}(S)$.

□

Here

$$\check{\Psi}_\ell : A^2(U)\widehat{\otimes}_\pi X' \to X'$$

denotes the local analytic functional calculus of the adjoint $T' \in \mathcal{B}(X')$ and

$$\check{J} : X' \to \left(W^m(U)\widehat{\otimes}_\pi X'\right) / \ker \check{\Psi}_\ell$$

is the corresponding continuous linear map given by

$$\check{J}(\varphi) := 1 \otimes \varphi + \ker \check{\Psi}_\ell \qquad \text{for all } \varphi \in X'.$$

Morever, the information in the last part of the theorem may be used in a proof of Eschmeier–Prunaru's result on the existence of closed invariant subspaces for operators with (β) and thick spectrum. There is more on this in Chapter 18.

As a consequence we have Theorem 23.1.5, that is, every operator $T \in \mathcal{B}(X)$ has one of the properties (β) or (δ) if and only if its adjoint operator $T' \in \mathcal{B}(X')$ has the other one.

References

Albrecht, E. and Eschmeier, J. (1997). Analytic functional models and local spectral theory, *Proc. London Math. Soc.* (3), **75**, 323–48.

Bishop, E. (1959). A duality theorem for an arbitrary operator, *Pacific J. Math.*, **9**, 379–97.

Bonsall, F. F. and Duncan, J. (1973). *Complete normed algebras*, Berlin–Heidelberg–New York, Springer-Verlag.

Clary, S. (1975). Equality of spectra of quasi-similar hyponormal operators, *Proc. American Math. Soc.*, **53**, 88–90.

Conway, J. B. (1991). *The theory of subnormal operators*, Mathematical Surveys and Monographs, **36**. Providence, Rhode Island, American Mathematical Society.

Dales, H. G. (2000) *Banach algebras and automatic continuity*, London Mathematical Society Monograph **24**, Oxford, Clarendon Press.

Davis, C. and Rosenthal, P. (1974). Solving linear operator equations, *Canadian J. Math.*, **26**, 1384–9.

Douglas, R. G. (1969). On extending commutative semigroups of isometries, *Bull. London Math. Soc.*, **1**, 157–9.

Dunford, N. and Schwartz, J. T. (1958, 1963, 1971). *Linear operators, Vols. I, II, III*. New York, Wiley-Interscience.

Engelking, R. (1977). *General topology*, Warsaw, Polish Scientific Publishers.

Eschmeier, J. (2000). On the essential spectrum of Banach-space operators, *Proc. Edinburgh Math. Soc.*, **43**, 511–28.

Eschmeier, J. and Prunaru, B. (1990). Invariant subspaces for operators with Bishop's property (β) and thick spectrum, *J. Functional Analysis*, **94**, 196–222.

Fialkow, L. A. (1977). A note on quasisimilarity of operators, *Acta Sci. Math. (Szeged)*, **39**, 67–85.

Frunză, Şt. (1971). A duality theorem for decomposable operators, *Rev. Roumaine Math. Pures Appl.*, **16**, 1055–8.

Gleason, A. M. (1962). The abstract theorem of Cauchy–Weil, *Pacific J. Math.*, **12**, 511–25.

Grothendieck, A. (1953). Sur certains espaces de fonctions holomorphes, I and II, *J. Reine Angew. Math.*, **192**, 35–64 and 77–95.

Helson, H. (1953). Isomorphisms of abelian group algebras, *Ark. Math.*, **2**, 475–487.

Herrero, D. A. (1988). On the essential spectra of quasi-similar operators, *Canadian J. Math.*, **40**, 1436–57.

Hewitt, E. and Ross, K. A. (1963, 1970). *Abstract harmonic analysis, Vols. I, II*. Berlin, Springer-Verlag.

Johnson, B. E. and Sinclair, A. M. (1969). Continuity of linear operators commuting with continuous linear operators II, *Trans. American Math. Soc.*, **146**, 533–40.

Kato, T. (1958). Perturbation theory for nullity, deficiency and other quantities of linear operators, *J. Analysis Math.*, **6**, 261–322.

Laursen, K. B. and Neumann, M. M. (1994). Local spectral theory and spectral inclusions, *Glasgow Math. J.*, **36**, 331–43.

Laursen, K. B. and Neumann, M. M. (2000). *An introduction to local spectral theory*, London Mathematical Society Monograph **20**, Oxford, Clarendon Press.

Lumer, G. and Rosenblum, M. (1959). Linear operator equations, *Proc. American Math. Soc.*, **10**, 32–41.

Miller, T. L., Miller, V. G., and Smith, R. C. (1998). Bishop's property (β) and the Cesàro operator, *J. London Math. Soc.* (2), **58**, 197–207.

Putinar, M. (1984). Hyponormal operators are subscalar, *J. Operator Theory*, **12**, 385–95.

Putinar, M. (1992). Quasi-similarity of tuples with Bishop's property (β), *Integral Equations Operator Theory*, **15**, 1047–52.

Rosenblum, M. (1956). On the operator equation $BX - XA = Q$, *Duke Math. J.*, **23**, 263–9.

Rudin, W. (1991). *Functional analysis* (2nd edn), New York, McGraw-Hill.

Trèves, F. (1967). *Topological vector spaces, distributions and kernels*, New York, Academic Press.

Vasilescu, F.-H. (1982). *Analytic functional calculus and spectral decompositions*, Bucharest and Dordrecht, Editura Academiei and D. Reidel.

Wendel, J. G. (1952). Left centralizers and isomorphisms of group algebras, *Pacific J. Math.*, **2**, 251–61.

West, T. T. (1990). Removing the jump – Kato's decomposition, *Rocky Mountain J. Math.*, **20**, 603–12.

Yang, L. (1993). Quasisimilarity of hyponormal and subdecomposable operators, *J. Functional Analysis*, **112**, 204–17.

Part V

Single-valued extension property and Fredholm theory

PIETRO AIENA
Università degli Studi, Palermo, Italy

26
Semi-regular operators

Let T be a bounded operator on a Banach space X. It has been observed in Chapter 21 that, if the spectrum $\sigma(T)$ is totally disconnected, then T is decomposable. In particular, every compact operator K is decomposable, because $\sigma(K)$ is either finite or a sequence which converges to 0. The classical Riesz–Schauder theory for compact operators shows that, for every $\lambda \neq 0$, the space $\lambda I - T$ has a finite-dimensional kernel and a finite-codimensional range, that is, $\lambda I - T$ is a *Fredholm operator*, see Part I, Theorem 2.2.5 or Heuser (1982). Similarly, the class of normal operators on a Hilbert space consists of decomposable operators (see Part IV, Chapter 21), and has many remarkable properties from the point of view of Fredholm theory. For instance, for normal operators some of the spectra originating from Fredholm theory coincide, as in the case of compact operators.

A natural question is to what extent the results which hold for compact and normal operators in Fredholm theory may be extended to decomposable operators. In this part of the book we shall give some answers to this question, and establish important connections between Fredholm theory and the *single-valued extension property*, a property which plays a leading role in the investigation of decomposable operators.

26.1 Definitions

Among the various concepts of regularity originating from Fredholm theory, the concept of semi-regularity, which will be introduced in this chapter, seems to be the most appropriate to investigate some important aspects of local spectral

H.G. Dales, P. Aiena, J. Eschmeier, K.B. Laursen, and G. A. Willis, *Introduction to Banach Algebras, Operators, and Harmonic Analysis*. Published by Cambridge University Press.

theory. The concept of semi-regularity originates in the classical Kato's treatment (1958) of perturbation theory of Fredholm operators.

To introduce the class of semi-regular operators, we need first to establish some connections between the kernels and the ranges of the iterates T^n of an operator T on a vector space X.

Proposition 26.1.1 *For a linear operator T on a vector space X, the following statements are equivalent:*

(i) $\ker T \subseteq T^m(X)$ *for each* $m \in \mathbb{N}$*;*
(ii) $\ker T^n \subseteq T(X)$ *for each* $n \in \mathbb{N}$*;*
(iii) $\ker T^n \subset T^m(X)$ *for each* $n \in \mathbb{N}$ *and each* $m \in \mathbb{N}$*;*
(iv) $\ker T^n = T^m(\ker T^{m+n})$ *for each* $n \in \mathbb{N}$ *and each* $m \in \mathbb{N}$*.*

Proof (i)⇒(ii) We proceed by induction. The case $n = 1$ is obvious from the assumption (i). Assume that (ii) holds for $n = k$, and let $x \in \ker T^{k+1}$. Then

$$T^k x \in \ker T \subseteq T^{k+1}(X),$$

by assumption (i). Hence there exists $y \in X$ such that $T^k x = T^{k+1} y$. It is obvious that $T^k(x - Ty) = 0$. Hence, if $z := x - Ty$, then $z \in \ker T^k \subseteq T(X)$ by the inductive assumption. Therefore $x = z + Ty \in T(X)$, so that the inclusion $\ker\ T^{k+1} \subseteq T(X)$ is proved.

(ii)⇒(iii) We proceed by induction on m. The case $m = 1$ is true by assumption. Assume that (iii) is valid for $m = k$ and each $n \in \mathbb{N}$, that is, $\ker T^n \subseteq T^k(X)$ for every $n \in \mathbb{N}$. Let $x \in \ker T^n$, where $n \in \mathbb{N}$ is arbitrary. Since $\ker T^n \subseteq T(X)$, we have $x = Ty$ for some $y \in X$. From $0 = T^n x = T^{n+1} y$, we obtain that $y \in \ker T^{n+1} \subseteq T^k(X)$, and this implies that $x = Ty \in T^{k+1}(X)$. Hence (iii) is proved.

(iii)⇒(iv) Clearly, if $x \in T^m(\ker T^{n+m})$, then $T^n x = 0$, so that

$$T^m(\ker T^{n+m}) \subseteq \ker T^n.$$

To prove the opposite inclusion, let $x \in \ker T^n$ for a fixed $n \in \mathbb{N}$. If $m \in \mathbb{N}$, then, by assumption (iii), there exists $y \in X$ such that $x = T^m y$. From the equality $0 = T^n x = T^{n+m} y$, we conclude that $y \in \ker T^{n+m}$; so (iv) is proved.

(iv)⇒ (i) It suffices to consider the equality (iv) in the case where $n = 1$. In fact, in this case, for each $m \in \mathbb{N}$, we have

$$\ker T = T^m(\ker T^{m+1}) \subseteq T^m(X).$$

□

As in Part I, Example 1.2(vii), $\mathcal{B}(X)$ denotes the Banach algebra of all bounded linear operators on a Banach space X.

Definition 26.1.2 *Given a Banach space X, a bounded operator $T \in \mathcal{B}(X)$ is* semi-regular *if $T(X)$ is closed and if T satisfies one of the equivalent conditions of Proposition 26.1.1.*

It is not difficult to show that T is semi-regular if and only if the dual operator T' is semi-regular.

Given a bounded operator $T \in \mathcal{B}(X, Y)$, where X, Y are Banach spaces, the property of $T(X)$ being closed may be characterized by means of the following quantity associated with T.

Definition 26.1.3 *If $T \in \mathcal{B}(X, Y)$, where X and Y are Banach spaces, the* minimal modulus *of T is*

$$\gamma(T) := \inf_{x \notin \ker T} \frac{\|Tx\|}{\operatorname{dist}(x, \ker T)},$$

with the convention that $\gamma(T) = \infty$ if $T = 0$.

A classical result from perturbation operator theory (which is easy to see directly) establishes that $\gamma(T) > 0$ if and only if $T(X)$ is closed. Moreover, if T' is the dual of T, then $\gamma(T) = \gamma(T')$ for every $T \in \mathcal{B}(X)$; see Kato (1966) or Goldberg (1966) for details. Of course, this implies that $T(X)$ is closed if and only if $T'(X')$ is closed.

The next result gives a useful condition which ensures that $T(X)$ is closed.

Proposition 26.1.4 *Let $T \in \mathcal{B}(X)$, where X is a Banach space, and suppose that there exists a closed subspace Y of X such that $T(X) \cap Y = \{0\}$ and $T(X) \oplus Y$ is closed. Then $T(X)$ is also closed.*

Proof Consider the product space $X \times Y$ under the norm

$$\|(x, y)\| := \|x\| + \|y\| \quad (x \in X, y \in Y).$$

Then $X \times Y$ is a Banach space, and the continuous map $S : X \times Y \to X$ defined by $S(x, y) := Tx + y$ has range $S(X \times Y) = T(X) \oplus Y$, which is closed by assumption. From this, it follows that the minimal modulus

$$\gamma(S) := \inf_{(x,y) \notin \ker S} \frac{\|S(x, y)\|}{\operatorname{dist}((x, y), \ker S)} > 0.$$

Moreover, $\ker S = \ker T \times \{0\}$, so $\operatorname{dist}((x,0), \ker S) = \operatorname{dist}(x, \ker T)$, and hence

$$\|Tx\| = \|S(x,0)\| \geq \gamma(S)\operatorname{dist}((x,0), \ker S)) = \gamma(S)\operatorname{dist}(x, \ker T).$$

This implies that $\gamma(T) \geq \gamma(S)$, and therefore T has closed range. □

Recall that every subspace M of a vector space X admits at least one *algebraic complement* N, that is, $X = M + N$ and $M \cap N = \{0\}$. The *codimension* of a subspace M of X is the dimension of each algebraic complement N of M. An immediate consequence of Proposition 26.1.4 is that, if Y is a finite-dimensional subspace of X and $T(X) + Y$ is closed, then $T(X)$ is closed. In particular, it follows from Proposition 26.1.4 that, if $\operatorname{codim} T(X) < \infty$, then $T(X)$ is closed.

The following T-invariant subspace for a linear operator T on a vector space has been introduced by Saphar (1964).

Definition 26.1.5 *Let T be a linear operator on a vector space X. The* algebraic core $C(T)$ *is defined to be the greatest subspace M of X for which $T(M) = M$.*

Evidently, $C(T)$ coincides with the *algebraic spectral subspace* $E_T(\mathbb{C} \setminus \{0\})$ defined in Chapter 22.

It is not difficult to show that $C(T)$ may be characterized in terms of sequences. In fact, a recursive argument proves that $x \in C(T)$ if and only if there exists a sequence $(u_n)_{n \in \mathbb{Z}^+} \subset X$ such that $x = u_0$ and $Tu_{n+1} = u_n \ (n \in \mathbb{Z}^+)$.

The subspace $C(T)$ is defined in purely algebraic terms. The second subspace that we shall consider is, in a certain sense, the analytic counterpart of $C(T)$.

Definition 26.1.6 *Let X be a Banach space. The* analytical core *of $T \in \mathcal{B}(X)$ is the set $K(T)$ of all $x \in X$ for which there is a sequence (u_n) and a $\delta > 0$ such that:*

(i) $x = u_0$ *and* $Tu_{n+1} = u_n \ (n \in \mathbb{Z}^+)$;
(ii) $\|u_n\| \leq \delta^n \|x\| \ (n \in \mathbb{Z}^+)$.

It is easily seen from definition that $K(T)$ is a linear subspace of X contained in $C(T)$ and that $T(K(T)) = K(T)$. We shall see in Theorem 27.2.2 that $K(T) = X_T(\mathbb{C} \setminus \{0\})$, where X_T denotes the local analytic subspace, as defined in Definition 22.1.1. Observe that in general neither $K(T)$ nor $C(T)$ is closed. The next result shows that, under the assumption that $C(T)$ is closed, these two subspaces coincide.

Proposition 26.1.7 *Let $T \in \mathcal{B}(X)$, where X is a Banach space, and let F be a closed subspace of X such that $T(F) = F$. Then $F \subseteq K(T)$. In particular, if $C(T)$ is closed, then $C(T) = K(T)$.*

Proof Let $T_0 : F \to F$ denote the restriction of T to F. By assumption, F is a Banach space and $T(F) = F$, and so, by the open mapping theorem, T_0 is open. This means that there is a constant $\delta > 0$ with the property that, for every $x \in F$, there exists $u \in F$ such that $Tu = x$ and $\|u\| \leq \delta\|x\|$. Let $x \in F$ be arbitrarily given, let $u_0 = x$, and consider an element $u_1 \in F$ such that $Tu_1 = u_0$ and $\|u_1\| \leq \delta\|u_0\|$. By repeating this procedure, we can find an element $u_n \in F$ for every $n \in \mathbb{Z}^+$ such that $Tu_{n+1} = u_n$ and $\|u_n\| \leq \delta\|u_{n-1}\|$. The last inequality gives the estimate $\|u_n\| \leq \delta^n\|u_0\| = \delta^n\|x\|$ for $n \in \mathbb{Z}^+$. Thus $x \in K(T)$, and hence $F \subseteq K(T)$.

The last assertion is clear. □

Definition 26.1.8 *Let T be a linear operator on a vector space X. The* generalized range *and* generalized kernel *of T are*

$$T^\infty(X) := \bigcap_{n\in\mathbb{N}} T^n(X) \quad \text{and} \quad \mathcal{N}^\infty(T) := \bigcup_{n\in\mathbb{N}} \ker T^n .$$

A simple inductive argument shows that the inclusion $C(T) \subseteq T^n(X)$ holds for all $n \in \mathbb{N}$. From this it follows that $C(T) \subseteq T^\infty(X)$. By Proposition 26.1.1, we easily obtain that $T \in \mathcal{B}(X)$ is semi-regular if and only if $T(X)$ is closed and $\mathcal{N}^\infty(T) \subseteq T^\infty(X)$.

The next lemma shows that under certain conditions the algebraic core and the generalized range of an operator coincide.

Lemma 26.1.9 *Let T be a linear operator on a vector space X. Suppose that the equality*

$$\ker T \cap T^m(X) = \ker T \cap T^{m+k}(X) \quad (k \in \mathbb{Z}^+), \tag{26.1.1}$$

holds for some $m \in \mathbb{N}$. Then $C(T) = T^\infty(X)$.

Proof We need only to prove the inclusion $T^\infty(X) \subseteq C(T)$, that is, $T(T^\infty(X)) = T^\infty(X)$. The inclusion $T(T^\infty(X)) \subseteq T^\infty(X)$ is obvious for every linear operator, so we need only to prove the opposite inclusion.

Suppose that (26.1.1) holds for $m \in \mathbb{N}$, and consider $D := \ker T \cap T^m(X)$. Clearly, $D = \ker T \cap T^\infty(X)$. Now, let y be an arbitrary element of $T^\infty(X)$. Since $y \in T^n(X)$ $(n \in \mathbb{N})$, there exists $x_k \in X$ such that $y = T^{m+k}x_k$. We now set

$$z_k := T^m x_1 - T^{m+k-1}x_k \quad (k \in \mathbb{N}).$$

Then $z_k \in T^m(X)$, and from the fact that $Tz_k = T^{m+1}x_1 - T^{m+k}x_k = 0$ we obtain that $z_k \in \ker T$. Hence $z_k \in D$. The inclusion

$$D = \ker T \cap T^{m+k}(X) \subseteq \ker T \cap T^{m+k-1}(X)$$

now entails that $z_k \in T^{m+k-1}(X)$. This implies that

$$T^m x_1 = z_k + T^{m+k-1}x_k \in T^{m+k-1}(X) \quad (k \in \mathbb{N}),$$

and therefore $T^m x_1 \in T^\infty(X)$. From the equalities $T(T^m x_1) = T^{m+1}x_1 = y$, we finally conclude that $y \in T(T^\infty(X))$, and therefore $T^\infty(X) \subseteq T(T^\infty(X))$, as required. □

Proposition 26.1.10 *Let T be a linear operator on a vector space X. Suppose that one of the following conditions holds:*

(i) $\dim \ker T < \infty$;
(ii) $\operatorname{codim} T(X) < \infty$;
(iii) $\ker T \subseteq T^n(X)$ $(n \in \mathbb{N})$.

Then $C(T) = T^\infty(X)$.

Proof (i) It is evident that, if $\ker T$ is finite dimensional, then there exists a positive integer $m \in \mathbb{N}$ such that the equality (26.1.1) holds. Hence Lemma 26.1.9 applies.

(ii) Assume that $X = F \oplus T(X)$ with $\dim F < \infty$. Clearly, if

$$D_n := \ker T \cap T^n(X),$$

then $D_n \supseteq D_{n+1}$ for all $n \in \mathbb{N}$. Suppose that there exist k distinct subspaces D_n. There is no loss of generality if we assume that $D_j \neq D_{j+1}$ for $j = 1, 2, \dots, k$. Then for each one of these j, there exists $w_j \in X$ such that $T_j w_j \in D_j$ and $T^j w_j \notin D_{j+1}$. By means of the decomposition $X = F \oplus T(X)$, we can find $u_j \in F$ and $v_j \in T(X)$ such that $w_j = u_j + v_j$. We claim that the set $\{u_1, \dots, u_k\}$ is linearly independent.

To see this, let us suppose that $\sum_{j=1}^{k} \lambda_j u_j = 0$. Then $\sum_{j=1}^{k} \lambda_j w_j = \sum_{j=1}^{k} \lambda_j v_j$ and hence, since $T^k w_1 = \cdots = T^k w_{k-1} = 0$, we have

$$T^k \left(\sum_{j=1}^{k} \lambda_j w_j \right) = \lambda_k T^k w_k = T^k \left(\sum_{j=1}^{k} \lambda_j v_j \right) \in T^k(T(X)) = T^{k+1}(X).$$

From $T^k w_k \in \ker T$, we obtain $\lambda_k T^k w_k \in D_{k+1}$ and, since $T^k w_k \notin D_{k+1}$, this is possible only if $\lambda_k = 0$. Analogously we have $\lambda_{k-1} = \ldots = \lambda_1 = 0$, and consequently the set $\{u_1, \ldots, u_k\}$ is linearly independent. From this, we conclude that $k \leq \dim F$. But then, for a sufficiently large m, we have

$$\ker T \cap T^m(X) = \ker T \cap T^{m+j}(X) \quad (j \in \mathbb{Z}^+).$$

So we are again in the situation of Lemma 26.1.9, and so (ii) implies that $C(T) = T^\infty(X)$.

(iii) Obviously, if $\ker T \subseteq T^n(X)$ for all $n \in \mathbb{N}$, then

$$\ker T \cap T^n(X) = \ker T \cap T^{n+k}(X) = \ker\, T \quad (k \in \mathbb{Z}^+).$$

Hence also in this case we can apply Lemma 26.1.9. □

Lemma 26.1.11 *Suppose that $T \in \mathcal{B}(X)$, where X is a Banach space, and that T has closed range $T(X)$. For each (not necessarily closed) subspace $Y \subseteq X$, $T(Y)$ is closed whenever $Y + \ker T$ is closed.*

Proof Let $\overline{x}$ be the equivalence class $x + \ker\, T$ in the quotient space $X/\ker T$, and denote by $\overline{T} : X/\ker T \to X$ the canonical injection which is defined by $\overline{T}(\overline{x}) := Tx$, where $x \in \overline{x}$. By assumption, $T(X)$ is closed, so that $\overline{T}$ has a bounded inverse $\overline{T}^{-1} : T(X) \to X/\ker\, T$. Let $\overline{Y} := \{\overline{y} : y \in Y\}$. It is evident that $T(Y) = \overline{T}(\overline{Y})$ is the inverse image of $\overline{Y}$ under the continuous map $\overline{T}^{-1}$. Thus $T(Y)$ is closed whenever $\overline{Y}$ is closed.

It remains to show that $\overline{Y}$ is closed whenever $Y + \ker\, T$ is closed. Let $(\overline{x}_n)$ be a sequence of $\overline{Y}$ which converges to $\overline{x} \in X/\ker T$. Then there exists a sequence (x_n) with $x_n \in \overline{x}_n$ such that $\text{dist}(x_n - x, \ker T)$ converges to zero, and so there exists a sequence $(z_n) \subset \ker T$ with $x_n - x - z_n \to 0$ as $n \to \infty$. Evidently, the sequence $(x_n - z_n) \subset Y + \ker T$ converges to x, and since, by assumption, $Y + \ker T$ is closed, we conclude that $x \in Y + \ker T$. This implies $\overline{x} \in \overline{Y}$, and hence $\overline{Y}$ is closed. □

Proposition 26.1.12 *Let X be a Banach space, and let $T \in \mathcal{B}(X)$ be semi-regular. Then:*

(i) *the subspace $T^n(X)$ is closed for each $n \in \mathbb{N}$;*
(ii) *$C(T)$ is closed and $C(T) = K(T) = T^\infty(X)$.*

Proof (i) We proceed by induction. The case $n = 1$ is obvious because, by assumption, $Y := T(X)$ is closed. Assume that $T^n(X)$ is closed for some $n \in \mathbb{N}$. Since T is semi-regular, $\ker T^n \subseteq T(X) = Y$. From this, it follows that the sum $Y + \ker T^n = Y$ is closed, and hence, by Lemma 26.1.11, $T^n(Y) = T^{n+1}(X)$ is closed.

(ii) The operator T is semi-regular, and so, by Proposition 26.1.10,

$$C(T) = T^\infty(X) = \bigcap_{n=1}^{\infty} T^n(X),$$

and therefore $C(T)$ is closed by part (i). From Proposition 26.1.7, we now conclude that $C(T) = K(T)$. □

26.2 The Kato spectrum

Among the many concepts dealt with in Kato's extensive treatment of perturbation theory (Kato 1958), there is a very important part of the spectrum defined as follows.

Definition 26.2.1 *Let X be a Banach space, and let $T \in \mathcal{B}(X)$. The* Kato resolvent set *of T is*

$$\rho_K(T) := \{\lambda \in \mathbb{C} : \lambda I - T \text{ is semi-regular}\}.$$

The Kato spectrum *of T is the set $\sigma_K(T) := \mathbb{C} \setminus \rho_K(T)$.*

It is clear that $\sigma_K(T) \subseteq \sigma(T)$ and $\rho(T) \subseteq \rho_K(T)$. Later we shall prove that $\sigma_K(T)$ is a non-empty, compact subset of $\mathbb{C}$.

Recall that an operator $T \in \mathcal{B}(X)$ is *bounded below* if T is injective and has closed range. This is equivalent to saying that there exists $K > 0$ such that $\|Tx\| \geq K\|x\|$ for all $x \in X$. Clearly, if T is bounded below or surjective, then T is semi-regular. A standard result of duality theory shows that T is bounded below (respectively, surjective) if and only if T' is surjective (respectively, bounded below).

Lemma 26.2.2 *Let X be a Banach space, and let $T \in \mathcal{B}(X)$. Then T is bounded below (respectively, surjective) if and only if $\lambda I - T$ is bounded below (respectively, surjective) for every $|\lambda| < \gamma(T)$.*

Proof Suppose T bounded below. Since $T(X)$ is closed, $\gamma(T) > 0$ and

$$\gamma(T) \cdot \operatorname{dist}(x, \ker T) = \gamma(T)\|x\| \leq \|Tx\| \quad (x \in X).$$

Take $|\lambda| < \gamma(T)$. Then

$$\|(\lambda I - T)x\| \geq \|Tx\| - |\lambda|\|x\| \geq (\gamma(T) - |\lambda|)\|x\|,$$

and so $\lambda I - T$ is bounded below.

The case where T is surjective now follows easily by considering the dual T' of T. □

Theorem 26.2.3 *Let X be a Banach space, and let $T \in \mathcal{B}(X)$ be semi-regular. Then $\lambda I - T$ is semi-regular for $|\lambda| < \gamma(T)$. Consequently, $\sigma_K(T)$ is closed.*

Proof First we show that $C(T) \subseteq C(\lambda I - T)$ for all $|\lambda| < \gamma(T)$. Indeed, let $T_0 : C(T) \to C(T)$ denote the restriction of T to $C(T)$. By Proposition 26.1.12, $C(T)$ is closed and T_0 is surjective. Thus, by Lemma 26.2.2, the equalities

$$(\lambda I - T_0)(C(T)) = (\lambda I - T)(C(T)) = C(T)$$

hold for all $|\lambda| < \gamma(T_0)$.

On the other hand, T is semi-regular so, by Proposition 26.1.10, we have $\ker T \subseteq T^\infty(X) = C(T)$. From this it easily follows that $\gamma(T_0) \geq \gamma(T)$ and, consequently, we see that $(\lambda I - T)(C(T)) = C(T)$ for all $|\lambda| < \gamma(T)$. Note that this last equality implies that

$$C(T) \subseteq C(\lambda I - T) \quad (\lambda \in \mathbb{D}(0, \gamma(T))). \tag{26.2.1}$$

Moreover, for every $\lambda \neq 0$, we have $T(\ker(\lambda I - T)) = \ker(\lambda I - T)$, and so, by Proposition 26.1.12 and Proposition 26.1.7, we have $\ker(\lambda I - T) \subseteq C(T)$ for $\lambda \neq 0$. From the inclusion (26.2.1), we now obtain that the inclusions

$$\ker(\lambda I - T) \subseteq C(\lambda I - T) \subseteq (\lambda I - T)^n(X) \tag{26.2.2}$$

hold for all $|\lambda| < \gamma(T)$, $\lambda \neq 0$ and $n \in \mathbb{N}$. This is still true for $\lambda = 0$ since T is semi-regular, so that (26.2.2) is valid for all $|\lambda| < \gamma(T)$.

To show that $\lambda I - T$ is semi-regular for all $|\lambda| < \gamma(T)$, it only remains to prove that $(\lambda I - T)(X)$ is closed for all $|\lambda| < \gamma(T)$. Observe that, as a consequence of Lemma 26.2.2, we can limit ourselves to considering only the case where $\{0\} \neq C(T) \neq \{X\}$. Indeed, if $C(T) = \{0\}$, then we have ker $T \subseteq C(T) = \{0\}$, and hence T is injective, while, if $C(T) = X$, the operator T is surjective.

Let $\overline{X} := X/C(T)$, and let $\overline{T} : \overline{X} \to \overline{X}$ be the continuous quotient map, defined by $\overline{T}\,(\overline{x}) := \overline{Tx}$ where $x \in \overline{x}$. The operator $\overline{T}$ is injective, since from $\overline{T}\,\overline{x} = \overline{Tx} = \overline{0}$ we have $Tx \in C(T)$ and this easily implies, since T is semi-regular, that $x \in C(T)$, which yields $\overline{x} = \overline{0}$.

Next we prove that $\overline{T}$ is bounded below. Evidently, we only need to prove that $\overline{T}$ has closed range. To see this we show the inequality $\gamma(\overline{T}) \geq \gamma(T)$. In fact, for $x \in X$ and $u \in C(T)$, we have

$$\begin{aligned}\|\overline{x}\| &= \text{dist}(x, C(T)) = \text{dist}(x - u, C(T)) \\ &\leq \text{dist}(x - u, \ker T) \leq \|Tx - Tu\|/\gamma(T).\end{aligned}$$

From the equality $C(T) = T(C(T))$, we obtain $\|\overline{x}\| = \|\overline{T}\,\overline{x}\|/\gamma(T)$ and, consequently, $\gamma(\overline{T}) \geq \gamma(T)$. Therefore $\overline{T}$ is bounded below. By Lemma 26.2.2, $\lambda\overline{I} - \overline{T}$ is bounded below for all $|\lambda| < \gamma(\overline{T})$ and a fortiori for all $|\lambda| < \gamma(T)$.

Finally, to show that $(\lambda I - T)(X)$ is closed for all $|\lambda| < \gamma(T)$, let us consider a sequence (u_n) of $(\lambda I - T)(X)$ which converges to $x \in X$. Clearly, the sequence $(\overline{x}_n)$ converges to $\overline{x}$ and $\overline{x}_n \in (\lambda\overline{I} - \overline{T})(\overline{X})$. The last space is closed for all $|\lambda| < \gamma(T)$, and hence $\overline{x} \in (\lambda\overline{I} - \overline{T})(\overline{X})$. Let $\overline{x} = (\lambda\overline{I} - \overline{T})\overline{v}$ and $v \in \overline{v}$. Then

$$x - (\lambda I - T)v \in C(T) = (\lambda I - T)(C(T)) \quad (|\lambda| < \gamma(T)),$$

and so there exists $u \in C(T)$ such that $x = (\lambda I - T)(v + u)$, that is, $x \in (\lambda I - T)(X)$, for each $|\lambda| < \gamma(T)$. Therefore $(\lambda I - T)(X)$ is closed for $|\lambda| < \gamma(T)$, and consequently $\lambda I - T$ is semi-regular for $|\lambda| < \gamma(T)$, that is, $\rho_K(T)$ is an open subset of $\mathbb{C}$. □

The set $\rho_K(T)$ is open, and therefore it can be decomposed into open, connected, pairwise disjoint, non-empty components. We want to prove that $C(\lambda I - T)$ is constant on each component Ω of $\rho_K(T)$. To show this, we first need some preliminary results on gap theory.

Let M, N be two closed linear subspaces of a Banach space X, and define

$$\delta(M, N) := \sup\{\text{dist}(u, N) : u \in M, \|u\| = 1\}.$$

Lemma 26.2.4 *Let M and N be two closed subspaces of a Banach space X. For every $x \in X$ and $0 < \varepsilon < 1$, there exists $x_0 \in X$ such that $x - x_0 \in M$ and*

$$\operatorname{dist}(x_0, N) \geq (1 - \varepsilon)\frac{1 - \delta(M, N)}{1 + \delta(M, N)}\|x_0\| . \tag{26.2.3}$$

Proof The case $x \in M$ is obvious: it suffices to take $x_0 = 0$. Hence assume that $x \notin M$. Let $\overline{X} := X/M$ be the quotient space, and set $\overline{x} := x + M$. Then $\|\overline{x}\| = \inf_{z \in \overline{X}} \|z\| > 0$. We claim that there exists $x_0 \in X$ such that $\|\overline{x_0}\| = \operatorname{dist}(x_0, M) \geq (1 - \varepsilon)\|x_0\|$. Indeed, were this not so, then

$$\|\overline{x}\| = \|\overline{z}\| < (1 - \varepsilon)\|z\|$$

for every $z \in \overline{x}$ and, consequently, $\|\overline{x}\| \leq (1 - \varepsilon) \inf_{z \in \overline{x}} \|z\| = (1 - \varepsilon)\|\overline{x}\|$. This is impossible because $\|\overline{x}\| > 0$.

Let $\mu := \operatorname{dist}(x_0, N) = \inf_{u \in N} \|x_0 - u\|$. We know that there exists $y \in N$ such that $\|x_0 - y\| \leq \mu + \varepsilon\|x_0\|$. From this we obtain

$$\|y\| \leq (1 + \varepsilon)\|x_0\| + \mu .$$

On the other hand we have $\operatorname{dist}(y, M) = \delta(N, M) \cdot \|y\|$, and so

$$\begin{aligned}(1 - \varepsilon)\|\overline{x_0}\| &\leq \operatorname{dist}(x_0, M) \leq \|x_0 - y\| + \operatorname{dist}(y, M)\\ &\leq \mu + \varepsilon\|x_0\| + \delta(N, M) \cdot \|y\|\\ &\leq \mu + \varepsilon\|x_0\| + \delta(N, M)[(1 + \varepsilon)\|x_0\| + \mu] ,\end{aligned}$$

and therefore

$$\mu \geq \left(\frac{1 - \varepsilon - \delta(N, M)}{1 + \delta(N, M)} - \varepsilon\right) \|x_0\| .$$

Since $\varepsilon > 0$ is arbitrary, this implies the inequality (26.2.3). □

Lemma 26.2.5 *Let X be a Banach space, and let $T \in \mathcal{B}(X)$ be semi-regular. Then*

$$\gamma(\lambda I - T) \geq \gamma(T) - 3|\lambda| \quad (\lambda \in \mathbb{C}) . \tag{26.2.4}$$

Proof Of course, for every $T \in \mathcal{B}(X)$ and $|\lambda| \geq \gamma(T)$, we have

$$\gamma(\lambda I - T) \geq 0 \geq \gamma(T) - 3|\lambda| ,$$

hence we need only to prove the inequality (26.2.4) in the case where $|\lambda| < \gamma(T)$.

If $C(T) = \{0\}$, from the assumption that $\ker T \subseteq T^\infty(X) = C(T)$, we conclude that $\ker T = \{0\}$, that is, T is bounded below. From an inspection of the proof of Lemma 26.2.2, we obtain that $\gamma(\lambda I - T) \geq \gamma(T) - |\lambda| \geq \gamma(T) - 3|\lambda|$ for all $|\lambda| < \gamma(T)$. Also the case where $C(T) = X$ is trivial, because in this case T is surjective and hence T' is bounded below, so that

$$\gamma(\lambda I - T) = \gamma(\lambda I' - T') \geq \gamma(T') - 3|\lambda| = \gamma(T) - 3|\lambda| .$$

It remains to prove the inequality (26.2.4) in the case where $C(T) \neq \{0\}$ and $C(T) \neq X$. Suppose that $|\lambda| < \gamma(T)$, and let $x \in C(T) = T(C(T))$. Then there exists $u \in C(T)$ such that $x = Tu$, and hence

$$\operatorname{dist}(u, \ker T) \leq \|Tu\|/\gamma(T) = \|x\|/\gamma(T) .$$

Let $\varepsilon > 0$ be arbitrary, and choose $w \in \ker T$ such that

$$\|u - w\| \leq [(1-\varepsilon)\|x\|/\gamma(T).$$

Let $u_1 := u - w$, and $\mu := (1-\varepsilon)\gamma(T)$. Clearly, $u_1 \in C(T)$, $Tu_1 = x$, and $\|u_1\| \leq \mu^{-1}\|x\|$. Since $u_1 \in C(T)$, by repeating the same procedure, we obtain a sequence (u_n), where $u_0 := x$, such that $u_n \in C(T)$, $Tu_{n+1} = u_n$, and $\|u_n\| \leq \mu^{-n}\|x\|$.

Let us consider the function $f : \mathbb{D}(0, \mu) \to X$ defined by

$$f(\lambda) := u_0 + \sum_{n=1}^{\infty} \lambda^n u_n .$$

Obviously, $f(0) = x$ and $f(\lambda) \in \ker(\lambda I - T)$ for all $|\lambda| < \mu$. Moreover,

$$\|x - f(\lambda)\| = \left\| \sum_{n=1}^{\infty} \lambda^n u_n \right\| \leq \frac{|\lambda|}{\mu - |\lambda|} .$$

From this we obtain

$$\operatorname{dist}(x, \ker(\lambda I - T)) \leq \frac{|\lambda|}{\mu - |\lambda|} ,$$

which implies that

$$\delta(\ker T, \ker(\lambda I - T)) \leq \frac{|\lambda|}{\mu - |\lambda|} = \frac{|\lambda|}{(1-\varepsilon)\gamma(T) - |\lambda|} \quad (|\lambda| < \mu) .$$

Since $\varepsilon > 0$ is arbitrary, we then conclude that

$$\delta(\ker T, \ker(\lambda I - T)) \leq \frac{|\lambda|}{\gamma(T) - |\lambda|} \quad (|\lambda| < \gamma(T)) . \tag{26.2.5}$$

Now, let $\delta := \delta(\ker T, \ker(\lambda I - T))$. By Lemma 26.2.4, to the element u and $\varepsilon > 0$ there corresponds an element $v \in X$ such that $z := u - v \in \ker(\lambda I - T)$

and

$$\operatorname{dist}(v, \ker T) \geq \frac{1-\delta}{1+\delta}(1-\varepsilon)\|v\|.$$

From this it follows that

$$\begin{aligned}\|(\lambda I - T)u\| = \|(\lambda I - T)v\| &\geq \|Tv\| - |\lambda|\|v\| \\ &\geq \gamma(T) \cdot \operatorname{dist}(v, \ker T) - |\lambda|\|v\| \\ &\geq \gamma(T)\frac{1-\delta}{1+\delta}(1-\varepsilon)\|v\| - |\lambda|\|v\|.\end{aligned}$$

Then, by using the inequality (26.2.5), we obtain

$$\begin{aligned}\|(\lambda I - T)u\| &\geq [(1-\varepsilon)(\gamma(T) - 2|\lambda|) - |\lambda|]\|v\| \\ &\geq [(1-\varepsilon)(\gamma(T) - 2|\lambda|) - |\lambda|]\|u - z\| \\ &\geq [(1-\varepsilon)(\gamma(T) - 2|\lambda|) - |\lambda|] \cdot \operatorname{dist}(u, \ker(\lambda I - T)).\end{aligned}$$

From the last inequality, it easily follows that

$$\gamma(\lambda I - T) \geq (1-\varepsilon)(\gamma(T) - 2|\lambda|) - |\lambda|,$$

and, since ε is arbitrary, we conclude that $\gamma(\lambda I - T) \geq \gamma(T) - 3|\lambda|$. □

Theorem 26.2.6 *Let $T \in \mathcal{B}(X)$, where X is a Banach space, and consider a component Ω of $\rho_K(T)$. If $\lambda_0 \in \Omega$, then $C(\lambda I - T) = C(\lambda_0 I - T)$ for all $\lambda \in \Omega$, that is, the subspaces $C(\lambda I - T) = K(\lambda I - T)$ are constant on Ω.*

Proof Observe first that, by the first part of the proof of Theorem 26.2.3, we have $C(T) \subseteq C(\delta I - T)$ for every $|\delta| < \gamma(T)$. Now take $|\delta| < \gamma(T)/4$ and define $S = \delta I - T$. From Lemma 26.2.5, we obtain

$$\gamma(S) = \gamma(\delta I - T) \geq \gamma(T) - 3|\delta| > |\delta|,$$

and hence, again by the observation above,

$$C(S) = C(\delta I - T) \subseteq C(\delta I - (\delta I - T)) = C(T).$$

This shows that $C(\delta I - T) = C(T)$ for δ sufficiently small.

Assume now that $\lambda, \mu \in \Omega$. Then $\lambda I - T = (\lambda - \mu)I - (T - \mu I)$, and the previous argument shows that, if we take λ, μ sufficiently close to each other, then

$$C(\lambda I - T) = C((\lambda - \mu)I - (T - \mu I)) = C(\mu I - T).$$

A compactness argument shows that we have $C(\lambda I - T) = C(\mu I - T)$ for all $\lambda, \mu \in \Omega$. □

In the next definition we introduce another important T-invariant subspace for a bounded operator T.

Definition 26.2.7 *Let $T \in \mathcal{B}(X)$, where X is a Banach space. The* quasi-nilpotent part *of T is the set*

$$H_0(T) := \{x \in X : \lim_{n\to\infty} \|T^n x\|^{1/n} = 0\}.$$

As in Definition 2.1.2, the operator $T \in \mathcal{B}(X)$ is said to be *quasi-nilpotent* if its *spectral radius* $\nu(T) := \inf\{\|T^n\|^{1/n}\} = \lim_{n\to\infty} \|T^n\|^{1/n}$ is zero.

Clearly, $H_0(T)$ is a linear subspace of X, generally not closed.

Lemma 26.2.8 *Let X be a Banach space, and let $T \in \mathcal{B}(X)$ be semi-regular. Then $\gamma(T^n) \geq \gamma(T)^n$ $(n \in \mathbb{N})$.*

Proof We proceed by induction. The case $n = 1$ is trivial. Assume inductively that $\gamma(T^n) \geq \gamma(T)^n$. For every element $x \in X$ and $u \in \ker T^{n+1}$, we have

$$\operatorname{dist}(x, \ker T^{n+1}) = \operatorname{dist}(x - u, \ker T^{n+1}) \leq \operatorname{dist}(x - u, \ker T).$$

Since T is semi-regular, Lemma 26.1.1 implies that $\ker T = T^n(\ker T^{n+1})$. Therefore

$$\begin{aligned}\operatorname{dist}(T^n x, \ker T) = \operatorname{dist}(T^n x, T^n(\ker T^{n+1})) &= \inf_{u\in\ker T^{n+1}} \|T^n(x-u)\| \\ &\geq \gamma(T^n) \cdot \inf_{u\in\ker T^{n+1}} \operatorname{dist}(x-u, \ker T^n) \\ &\geq \gamma(T^n) \operatorname{dist}(x, \ker T^{n+1}).\end{aligned}$$

From this we obtain

$$\|T^{n+1}x\| \geq \gamma(T) \operatorname{dist}(T^n x, \ker T) \geq \gamma(T)\,\gamma(T^n) \cdot \operatorname{dist}(x, \ker T^{n+1}).$$

Consequently, $\gamma(T^{n+1}) \geq \gamma(T)\gamma(T)^n = \gamma(T)^{n+1}$, so the proof is complete. □

Let $M^\perp$ denote the *annihilator* of $M \subseteq X$, and ${}^\perp N$ the pre-annihilator of $N \subseteq X'$.

Proposition 26.2.9 *For every bounded operator $T \in \mathcal{B}(X)$, where X is a Banach space, we have $H_0(T) \subseteq {}^\perp K(T')$ and $K(T) \subseteq {}^\perp H_0(T')$. Moreover, if T is semi-regular, then $\overline{H_0(T)} = \overline{\mathcal{N}^\infty(T)} = {}^\perp K(T')$ and $K(T) = {}^\perp H_0(T')$.*

Proof Consider an element $u \in H_0(T)$ and $f \in K(T')$. Then, according to the definition of $K(T')$, there exists $\delta > 0$ and a sequence $(g_n) \subset X'$ such that $g_0 = f$, $T'g_{n+1} = g_n$, and $\|g_n\| \le \delta^n \|f\|$ for every $n \in \mathbb{Z}^+$. These equalities entail that $f = (T')^n g_n$, and hence that $f(u) = (T')^n g_n(u) = g_n(T^n u)$, for $n \in \mathbb{Z}^+$. From this it follows that $|f(u)| \le \|T^n u\| \|g_n\|$ for $n \in \mathbb{Z}^+$, and therefore

$$|f(u)| \le \delta^n \|f\| \|T^n u\| \quad (n \in \mathbb{Z}^+). \tag{26.2.6}$$

Now, from $u \in H_0(T)$, we obtain that $\lim_{n\to\infty} \|T^n u\|^{1/n} = 0$, and hence, by taking the nth root in (26.2.6), we conclude that $f(u) = 0$. Therefore, $H_0(T) \subseteq {}^{\perp}K(T')$.

The inclusion $K(T) \subseteq {}^{\perp}H_0(T')$ is proved in a similar way.

Now, suppose that T is semi-regular. The inclusion $\overline{\mathcal{N}^\infty(T)} \subseteq \overline{H_0(T)}$ is obvious for every operator. To show the reverse inclusion, let suppose that $x \notin \overline{\mathcal{N}^\infty(T)}$ and take

$$\delta := \operatorname{dist}(x, \overline{\mathcal{N}^\infty(T)}) = \operatorname{dist}\left(x, \overline{\bigcup_{n=1}^{\infty} \ker T^n}\right).$$

Obviously, $\delta > 0$ and, by Lemma 26.2.8, we have

$$\|T^n x\| \ge \gamma(T^n) \operatorname{dist}(x, \ker T^n)) \ge \gamma(T^n)\, \delta \ge \gamma(T)^n\, \delta,$$

so that $\lim_{n\to\infty} \|T^n x\|^{1/n} \ge \gamma(T) > 0$, and hence $x \notin H_0(T)$. This shows the reverse inclusion, so that we have $\overline{H_0(T)} = \overline{\mathcal{N}^\infty(T)}$.

Finally, we show the equality $\overline{H_0(T)} = {}^{\perp}K(T')$. From the first part, it is enough to show the inclusion ${}^{\perp}K(T') \subseteq \overline{H_0(T)}$. For $T \in \mathcal{B}(X)$ and $n \in \mathbb{N}$, we have $\ker T^n \subseteq H_0(T)$, and hence

$$H_0(T)^{\perp} \subseteq \ker T^{n\perp} = (T')^n(X'),$$

where the last equality holds since T' is semi-regular and therefore, by Proposition 26.1.12(i), $(T')^n(X')$ is closed for $n \in \mathbb{N}$. This easily implies that

$$\overline{H_0(T)}^{\perp} = H_0(T)^{\perp} \subseteq (T')^{\infty}(X') = K(T'),$$

again by Proposition 26.1.12. Consequently, ${}^{\perp}K(T') \subseteq \overline{H_0(T)}$.

The equality $K(T) = {}^{\perp} H_0(T')$ is proved in a similar way. □

Theorem 26.2.10 *Let $T \in \mathcal{B}(X)$, where X is a Banach space, and let $\Omega \subset \mathbb{C}$ be a connected component of $\rho_K(T)$. If $\lambda_0 \in \Omega$, then*

$$\overline{H_0(\lambda I - T)} = \overline{H_0(\lambda_0 I - T)} \quad (\lambda \in \Omega).$$

Proof We know that $\rho_K(T) = \rho_K(T')$. By Theorem 26.2.6, it follows that $K(\lambda I' - T') = K(\lambda_0 I' - T')$ for all $\lambda \in \Omega$. From Proposition 26.2.9, we then obtain that

$$\overline{H_0(\lambda I - T)} = {}^{\perp}K(\lambda I' - T') = {}^{\perp}K(\lambda_0 I' - T') = \overline{H_0(\lambda_0 I - T)},$$

for all $\lambda \in \Omega$. □

Theorem 26.2.11 *Let $T \in \mathcal{B}(X)$, and denote by Λ a connected component of $\sigma(T)$. Then the topological boundary $\partial\Lambda$ is contained in $\sigma_K(T)$. In particular, if $X \neq \{0\}$, the Kato spectrum $\sigma_K(T)$ is a non-empty, compact subset of $\mathbb{C}$ containing $\partial\sigma(T)$.*

Proof Let $\lambda_0 \in \partial\Lambda$, and suppose that $\lambda_0 \in \rho_K(T)$. Let Ω denote the component of $\rho_K(T)$ containing λ_0. The set Ω is open, so there exists a neighbourhood U of λ_0 contained in Ω and, since $\lambda_0 \in \partial\sigma(T)$, U also contains points of $\rho(T)$. Hence $\Omega \cap \rho(T) \neq \emptyset$.

Consider a point $\lambda_1 \in \Omega \cap \rho(T)$. Clearly, $\mathcal{N}^\infty(\lambda_1 I - T) = \{0\}$, and hence, by Theorem 26.2.10 and Proposition 26.2.9, we have

$$\overline{H_0(\lambda I - T)} = \overline{H_0(\lambda_1 I - T)} = \overline{\mathcal{N}^\infty(\lambda_1 I - T)} = \{0\} \quad (\lambda \in \Omega).$$

This shows that $\ker(\lambda I - T) = \{0\}$ for every $\lambda \in \Omega$, so that $\lambda I - T$ is injective.

On the other hand, from Theorem 26.2.6, we know that

$$K(\lambda I - T) = K(\lambda_1 I - T) = X \; (\lambda \in \Omega),$$

so that $\lambda I - T$ is onto for every $\lambda \in \Omega$. In particular, since $\lambda_0 \in \Omega$, we conclude that $\lambda_0 \in \rho(T)$, a contradiction. Hence the first assertion is proved.

The second assertion is clear: $\sigma_K(T)$ is compact by Theorem 26.2.3 and $\partial\sigma(T) \subseteq \sigma_K(T)$ by the first part. □

26.3 Exercises

1. Show that the equality $(\lambda I + T)(\mathcal{N}^\infty(T)) = \mathcal{N}^\infty(T)$ holds for all $\lambda \neq 0$. Show that $\mathcal{N}^\infty(\lambda I + T) \subseteq (\mu I + T)^\infty(X)$ for $\lambda \neq \mu$.
2. Show that $T \in \mathcal{B}(X)$ is semi-regular if and only if T' is semi-regular. *Hint*: show first that $(\ker T^n)^\perp = (T')^n(X')$ and ${}^\perp(\ker (T')^n) = T^n(X)$ for every $n \in \mathbb{N}$.

3. Show that $x \in C(T)$ if and only if there exists a sequence $(u_n)_{n \in \mathbb{Z}^+} \subset X$ such that $x = u_0$ and $Tu_{n+1} = u_n$. Check that, for each semi-regular operator, the absorbency property $x \in C(T) \Leftrightarrow Tx \in C(T)$.
4. Show that T is quasi-nilpotent if and only if $H_0(T) = X$. *Hint*: for every $x \in X$ and $\lambda \neq 0$, the series $\sum_{n=0}^{\infty} T^n x/\lambda^{n+1}$ converges to some y for which $(\lambda I - T)x = y$.
5. Let T be a multiplier of a semisimple Banach algebra A, see Chapter 25. Show that $H_0(T) = \ker T$. *Hint*: show first that, if $x \in H_0(T)$, then Tx belongs to the radical rad A, for every $a \in A$. See Part I, §2.2 for definition and basic properties.
6. Show that the operator $T \in \mathcal{B}(X)$ is semi-regular if and only if the mapping $\lambda \mapsto \gamma(\lambda I - T)$ is continuous at 0. *Hint*: use Lemma 26.2.5.
7. Show that $T \in \mathcal{B}(X)$ is semi-regular if and only if there exists a closed subspace M of X such that $T(M) = M$ and the operator $\widetilde{T} : X/M \to X/M$, induced by T, is injecive and has closed range.
8. Show that, if $T, S \in \mathcal{B}(X)$ commute and if TS is semi-regular, then both T and S are semi-regular. Find an example which shows that the product of two semi-regular operators, also commuting semi-regular operators, need not be semi-regular. *Hint*: let H be a Hilbert space with an orthonormal basis $(e_{i,j})$ where i, j are integers for which $ij \leq 0$. Let $T, S \in \mathcal{B}(H)$ be defined by the assignments:

$$Te_{i,j} := \begin{cases} 0 & \text{if } i = 0,\ j > 0, \\ e_{i+1,j} & \text{otherwise}, \end{cases} \quad \text{and } Se_{i,j} := \begin{cases} 0 & \text{if } j = 0,\ i > 0, \\ e_{i,j+1} & \text{otherwise}. \end{cases}$$

Show that T and S are semi-regular and that TS is not semi-regular.
9. Check that the set of all semi-regular operators need not be an open set of $\mathcal{B}(X)$. *Hint*: let H be a Hilbert space with an orthonormal basis $(e_{i,j})$ where i, j are integers and $i \geq 1$. Let $T \in \mathcal{B}(H)$ be defined by the assignment:

$$Te_{i,j} := \begin{cases} e_{i,j+1} & \text{if } j \neq 0, \\ 0 & \text{if } j = 0. \end{cases}$$

Show first that T is semi-regular.
Now, let $\varepsilon > 0$ be arbitrarily given and define $S \in \mathcal{B}(H)$ by

$$Se_{i,j} := \begin{cases} (\varepsilon/i)e_{i,0} & \text{if } j = 0, \\ 0 & \text{if } j \neq 0. \end{cases}$$

Note first that $\|S\| = \varepsilon$. Show that $(T + S)(H)$ is not closed and hence $T + S$ is not semi-regular; see Müller (1994).

26.4 Additional notes

The concept of semi-regularity of an operator $T \in \mathcal{B}(X)$, where X is a Banach space, arose from the treatment of perturbation theory due to Kato (1958), even if originally these operators were not given this name. Subsequently, this class of operators has received a lot of attention from several other authors (see, for instance, Mbekhta 1987, 1990; Mbekhta and Ouahab 1994; Schmoeger 1990; Müller 1994). The Kato spectrum, also known as the *Apostol spectrum* in the literature, was first introduced by Apostol (1984) for Hilbert space operators and was defined as the set of all complex λ such that either $\lambda I - T$ is not closed or λ is a discontinuity point for the function $\lambda \to (\lambda I - T)^{-1}$. Subsequently, the spectrum $\sigma_K(T)$ has been studied by different authors in the more general framework of Banach spaces.

The local constancy of the analytic core on the components of the Kato resolvent has been established by several authors (see, for instance, Förster 1966; Ó Searcóid and West 1989), but the methods adopted in this notes are inspired by the paper of Mbekhta and Ouahab (1994). A different approach to this result may be found in Section 3.7 of Laursen and Neumann (2000). The section on the quasi-nilpotent part of an operator is modelled after Mbekhta (1990).

The concept of algebraic core of an operator has been introduced by Saphar (1964), while the analytical core has been introduced by Vrbová (1973) and, subsequently, studied by Mbekhta (1990), Mbekhta and Ouahab (1994). Lemma 26.1.9 and Proposition 26.1.10 are taken from Aiena and Monsalve (2000). Proposition 26.1.12 is due to Schmoeger (1990), while the subsequent part, except Lemma 26.2.4 due to Kato (1958), can be found in Mbekhta (1990). Theorem 26.2.11 is due to Schmoeger (1990), who also showed that a spectral mapping theorem holds for $\sigma_K(T)$.

27

The single-valued extension property

The basic role of the single-valued extension property (SVEP) arises in the spectral decomposition theory, since every decomposable operator T enjoys this property, as does its dual T'. Indeed, in part IV, Chapter 21 it has been shown that the decomposability of an operator may be viewed as the union of two properties, the so-called *Bishop's property* (β) and the *property* (δ). Property (β) for T implies the SVEP for T (see part IV, Chapter 21) and, as observed in part IV, Chapter 23, properties (β) and (δ) have a complete duality, so that, if T has (δ), then the dual T' has (β) and therefore SVEP.

The main goal of this chapter is to investigate in detail a localized version of SVEP. First we shall show that local spectral theory provides a suitable frame for some characterizations of the analytical core and of the quasi-nilpotent part. Then we shall use these characterizations to describe the localized SVEP by means of a variety of conditions that involve the analytical core and the quasi-nilpotent part of an operator, as well as the generalized range and the generalized kernel.

27.1 The SVEP at a point

To explain the role of SVEP in local spectral theory we begin with some preliminary and well-known facts from operator theory.

The resolvent function $R(\lambda, T) := (\lambda I - T)^{-1}$ of $T \in \mathcal{B}(X)$, where X is a complex Banach space, is an analytic operator-valued function defined on the resolvent set $\rho(T)$. Setting $f_x(\lambda) := R(\lambda, T)x$ for each $x \in X$, we obtain a vector-valued analytic function $f_x : \rho(T) \to X$ which satisfies

$$(\lambda I - T)f_x(\lambda) = x \quad (\lambda \in \rho(T)). \tag{27.1.1}$$

However, it is possible to find analytic solutions to the equation

$(\lambda I - T)f_x(\lambda) = x$ for some (sometimes even for all) values of λ that are in the spectrum of T, as the following example shows.

Let $T \in \mathcal{B}(X)$ be a bounded operator on a Banach space X such that the spectrum $\sigma(T)$ has a non-empty spectral subset $\sigma \neq \sigma(T)$, that is, σ and $\sigma(T) \setminus \sigma$ are closed. From the functional calculus, if $P := P(\sigma, T)$ denotes the spectral projection of T associated with σ, we know that $P(X)$ is a closed T-invariant subspace and $\sigma(T \mid P(X)) = \sigma$, so the restriction $(\lambda I - T) \mid P(X)$ is invertible for all $\lambda \notin \sigma$ (see part I, Theorem 4.3.1). Let $x \in P(X)$. Then the equation (27.1.1) has the analytic solution $f_x(\lambda) := ((\lambda I - T) \mid P(X))^{-1}x$ for all $\lambda \in \mathbb{C} \setminus \sigma$.

These considerations lead in a natural way to the concepts of *local resolvent* $\rho_T(x)$ at x of a bounded operator T on a Banach space X. This is defined as the set of all $\lambda \in \mathbb{C}$ for which there is an open disc $\mathbb{D}(\lambda, \varepsilon)$ and an analytic function $f : \mathbb{D}(\lambda, \varepsilon) \to X$ such that the equation

$$(\mu I - T)f(\mu) = x \quad (\mu \in \mathbb{D}(\lambda, \varepsilon))$$

holds. See Part IV, Definition 22.1.1. The *local spectrum* $\sigma_T(x)$ at x is defined $\sigma_T(x) := \mathbb{C} \setminus \rho_T(x)$. Recall that, if $\lambda \in \rho_T(x)$ and $f : \mathbb{D}(\lambda, \varepsilon) \to X$ is an analytic function on $\mathbb{D}(\lambda, \varepsilon)$ which satisfies the equation $(\mu I - T)f(\mu) = x$ $(\mu \in \mathbb{D}(\lambda, \varepsilon))$, then $\sigma_T(f(\mu)) = \sigma_T(x)$ for all $\mu \in \mathbb{D}(\lambda)$; for this, see Part IV, Lemma 22.1.4.

The single-valued extension property has been defined in Part IV, Chapter 21. In this chapter we shall consider a localized version of this property.

Let X be a complex Banach space, and let $T \in \mathcal{B}(X)$. The operator T has *the single-valued extension property at* $\lambda_0 \in \mathbb{C}$, abbreviated 'T has SVEP at λ_0', if, for every open disc $\mathbb{D}(\lambda_0)$ centred at λ_0, the only analytic function $f : \mathbb{D}(\lambda_0) \to X$ for which the equation $(\lambda I - T)f(\lambda) = 0$ holds is the constant function $f \equiv 0$. Obviously, if T has SVEP for every $\lambda \in \mathbb{C}$, then T has SVEP.

Remark 27.1.1

(i) If $x \in X$ and T has SVEP at $\lambda_0 \in \rho_T(x)$, then there exist an open disc $\mathbb{D}(\lambda_0, \varepsilon)$ and a *unique* analytic function $f : \mathbb{D}(\lambda_0) \to X$ satisfying the equation $(\lambda I - T)f(\lambda) = x$ for all $\lambda \in \mathbb{D}(\lambda_0, \varepsilon)$. Consequently, SVEP implies the existence of a maximal analytic extension $\widetilde{f}$ of

$$R(\lambda, T)x := (\lambda I - T)^{-1}x$$

to the set $\rho_T(x)$ for every $x \in X$. It is evident that this function identically satisfies the equation

$$(\lambda I - T)\widetilde{f}(\lambda) = x \quad (\lambda \in \rho_T(x)),$$

and that $\widetilde{f}(\lambda) = (\lambda I - T)^{-1}x$ for every $\lambda \in \rho(T)$.

(ii) Trivially, a bounded operator $T \in \mathcal{B}(X)$ has SVEP at every point of the resolvent $\rho(T)$. Moreover, from the identity theorem for analytic functions, it easily follows that $T \in \mathcal{B}(X)$ has SVEP at every point of the boundary $\partial\sigma(T)$ of the spectrum. In particular, every operator having discrete spectrum has SVEP.

27.2 Local analytic subspace

For every subset Ω of $\mathbb{C}$, let $X_T(\Omega)$ denote the *local analytical subspace* of T associated with Ω, as in Definition 22.1.1, so that

$$X_T(\Omega) := \{x \in X : \sigma_T(x) \subseteq \Omega\}.$$

The next result gives a precise description of the analytical core $K(T)$ by means of the local spectrum $\sigma_T(x)$.

Theorem 27.2.1 *Let $T \in \mathcal{B}(X)$ be a bounded operator on a Banach space X. Then*

$$K(T) = X_T(\mathbb{C} \setminus \{0\}) = \{x \in X : 0 \notin \sigma_T(x)\}.$$

Proof Let $x \in K(T)$ and, according to the definition of $K(T)$, let $\delta > 0$ and $(u_n) \subset X$ a sequence for which $x = u_0$, $Tu_{n+1} = u_n$, $\|u_n\| \leq \delta^n \|x\|$ for every $n \in \mathbb{Z}^+$. The function $f : \mathbb{D}(0, 1/\delta) \to X$, where $\mathbb{D}(0, 1/\delta)$ is the open disc centred at 0 and with radius $1/\delta$, defined by $f(\lambda) := -\sum_{n=1}^{\infty} \lambda^{n-1} u_n$, is an analytic vector-valued function which satisfies the equation $(\lambda I - T)f(\lambda) = x$ for every $\lambda \in \mathbb{D}(0, 1/\delta)$. This means that $0 \in \rho_T(x)$, and hence $\sigma_T(x) \subseteq \mathbb{C} \setminus \{0\}$.

Conversely, if $\sigma_T(x) \subseteq \mathbb{C} \setminus \{0\}$, then $0 \in \rho_T(x)$, and hence there is an open disc $\mathbb{D}(0, \varepsilon)$ and an analytic function $f : \mathbb{D}(0, \varepsilon) \to X$ such that

$$(\lambda I - T)f(\lambda) = x \quad (\lambda \in \mathbb{D}(0, \varepsilon)). \tag{27.2.1}$$

Since f is analytic on $\mathbb{D}(0, \varepsilon)$, there exists a sequence $(u_n) \subset X$ such that

$$f(\lambda) := -\sum_{n=1}^{\infty} \lambda^{n-1} u_n \quad (\lambda \in \mathbb{D}(0, \varepsilon)). \tag{27.2.2}$$

Obviously, $f(0) = -u_1$ and, taking $\lambda = 0$ in (27.2.1), we obtain the equation $Tu_1 = -T(f(0)) = x$.

On the other hand, we have

$$x = (\lambda I - T)f(\lambda) = Tu_1 + \lambda(Tu_2 - u_1) + \lambda^2(Tu_3 - u_2) + \cdots \quad (\lambda \in \mathbb{D}(0, \varepsilon)).$$

From $x = Tu_1$ we obtain $Tu_{n+1} = u_n$ for $n \in \mathbb{N}$. Letting $u_0 = x$, the sequence (u_n) satisfies the condition (1) of the definition of $K(T)$, so it remains to prove the condition $\|u - n\| \leq \delta^n \|x\|$ for a suitable $\delta > 0$ and all $n \in \mathbb{Z}^+$. Take $\mu > 1/\varepsilon$. Then there exists $c > 0$ such that

$$\|u_n\| \leq c\,\mu^{n-1} \quad (n \in \mathbb{N}). \tag{27.2.3}$$

From (27.2.1) and (27.2.2) we easily obtain $Tu_1 = u_0$ and $Tu_n = u_{n-1}$ for every $n = 2, 3, \ldots$. Obviously, if $x = 0$, then $x \in K(T)$. Suppose that $x \neq 0$. From the estimates (27.2.3), it then follows that $\|u_n\| \leq \|x\|(\mu + c/\|x\|)^n$, and hence $x \in K(T)$. □

A certain variant of the local spectral subspaces which is better suited for operators without SVEP is given, for every closed subset Ω of $\mathbb{C}$, by the subspace $\mathcal{X}_T(\Omega)$ of all $x \in X$ such that there is an analytic function $f : \mathbb{C} \setminus \Omega \to X$ such that $(\lambda I - T)f(\lambda) = x$ for all $\lambda \in \mathbb{C} \setminus \Omega$. Clearly, $\mathcal{X}_T(\Omega) \subseteq X_T(\Omega)$ for every closed subset $\Omega \subseteq \mathbb{C}$. The set $\mathcal{X}_T(\Omega)$ is called the *glocal spectral subspace* of T associated with Ω in Definition 21.2.3.

Note that $X_T(\Omega) = \mathcal{X}_T(\Omega)$ for every closed $\Omega \subseteq \mathbb{C}$ if and only if T has SVEP, and this happens if and only if, for every $x \neq 0$, the local spectrum $\sigma_T(x)$ is non-empty, that is, $X_T(\emptyset) = \{0\}$. See Proposition 22.1.2.

Theorem 27.2.2 *For every operator $T \in \mathcal{B}(X)$ on a Banach space X and every $\varepsilon \geq 0$, we have*

$$\mathcal{X}_T(\overline{\mathbb{D}(0, \varepsilon)}) = \{x \in X : \limsup_{n\to\infty} \|T^n x\|^{1/n} \leq \varepsilon\}. \tag{27.2.4}$$

In particular, $H_0(T) = \mathcal{X}_T(\{0\})$ and, if T has SVEP, then $H_0(T) = X_T(\{0\})$.

Proof Let $x \in X$ such that $\delta := \limsup_{n\to\infty} \|T^n x\|^{1/n} \leq \varepsilon$. The series

$$f(\lambda) := \sum_{n=1}^{\infty} \lambda^{-n} T^{n-1} x \quad (\lambda \in \mathbb{C} \setminus \overline{\mathbb{D}(0, \varepsilon)})$$

converges locally uniformly, so it defines an X-valued function on the open set $\mathbb{C} \setminus \overline{\mathbb{D}(0, \varepsilon)}$. Evidently, $(\lambda I - T)f(\lambda) = x$ for all $\lambda \in \mathbb{C} \setminus \overline{\mathbb{D}(0, \varepsilon)}$. Thus we see that $x \in \mathcal{X}_T\big(\overline{\mathbb{D}(0, \varepsilon)}\big)$.

Conversely, let us assume that $x \in \mathcal{X}_T(\overline{\mathbb{D}(0,\varepsilon)})$, and let us consider an analytic function $f : \mathbb{C} \setminus \overline{\mathbb{D}(0,\varepsilon)} \to X$ such that $(\lambda I - T)f(\lambda) = x$ holds for all $\lambda \in \mathbb{C} \setminus \overline{\mathbb{D}(0,\varepsilon)}$. For $|\lambda| > \max\{\varepsilon, \|T\|\}$, we obtain

$$f(\lambda) = (\lambda I - T)^{-1}x = \sum_{n}^{\infty} \lambda^{-n} T^{n-1} x,$$

and therefore $f(\lambda) \to 0$ as $|\lambda| \to \infty$. The analytic function $g : \mathbb{D}(0, 1/\varepsilon) \to X$ defined by

$$g(\mu) := \begin{cases} f(1/\lambda) & \text{if } 0 \neq \mu \in \mathbb{D}(0, 1/\varepsilon), \\ 0 & \text{if } \mu = 0, \end{cases}$$

satisfies the equality

$$g(\mu) = \sum_{n=1}^{\infty} \mu^n T^{n-1} x \quad (|\mu| < 1/\max\{\varepsilon, \|T\|\}). \tag{27.2.5}$$

Since g is analytic on $\mathbb{D}(0, 1/\varepsilon)$, it follows, exactly as in the scalar setting from Cauchy's integral formula, that the equality (27.2.5) holds even for all $\mu \in \mathbb{D}(0, 1/\varepsilon)$. This shows that the radius of convergence of the power series representing $g(\mu)$ is greater than $1/\varepsilon$. The standard formula for the radius of convergence of a vector-valued power series then implies that $\delta < \varepsilon$. Therefore the equality (27.2.4) holds.

The equality $H_0(T) = \mathcal{X}_T(\{0\})$ is clear, taking $\varepsilon = 0$ in (27.2.4). Finally, if T has SVEP, then $X_T(\Omega) = \mathcal{X}_T(\Omega)$ for every closed $\Omega \subseteq \mathbb{C}$, in particular $\mathcal{X}_T(\{0\}) = X_T(\{0\})$. □

We know that T does not have SVEP if and only if there exists an element $0 \neq x \in X$ such that $\sigma_T(x) = \emptyset$. The next result shows a localized version of this fact.

Theorem 27.2.3 *Suppose that $T \in \mathcal{B}(X)$, where X is a Banach space. Then the following conditions are equivalent:*

(i) *T has SVEP at λ_0;*
(ii) $\ker(\lambda_0 I - T) \cap X_T(\emptyset) = \{0\}$*;*
(iii) $\ker(\lambda_0 I - T) \cap K(\lambda_0 I - T) = \{0\}$.

Proof (i) $\Leftrightarrow$ (ii) We can suppose that $\lambda_0 = 0$. Assume that, for some $x \in \ker T$, we have $\sigma_T(x) = \emptyset$. Then $0 \in \rho_T(x)$, so there is an open disc $\mathbb{D}(0,\varepsilon)$ and an analytic function $f : \mathbb{D}(0,\varepsilon) \to X$ such that $(\lambda I - T)f(\lambda) = x \;\; (\lambda \in \mathbb{D}(0,\varepsilon))$.

Then

$$T((\lambda I - T)f(\lambda)) = (\lambda I - T)T(f(\lambda)) = Tx = 0 \quad (\lambda \in \mathbb{D}(0, \varepsilon)).$$

Since T has SVEP at 0, then $Tf(\lambda) = 0$, and therefore $T(f(0)) = x = 0$.

Conversely, suppose that, for every $0 \neq x \in \ker T$, we have $\sigma_T(x) \neq \emptyset$. There is an analytic function $f : \mathbb{D}(0, \varepsilon) \to X$ such that

$$(\lambda I - T)f(\lambda) = 0 \ (\lambda \in \mathbb{D}(0, \varepsilon)).$$

Then $f(\lambda) = \sum_{n=0}^{\infty} \lambda^n u_n$ for a suitable sequence $(u_n) \subset X$. Clearly,

$$Tu_0 = T(f(0)) = 0,$$

so that $u_0 \in \ker T$. Moreover, from the equalities

$$\sigma_T(f(\lambda)) = \sigma_T(0) = \emptyset \ (\lambda \in \mathbb{D}(0, \varepsilon)),$$

we obtain $\sigma_T(f(0)) = \sigma_T(u_0) = \emptyset$, and therefore, from the assumption, we conclude that $u_0 = 0$. For all $0 \neq \lambda \in \mathbb{D}(0, \varepsilon)$, we have

$$0 = (\lambda I - T)f(\lambda) = (\lambda I - T)\sum_{n=1}^{\infty} \lambda^n u_n = \lambda(\lambda I - T)\sum_{n=1}^{\infty} \lambda^n u_{n+1},$$

and therefore $0 = (\lambda I - T)(\sum_{n=0}^{\infty} \lambda^n u_{n+1})$ for every $0 \neq \lambda \in \mathbb{D}(0, \varepsilon)$. By continuity this is still true for every $\lambda \in \mathbb{D}(0, \varepsilon)$. At this point, by using the same argument as in the first part of the proof, it is possible to show that $u_1 = 0$ and, by iterating this procedure, we conclude that $u_2 = u_3 = \cdots = 0$. This shows that $f \equiv 0$ on $\mathbb{D}(0, \varepsilon)$, and therefore T has SVEP at 0.

To show the equivalence (ii) $\Leftrightarrow$ (iii), it suffices to prove the equality

$$\ker T \cap K(T) = \ker T \cap X_T(\emptyset).$$

From Theorem 27.2.1 we have

$$\ker T \cap K(T) = \ker T \cap X_T(\mathbb{C} \setminus \{0\}) \subseteq X_T(\{0\}) \cap X_T(\mathbb{C} \setminus \{0\}) = X_T(\emptyset),$$

so that $\ker T \cap K(T) = \ker T \cap K(T) \cap X_T(\emptyset) = \ker T \cap X_T(\emptyset)$, as required. □

Let $\sigma_p(T)$ denote the *point spectrum* of $T \in \mathcal{B}(X)$, that is ,

$$\sigma_p(T) := \{\lambda \in \mathbb{C} : \lambda \text{ is an eigenvalue of } T\}.$$

Corollary 27.2.4 *Let $T \in \mathcal{B}(X)$, where X is a Banach space. Then T does not have SVEP if and only if there exists $\lambda_0 \in \sigma_p(T)$ and a corresponding eigenvector x_0 ($\neq 0$) such that $\sigma_T(x_0) = \emptyset$. In such a case, T does not have SVEP at λ_0.* □

It is clear from the definition of SVEP, that, if the set of eigenvalues of operator T has empty interior, then T has this property. Consequently, all operators $T \in \mathcal{B}(X)$ with real spectrum have SVEP. But not every bounded operator enjoys SVEP. In fact, from Theorem 27.2.3, T has SVEP at 0 precisely when $\ker T \cap K(X) = \{0\}$, so that every non-injective, surjective operator does not have SVEP at 0.

Corollary 27.2.5 *Let $T \in \mathcal{B}(X)$, where X is a Banach space, and let $\lambda_0 \in \mathbb{C}$.*

(i) *If either*

$$K(\lambda_0 I - T) \cap H_0(\lambda_0 I - T) = \{0\} \quad \text{or}$$

$$\mathcal{N}^\infty(\lambda_0 I - T) \cap (\lambda_0 I - T)^\infty(X) = \{0\},$$

then T has SVEP at λ_0.

(ii) *T has SVEP $\Leftrightarrow K(\lambda I - T) \cap H_0(\lambda I - T) = \{0\}$ for every $\lambda \in \mathbb{C}$.*

Proof Part (i) is an obvious consequence of Theorem 27.2.3, since both conditions imply that $\ker(\lambda_0 I - T) \cap K(\lambda_0 I - T) = \{0\}$.

Suppose that T has SVEP and $x \in K(\lambda I - T) \cap H_0(\lambda I - T)$. By Theorem 27.2.1, we have $\sigma_{\lambda I - T}(x) \subseteq \{0\}$ and hence $\sigma_T(x) \subseteq \{\lambda\}$. On the other hand, we also have, by Theorem 27.2.2, $0 \notin \sigma_{\lambda I - T}(x)$, and hence $\lambda \notin \sigma_T(x)$. Consequently, $\sigma_T(x)$ is empty and therefore, since T has SVEP, $x = 0$. The reverse implication is clear from (i). □

Example 27.2.6 This example, based on the theory of weighted shifts, shows that SVEP at a point λ_0 does not necessarily imply that

$$H_0(\lambda_0 I - T) \cap K(\lambda_0 I - T) = \{0\}.$$

Let $\beta := (\beta_n)_{n \in \mathbb{Z}}$ be the sequence of real numbers defined by

$$\beta_n := \begin{cases} 1 + |n| & \text{if } n < 0, \\ e^{-n^2} & \text{if } n \geq 0. \end{cases}$$

Let $X := L_2(\beta)$ denote the Hilbert space of all formal Laurent series $\sum_{n=-\infty}^{\infty} a_n z^n$ for which $\sum_{n=-\infty}^{\infty} |\alpha_n|^2 \beta_n{}^2 < \infty$, endowed with the canonical norm $\|\cdot\|_\beta$. Let us consider the bilateral weighted right shift T defined by

$$T\left(\sum_{n=-\infty}^{\infty} a_n z^n\right) := \sum_{n=-\infty}^{\infty} a_n z^{n+1},$$

or equivalently, $Tz^n = z^{n+1}$ $(n \in \mathbb{Z})$. The operator T is bounded on $L_2(\beta)$ and

$$\|T\| = \sup\{\beta_{n+1}/\beta_n : n \in \mathbb{Z}\} = 1.$$

Clearly T is injective, and thus has SVEP at 0. The following argument shows that $H_0(T) \cap K(T) \neq \{0\}$. Since $\|z^n\|_\beta = \beta_n$ $(n \in \mathbb{Z})$, $\lim_{n\to\infty} \|z^{n-1}\|_\beta{}^{1/n} = 0$ and $\lim_{n\to\infty} \|z^{-n-1}\|_\beta{}^{1/n} = 1$. By the formula for the radius of convergence of a power series, we then conclude that the two series $f(\lambda) := \sum_{n=1}^{\infty} \lambda^{-n} z^{n-1}$ and $g(\lambda) := -\sum_{n=1}^{\infty} \lambda^n z^{-n-1}$ converge in $L_2(\beta)$ for all $|\lambda| > 0$ and $|\lambda| < 1$, respectively. Clearly, f is analytic on $\mathbb{C} \setminus \{0\}$, and

$$(\lambda I - T)f(\lambda) = -\sum_{n=1}^{\infty} \lambda^{-n} z^n - \sum_{n=1}^{\infty} \lambda^{1-n} z^{n-1} = 1 \quad (\lambda \in \mathbb{C} \setminus \{0\}),$$

while g is analytic on the open unit disc $\mathbb{D}$ and satisfies

$$(\lambda I - T)g(\lambda) = \sum_{n=0}^{\infty} \lambda^n z^{-n} - \sum_{n=0}^{\infty} \lambda^{1+n} z^{-n-1} = 1 \quad (\lambda \in \mathbb{D}).$$

This means that $1 \in \mathcal{X}_T(\{0\}) \cap \mathcal{X}_T(\mathbb{C} \setminus \mathbb{D}) = H_0(T) \cap K(T)$. □

An immediate consequence of part (ii) of Corollary 27.2.5 is that, for a quasi-nilpotent operator T, we have $K(T) = \{0\}$. In fact, T has SVEP and $H_0(T) = X$, so that

$$H_0(T) \cap K(T) = K(T) = \{0\}.$$

We shall use this fact in the proof of the following result.

Proposition 27.2.7 *Suppose that a bounded operator $T \in \mathcal{B}(X)$ on a Banach space X either has a closed quasi-nilpotent part $H(\lambda_0 I - T)$ or is such that that $H_0(\lambda_0 - T) \cap K(\lambda_0 I - T)$ is closed. Then*

$$H_0(\lambda_0 - T) \cap K(\lambda_0 I - T) = \{0\},$$

and hence T has SVEP at λ_0.

Proof Without loss of generality, we may consider $\lambda_0 = 0$.

Assume first that $H_0(T)$ is closed. Let $\widetilde{T}$ denote the restriction of T to the Banach space $H_0(T)$. It is easily seen that $H_0(T) \cap K(T) = K(\widetilde{T})$. On the other hand, we have $H_0(T) = H_0(\widetilde{T})$. Thus $\widetilde{T}$ is quasi-nilpotent, and hence $K(\widetilde{T}) = \{0\}$.

Suppose now that $Y := H_0(T) \cap K(T)$ is closed. Clearly, Y is invariant under T, so we can consider the restriction $S := T \mid Y$. If $y \in Y$, then

$$\|S^n y\|^{1/n} = \|T^n y\|^{1/n} \to 0 \quad \text{as } n \to \infty.$$

Thus $y \in H_0(S)$, and hence $H_0(S) = Y$. This shows that S is quasi-nilpotent, and hence has SVEP. This implies that $\sigma_S(y) = \{0\}$ for all $0 \neq y \in Y$.

On the other hand, we also have $K(S) = Y$. Indeed, take an element $y \in Y$. Then there is a sequence $(y_n) \subset X$ and a $\delta > 0$ such that

$$y_0 = y, \; Ty_n = y_{n-1} \quad \text{and} \quad \|y_n\| \leq \delta^n \|y\|$$

for all $n \in \mathbb{Z}^+$. From $y \in Y = H_0(T)$, we obtain that $y_n \in H_0(T)$ for each $n \in \mathbb{N}$. Moreover, since $y \in K(T) = X_T(\{0\})$, we easily obtain that $y_n \in K(T)$ $(n \in \mathbb{N})$. Thus $y_n \in Y$, and therefore $y \in K(S)$. This shows that $Y \subseteq K(S)$. The opposite inclusion is clear since $K(S) = K(T) \cap Y \subseteq Y$. Hence $Y = K(S)$.

Finally, from Theorem 27.2.1 we have

$$H_0(T) \cap K(T) = Y = X_S(\{0\}) = \{0\}. \qquad \square$$

Proposition 27.2.8 *Suppose that the sum $H_0(\lambda_0 I - T) + (\lambda_0 I - T)(X)$ is norm-dense in X. Then T' has SVEP at λ_0.*

Proof Also here we suppose that $\lambda_0 = 0$. From Proposition 26.2.9, we have the inclusion $K(T') \subseteq H_0(T)^\perp$. Now a standard duality argument shows that

$$\ker(T') \cap K(T') \subseteq T(X)^\perp \cap H_0(T)^\perp = (T(X) \cap H_0(T))^\perp .$$

Finally, if $H(T) + T(X)$ is norm-dense in X, then the last annihilator is zero. Thus $\ker T' \cap K(T') = \{0\}$, and hence, by Theorem 27.2.3, T' has SVEP at 0.

$\square$

Corollary 27.2.9 *Suppose for $T \in \mathcal{B}(X)$, where X is a Banach space, either that*

$$H_0(\lambda_0 I - T) + K(\lambda_0 I - T) \quad \text{or} \quad \mathcal{N}^\infty(\lambda_0 I - T) + (\lambda_0 I - T)^\infty(X)$$

is norm-dense in X. Then T' has SVEP at λ_0. $\square$

Let $T \in \mathcal{B}(X)$, where X is a Banach space, and let f be an analytic function on the open neighbourhood U of $\sigma(T)$. Let $f(T) \in \mathcal{B}(X)$ be the operator defined by the classical Riesz functional calculus, as explained in Part I, Chapter 4]. We have the following remarks.

(i) If the operator T has SVEP, then $f(T)$ also has SVEP: see Theorem 3.3.6 of Laursen and Neumann (2000). If f is non-constant on each of the connected components of U, then T has SVEP if and only if $f(T)$ does: see Theorem 3.3.9 of Laursen and Neumann (2000).

(ii) We have $f(\sigma_T(x)) \subseteq \sigma_{f(T)}(x)$ $(x \in X)$. If T has SVEP or if the function f is non-constant on each of the connected components of U, then $f(\sigma_T(x)) = \sigma_{f(T)}(x)$ $(x \in X)$: see Theorem 3.3.8 of Laursen and Neumann (2000).

For an arbitrary operator $T \in \mathcal{B}(X)$ on a Banach space X, let

$$\Xi(T) := \{\lambda \in \mathbb{C} : T \text{ does not have SVEP at } \lambda\} .$$

From the identity theorem for analytic functions it readily follows that $\Xi(T)$ is open and, consequently, is contained in the interior of the spectrum $\sigma(T)$. Clearly, $\Xi(T)$ is empty precisely when T has SVEP.

The next result shows that $\Xi(T)$ behaves canonically under the Riesz functional calculus.

Theorem 27.2.10 *Let $T \in \mathcal{B}(X)$, where X is a Banach space. Let $f : U \to \mathbb{C}$ be an analytic function on an open neighbourhood U of $\sigma(T)$, and suppose that f is non-constant on each of the connected components of U. Then $f(T)$ has SVEP at $\lambda \in \mathbb{C}$ if and only if T has SVEP at every point $\mu \in \sigma(T)$ for which $f(\mu) = \lambda$. Moreover, $f(\Xi(T)) = \Xi((f(T))$.*

Proof Suppose first that $f(T)$ has SVEP at $\lambda \in \mathbb{C}$. Let $\mu \in \sigma(T)$ be such that $f(\mu) = \lambda$. In order to show that T has SVEP at μ, it suffices, by Theorem 27.2.3, to show that

$$\ker(\mu I - T) \cap X_{\mu I - T}(\emptyset) = \{0\} .$$

Let $x \in \ker(\mu I - T) \cap X_{\mu I - T}(\emptyset)$ be given. Since $f(\mu) = \lambda$, there exists an analytic function g on U such that $\lambda - f = (\mu - Z)g$. The Riesz functional calculus preserves multiplication, and so

$$\lambda I - f(T) = (\mu I - T)g(T) ,$$

and therefore $x \in \ker(\lambda I - f(T))$. Moreover, from $\sigma_{\mu I - T}(x) = \emptyset$, we obtain $\sigma_T(x) = \emptyset$, and hence $\sigma_{f(T)}(x) = f(\sigma_T(x)) = \emptyset$. Consequently,

$$\ker(\mu I - T) \cap X_{\mu I - T}(\emptyset) \subseteq \ker(\lambda I - f(T)) \cap X_{\lambda I - f(T)}(\emptyset) ,$$

and hence, again by Theorem 27.2.3, T has SVEP at μ.

Conversely, let $\lambda \in \mathbb{C}$ and suppose that T has SVEP at every $\mu \in \sigma(T)$ for which $f(\mu) = \lambda$. From the classical spectral mapping theorem, $f(\sigma(T)) = \sigma(f(T))$, and so $\lambda \in \sigma(f(T))$. By assumption, f is non-constant on each of the connected components of U, and thus the function $f - \lambda$ has only finitely many zeros in $\sigma(T)$, and these zeros are of finite multiplicity. Hence there exist an analytic function g on U without zeros in $\sigma(T)$ and a polynomial p, of the form $p(z) = (z - \mu_1) \cdots (z - \mu_n)$, with not necessarily distinct elements $\mu_1, \dots, \mu_n \in \sigma(T)$ such that the factorization $f - \lambda = pg$ holds.

Now, assume that $x \in \ker(\lambda I - f(T)) \cap X_{\lambda I - f(T)}(\emptyset)$. In order to prove that $f(T)$ has SVEP at λ, it suffices, again by Theorem 27.2.3, to show that $x = 0$. From the classical spectral mapping theorem, we know that $g(T)$ is invertible. Thus the equality $f(T) - \lambda I = g(T) f(T)$ implies that $p(T)x = 0$, and hence $(\mu_1 I - T)y = 0$, where $y := q(T)x$ and $q(z) := (z - \mu_2) \cdots (z - \mu_n)$. On the other hand, $x \in X_{\lambda I - f(T)}(\emptyset)$ and f is non-constant on each of the connected components of U. By Remark (ii) after Corollary 27.2.9, we have $f(\sigma_T(x)) = \sigma_{f(T)}(x) = \emptyset$, and therefore, since T and $q(T)$ commute,

$$\sigma_T(y) = \sigma_T(q(T)x) \subseteq \sigma_T(x) = \emptyset .$$

But T has SVEP at μ_1, by assumption, so, again by Theorem 27.2.3, $y = 0$. Repetitions of this argument for $\mu_2, \dots, \mu_n$ then lead to the equality $x = 0$. Thus $f(T)$ has SVEP at λ.

The last claim is obvious, being nothing else than a reformulation of the equivalence proved above. □

Note that in the preceding result the additional condition on the analytic function is essential. In fact, if f is constant in U, then $\Xi((f(T))$ is certainly empty, while $f(\Xi(T))$ is empty only when T has SVEP.

An immediate consequence of Theorem 27.2.10 is that, in the characterization of SVEP at a point $\lambda_0 \in \mathbb{C}$ given in Theorem 27.2.3, the kernel $\ker(\lambda_0 I - T)$ may be replaced by the generalized kernel $\mathcal{N}^\infty(\lambda_0 I - T)$.

Corollary 27.2.11 *For every bounded operator T on a Banach space X the following properties are equivalent:*

(i) *T has SVEP at λ_0;*
(ii) *T^n has SVEP at λ_0 for each $n \in \mathbb{N}$;*
(iii) *$\mathcal{N}^\infty(\lambda_0 I - T) \cap X_T(\emptyset) = \{0\}$;*
(iv) *$\mathcal{N}^\infty(\lambda_0 I - T) \cap K(\lambda_0 I - T) = \{0\}$.*

Proof The equivalence (i) $\Leftrightarrow$ (ii) is obvious from Theorem 27.2.10. Combining this equivalence with Theorem 27.2.3, we then obtain that T has SVEP at λ_0 if and only if $\ker(\lambda_0 I - T)^n \cap X_T(\emptyset) = \{0\}$ for every $n \in \mathbb{N}$. This argument shows the equivalence (i) $\Leftrightarrow$ (iii). The equivalence (i) $\Leftrightarrow$ (iv) follows from Theorem 27.2.3 in a similar way. □

27.3 Exercises

1. Show that T has SVEP at λ_0 if and only if, for every $0 \neq x \in \ker(\lambda_0 I - T)$, we have $\sigma_T(x) = \{\lambda_0\}$.
2. Show that a semi-regular operator T has SVEP at 0 if and only if T is injective. Analogously, T' has SVEP at 0 if and only if T is surjective.
3. Show that if, for an operator $T \in \mathcal{B}(X)$, one of the sums

$$H_0(\lambda_0 I' - T') + (\lambda_0 I' - T')(X'), \quad H_0(\lambda_0 I' - T') + K(\lambda_0 I' - T'),$$

or

$$\mathcal{N}^\infty(\lambda_0 I' - T') + (\lambda_0 I' - T')^\infty(X')$$

is weak-$*$ dense in X', then T has SVEP at λ_0.
4. Let $X := \ell^2(\mathbb{N}) \oplus \ell^2(\mathbb{N}) \oplus \cdots$, provided with the norm $\|\cdot\|$ specified by $\|x\| := (\sum_{n=1}^\infty \|x_n\|^2)^{1/2}$ for $x := (x_n) \in X$, and define

$$T_n e_i := \begin{cases} e_{i+1} & \text{if } i = 1, \ldots, n, \\ e_{i+1}/(i-n) & \text{if } i > n. \end{cases}$$

Let $T := T_1 \oplus \cdots \oplus T_n \oplus \cdots$. Show that T has SVEP, and that the quasi-nilpotent part $H_0(T)$ is not closed. *Hint*: check that T^n is quasi-nilpotent, and find an element $x := (x_n) \in X$ which shows that $H_0(T) \neq X$. Show that $H_0(T)$ is norm-dense in X.
5. Find an example of operator for which $\mathcal{N}^\infty(T) + T^\infty(X)$ is not norm-dense in X, while T' has SVEP. *Hint*: consider the Volterra operator defined in Part IV, Example 22.2.6 and Part I, Exercise 1.5.6.
6. Let Ω be a component of $\rho_K(T)$. Show the following alternative: either T has SVEP at every $\lambda \in \Omega$ or T does not have SVEP at any $\lambda \in \Omega$. *Hint*: use Theorem 26.2.10.
7. Let K denote a compact subset of $\mathbb{C}$ with a non-empty interior, and let $X := \ell^\infty(K)$ denote the Banach space of all bounded, complex-valued functions on K, as in Chapter 1, and consider the operator $T : X \to X$ defined by the assignment $(Tf)(\lambda) := \lambda f(\lambda)$ for $f \in X$ and $\lambda \in K$. Show that T has SVEP, while $\sigma_p(T) = K$.
8. Let $1 \leq p < \infty$, and denote by $\omega := (\omega_n)$ a bounded sequence of strictly positive real numbers. The corresponding *unilateral weighted right shift* operator on the Banach space $\ell^p(\mathbb{N})$ is the operator defined by:

$$Tx = \sum_{n=1}^\infty \omega_n x_n e_{n+1} \quad (x = (x_n) \in \ell^p(\mathbb{N})).$$

Check that T has SVEP. Moreover, show that $H_0(T) + T(X)$ is norm-dense in $\ell^p(\mathbb{N})$ if and only if $\lim_{n\to\infty} \sup(\omega_1 \cdots \omega_n)^{1/n} = 0$. Show that T' has SVEP at 0 precisely when $\lim_{n\to\infty} \inf(\omega_1 \cdots \omega_n)^{1/n} = 0$.

9. Let $T \in \mathcal{B}(X)$, where X is a Banach space. Given an element $x \in X$, the quantity $\nu_T(x) := \limsup_{n\to\infty} \|T^n x\|^{1/n}$ is called the *local spectral radius* of T at x. Show that, if T has SVEP, then $\nu_T(x) = \max\{|\lambda| : \lambda \in \sigma_T(x)\}$. Further information on this local spectral radius may be found in Laursen and Neumann (2000).

10. Show that, if T is an isometric non-unitary operator on a Hilbert space H, then the adjoint T' does not have the SVEP at 0. See Example 1.7 of Colojoară and Foiaş (1968).

27.4 Additional notes

The single-valued extension property appeared first in Dunford (1952, 1954), and then received a systematic treatment in Dunford and Schwartz (1958, 1963, 1971). This property also forms a preliminary topic of the book by Colojoară and C. Foiaş (1968). The fundamental role of SVEP in local spectral theory is more clearly explained in the recent monograph Laursen and Neumann (2000).

The characterizations of the analytical core and the quasi-nilpotent part of an operator given in Theorem 27.2.1 and Theorem 27.2.2 are due to Vrbová (1973) and Mbekhta (1987, 1990); see also Proposition 2.2 of Laursen (1992). The localized SVEP at a point has been introduced by Finch (1975), who showed that a surjective operator T has the SVEP at 0 precisely when it is injective. The characterization of SVEP at a single point λ_0 given in Theorem 27.2.3 is taken from Aiena and Monsalve (2000).

Except for Corollary 27.2.5, which is due to Mbekhta (1990), the source of the results of the second section is essentially that of Aiena, Miller and Neumann (2001) and Aiena, Colasante and Gonzalez (2002). In particular, Example 27.2.6 and the local spectral mapping result of Theorem 27.2.10 are taken from Aiena, Miller and Neumann (2001). In this paper the reader may find applications to isometries, analytic Toeplitz operators, invertible composition operators on Hardy spaces, and weighted shifts.

28

SVEP for semi-Fredholm operators

In Chapter 27 we established many conditions which imply SVEP at a point. The main goal of this chapter is to show that all these implications become equivalences for an important class of operators, the class of all semi-Fredholm operators. It will be also shown that, for these operators, SVEP at a point $\lambda_0 \in \mathbb{C}$ is equivalent to the finiteness of two important quantities, the ascent and the descent of the operator $\lambda_0 I - T$. These equivalences also provide useful information on the fine structure of the spectrum. In particular, we shall show that many spectra originating from Fredholm theory coincide whenever T or T' have SVEP.

28.1 Ascent, descent, and semi-Fredholm operators

Let us recall the definition of some classical quantities associated with an operator. Given a linear operator T on a vector space X, it is easy to see that $\ker T^n \subseteq \ker T^{n+1}$ and $T^{n+1}(X) \subseteq T^n(X)$ for every $n \in \mathbb{N}$.

Definition 28.1.1 *The* ascent *of T is the smallest positive integer $p = p(T)$, whenever it exists, such that* $\ker T^p = \ker T^{p+1}$. *If such p does not exist, set $p = \infty$. Analogously, the* descent *of T is defined to be the smallest integer $q = q(T)$, whenever it exists, such that $T^{q+1}(X) = T^q(X)$. If such q does not exist, set $q = \infty$. If both $p(T)$ and $q(T)$ are finite, then T has finite chains.*

Note that $p(T) = 0$ means that T is injective and $q(T) = 0$ means that T is surjective. Clearly, T has finite ascent if and only if $\mathcal{N}^\infty(T) = \ker T^k$ for some $k \in \mathbb{N}$, and, analogously, T has finite descent if and only if $T^\infty(X) = T^k(X)$ for some $k \in \mathbb{N}$.

The relevant fact is that the finiteness of the ascent $p(T)$ may be expressed by means of some intersection properties between kernels and ranges of the iterates of T.

Lemma 28.1.2 *Let T be a linear operator on a vector space X, and let $m \in \mathbb{Z}^+$. Then*

(i) $p(T) \leq m < \infty$ *if and only if, for every $n \in \mathbb{N}$, $T^m(X) \cap \ker T^n = \{0\}$.*
(ii) $q(T) \leq m < \infty$ *if and only if, for every $n \in \mathbb{N}$, there exists a subspace $Y_n \subseteq \ker T^m$ such that $X = Y_n \oplus T^n(X)$.*

Proof (i) Suppose that $p(T) \leq m < \infty$ and that $n \in \mathbb{N}$. Let us consider an element $y \in T^m(X) \cap \ker T^n$. Then there exists $x \in X$ such that $y = T^m x$ and $T^n y = 0$. From this we obtain the equality $T^{m+n}x = T^n y = 0$, and therefore we have $x \in \ker\ T^{m+n} = \ker T^m$. Hence $y = T^m x = 0$.

Conversely, suppose that $T^m(X) \cap \ker T^n = \{0\}$ for some $m \in \mathbb{N}$, and let $x \in \ker T^{m+1}$. Then $T^m x \in \ker T$, and therefore

$$T^m x \in T^m(X) \cap \ker T \subseteq T^m(X) \cap \ker\ T^n = \{0\}.$$

Hence $x \in \ker T^m$. We have shown that $\ker\ T^{m+1} \subseteq \ker T^m$. Since the opposite inclusion is satisfied for all operators, we conclude that $\ker T^m = \ker\ T^{m+1}$.

(ii) Let $q := q(T) \leq m < \infty$, and let Y be a complementary subspace to $T^n(X)$ in X. Let $\{x_j : j \in J\}$ be a basis of Y. For every element x_j of the basis there exists an element $y_j \in X$ such that $T^q x_j = T^{q+n} y_j$. This follows because $T^q(Y) \subseteq T^q(X) = T^{q+n}(X)$. Set $z_j := x_j - T^n y_j$. Then clearly $T^q z_j = T^q x_j - T^{q+n} y_j = 0$. From this it follows that the linear subspace Y_n generated by the elements z_j is contained in $\ker T^q$ and *a fortiori* in $\ker T^m$. From the decomposition $X = Y \oplus T^n(X)$, we obtain for every $x \in X$ a representation of the form

$$x = \sum_{j \in J} \lambda_j x_j + T^n y = \sum_{j \in J} \lambda_j (z_j + T^n y_j) + T^n y = \sum_{j \in J} \lambda_j z_j + T^n z\,,$$

so that $X = Y_n + T^n(X)$. We show that this sum is direct. Indeed, choose an element $x \in Y_n \cap T^n(X)$. Then $x = \sum_{j \in J} \mu_j z_j = T^n v$ for some $v \in X$, and therefore

$$\sum_{j \in J} \mu_j x_j = \sum_{j \in J} \mu_j T^n y_j + T^n v \in T^n(X)\,.$$

From the decomposition $X = Y \oplus T^n(X)$, we then obtain that $\mu_j = 0$ for all $j \in J$, and hence that $x = 0$. Therefore Y_n is a complement of $T^n(X)$ contained in $\ker T^m$.

Conversely, if for $n \in \mathbb{N}$ the subspace $T^n(X)$ has a complement $Y_n \subseteq \ker T^m$, then $T^m(X) = T^m(Y_n) + T^{m+n}(X) = T^{m+n}(X)$, and therefore $q(T) \leq m$. □

Theorem 28.1.3 *If both $p(T)$ and $q(T)$ are finite, then $p(T) = q(T)$.*

Proof Set $p := p(T)$ and $q := q(T)$. Assume first that $p \leq q$, so that the inclusion $T^q(X) \subseteq T^p(X)$ holds. Obviously, we may suppose that $q > 0$. From part (ii) of Lemma 28.1.2, we have $X = \ker T^q + T^q(X)$, so every element $y := T^p(x) \in T^p(X)$ admits the decomposition $y = z + T^q w$, with $z \in \ker T^q$. From $z = T^p x - T^q w \in T^q(X)$, we then obtain that $z \in \ker T^q \cap T^q(X)$, and the last intersection is $\{0\}$ by part (i) of Lemma 28.1.2. Therefore, we have $y = T^q w \in T^q(X)$, and this shows the equality $T^p(X) = T^q(X)$, from which we obtain $p \geq q$, so that $p = q$.

Assume now that $q \leq p$ and $p > 0$, so that $\ker T^q \subseteq \ker T^p$. From part (ii) of Lemma 28.1.2, we have $X = \ker T^q + T^p(X)$, so that an arbitrary element x of $\ker T^p$ admits the representation $x = u + T^p v$, with $u \in \ker T^q$. From $T^p x = T^p u = 0$, it then follows that $T^{2p} v = 0$, so that $v \in \ker T^{2p} = \ker T^p$. Hence $T^p v = 0$, and consequently $x = u \in \ker T^q$. This shows that we have $\ker T^q = \ker T^p$, and hence $q \geq p$. Therefore $p = q$. □

If $p(T) < \infty$, then $C(T) = T^\infty(X)$. This is a consequence of Lemma 26.1.9 and part (i) of Lemma 28.1.2. The equality $C(T) = T^\infty(X)$ is still true when $q := q(T) < \infty$. Indeed, in this case, by definition,

$$\ker T \cap T^q(X) = \ker T \cap T^{q+k}(X) \quad \text{for } k \in \mathbb{Z}^+,$$

so that Lemma 26.1.9 again applies.

Two other important quantities associated with a linear operator T on a vector space X are the *nullity* $\alpha(T) := \dim \ker T$ and the *defect* $\beta(T) := \operatorname{codim} T(X)$.

Let $\Delta(X)$ denote the set of all linear operators on the vector space X for which $\alpha(T)$ and $\beta(T)$ are both finite. For every $T \in \Delta(X)$, the *index* of T, defined by

$$\operatorname{ind} T := \alpha(T) - \beta(T),$$

satisfies the basic *index theorem*:

$$\operatorname{ind}(TS) = \operatorname{ind} T + \operatorname{ind} S \quad (T, S \in \Delta(X)).$$

See Theorem 23.1 of Heuser (1982).

In the next theorem we establish the basic relationships between the quantities $\alpha(T)$, $\beta(T)$, $p(T)$, and $q(T)$.

Proposition 28.1.4 *Let T be a linear operator on a vector space X. Then the following properties hold.*

(i) *If $p(T) < \infty$ then $\alpha(T) \leq \beta(T)$.*
(ii) *If $q(T) < \infty$ then $\beta(T) \leq \alpha(T)$.*
(iii) *If $p(T) = q(T) < \infty$, then $\alpha(T) = \beta(T)$ (possibly infinite).*
(iv) *If $\alpha(T) = \beta(T) < \infty$, and, if one chain is finite, then $p(T) = q(T)$.*

Proof (i) Let $p := p(T) < \infty$. Obviously, if $\beta(T) = \infty$, there is nothing to prove. Suppose that $\beta(T) < \infty$. It is easy to check that also $\beta(T^n)$ is finite. By Lemma 28.1.2(i), we have $\ker T \cap T^p(X) = \{0\}$, and this implies the inequality $\alpha(T) < \infty$. From the index theorem, we obtain

$$n \cdot \text{ind } T = \text{ind } T^n = \alpha(T^p) - \beta(T^n) \quad (n \geq p).$$

Now, suppose that $q := q(T) < \infty$. We see that, for each $n \geq \max\{p, q\}$, the quantity $n \cdot \text{ind } T = \alpha(T^p) - \beta(T^p)$ is constant, so that $\text{ind } T = 0$, that is, $\alpha(T) = \beta(T)$.

Consider the other case, where $q = \infty$. Then $\beta(T^n) \to 0$, as $n \to \infty$, so $n \cdot \text{ind } T$ eventually becomes negative, and hence $\text{ind } T < 0$. Therefore in this case we have $\alpha(T) < \beta(T)$.

(ii) Let $q := q(T) < \infty$. Also here we can suppose that $\alpha(T) < \infty$, otherwise there is nothing to prove. Consequently, as is easy to check, we also have $\beta(T^n) < \infty$ and, by Lemma 28.1.2(ii), $X = Y \oplus T(X)$ with $Y \subseteq \ker T^q$. From this, it follows that

$$\beta(T) = \dim Y \leq \alpha(T^q) < \infty.$$

With appropriate changes, the index argument used in the proof of (i) shows that $\beta(T) = \alpha(T)$ if $p(T) < \infty$, and $\beta(T) < \alpha(T)$ if $p(T) = \infty$.

(iii) This is clear from (i) and (ii).

(iv) This is an immediate consequence of the equality

$$\alpha(T^n) - \beta(T^n) = \text{ind } T^n = n \cdot \text{ind } T = 0,$$

which is valid for every $n \in \mathbb{N}$. □

Proposition 28.1.5 *For a bounded operator T on a Banach space X, the following implications hold:*

(i) $p(\lambda_0 I - T) < \infty \Rightarrow \mathcal{N}^\infty(\lambda_0 I - T) \cap (\lambda_0 I - T)^\infty(X) = \{0\} \Rightarrow$ *T has SVEP at* λ_0*;*

(ii) $q(\lambda_0 I - T) < \infty \Rightarrow X = \mathcal{N}^\infty(\lambda_0 I - T) + (\lambda_0 I - T)^\infty(X) \Rightarrow$ T' *has SVEP at* λ_0.

Proof (i) There is no loss of generality in supposing that $\lambda_0 = 0$. We define $p := p(T) < \infty$. Then $\mathcal{N}^\infty(T) = \ker T^p$ and $T^\infty(X) \subseteq T^p(X)$. From Lemma 28.1.2 (i), we then conclude that $\mathcal{N}^\infty(T) \cap T^\infty(X) \subseteq \ker T^p \cap T^p(X) = \{0\}$. The second implication is a consequence of Theorem 27.2.3 because

$$\ker T \cap X_T(\emptyset) \subseteq \mathcal{N}^\infty(T) \cap T^\infty(X) = \{0\}.$$

(ii) Again we may suppose that $\lambda_0 = 0$. Suppose that $q := q(T) < \infty$. Then $T^\infty(X) = T^q(X)$ and $X = T^n(X) + \ker T^q$ $(n \in \mathbb{N})$ by Lemma 28.1.2(ii). From this it easily follows that $X = \mathcal{N}^\infty(T) + T^\infty(X)$.

In order to show that the second implication of (ii) holds, we first note that, if $X = \mathcal{N}^\infty(T) + T^\infty(X)$, then $\mathcal{N}^\infty(T)^\perp \cap T^\infty(X)^\perp = \{0\}$. Now take an element $x' \in \ker T' \cap X_{T'}(\emptyset)$. Then

$$x' \in \ker T' \subseteq \ker (T')^n = \ker (T^n)' = \overline{T^n(X)}^\perp = T^n(X)^\perp,$$

for every $n \in \mathbb{N}$ and therefore $x' \in T^\infty(X)^\perp$.

On the other hand, from $\sigma_{T'}(x') = \emptyset$ we obtain, by Theorem 27.2.1, that

$$x' \in K(T') \subseteq (T')^n(X') = (T^n)'(X') \subseteq (\ker T^n)^\perp$$

for every $n \in \mathbb{N}$. From this, it easily follows that $x' \in \mathcal{N}^\infty(T)^\perp$, and this implies that $x' \in \mathcal{N}^\infty(T)^\perp \cap T^\infty(X)^\perp$. Therefore $x' = 0$. From Theorem 27.2.3, we then conclude that T' has SVEP at 0. □

The preceding result is quite useful to establish SVEP for several classes of operators. In fact, there are several operators which satisfy the condition that $p(\lambda I - T) < \infty$ for all $\lambda \in \mathbb{C}$; see the Exercises.

28.2 Fredholm operators

Definition 28.2.1 *A bounded operator $T \in \mathcal{B}(X)$, where X is a Banach space, is* upper semi-Fredholm *if $\alpha(T) < \infty$ and $T(X)$ is closed. The operator T in $\mathcal{B}(X)$ is* lower semi-Fredholm *if $\beta(T) < \infty$. Denote by $\Phi_+(X)$ the class of all upper semi-Fredholm operators, and by $\Phi_-(X)$ the class of all lower semi-Fredholm operators. The class $\Phi(X)$ of all* Fredholm operators *is defined by $\Phi(X) := \Phi_+(X) \cap \Phi_-(X)$, while the class of all* semi-Fredholm operators *is defined by $\Phi_\pm(X) := \Phi_+(X) \cup \Phi_-(X)$.*

Note that, if $\beta(T) < \infty$, then $T(X)$ is closed (see Chapter 26). If $T \in \Phi_\pm(X)$, we can define the *index* of T as above. Obviously, the index of a semi-Fredholm operator is an integer or $\pm\infty$.

The three sets $\Phi_+(X)$, $\Phi_-(X)$, and $\Phi(X)$ are semi-groups, in the sense that, if T, S belong to one of these sets, then also the products TS and ST lie in the same set. Moreover, these three sets are each open, $T \in \Phi_+(X)$ if and only if $T' \in \Phi_-(X')$, and dually, $T \in \Phi_-(X)$ if and only if $T' \in \Phi_+(X')$; the reader can find these classical results on Fredholm theory in Caradus, Pfaffenberger and Yood (1974).

Note that, if $T \in \Phi_{\pm}(X)$, then $T^n \in \Phi_{\pm}(X$ $(n \in \mathbb{N})$, and this implies that $T^{\infty}(X)$ is closed because the subspaces $T^n(X)$ are closed for each $n \in \mathbb{N}$. From Proposition 26.1.10, we then have $T(T^{\infty}(X)) = T^{\infty}(X)$, and hence it follows from Proposition 26.1.7 that $T^{\infty}(X) \subseteq K(T)$. But the opposite inclusion $K(T) \subseteq T^{\infty}(X)$ is true for every operator $T \in \mathcal{B}(X)$, so that $K(T) = T^{\infty}(X)$ for every $T \in \Phi_{\pm}(X)$.

To explore SVEP at a point in the case of semi-Fredholm operators, we need a decomposition property for these operators due to Kato (1958). We state this classical result without proving it (the proof of this decomposition is rather involved; the interested reader will find a clear proof in West (1987)).

Theorem 28.2.2 *Let $T \in \Phi_{\pm}(X)$, where X is a Banach space. Then there exist closed T-invariant subspaces M, N of X such that $X = M \oplus N$, $T \mid M$ is semi-regular, $T \mid N$ is nilpotent, and N is finite-dimensional.*

For an arbitrary bounded operator $T \in \mathcal{B}(X)$, a pair (M, N) of closed T-invariant subspaces is called a *Kato decomposition* for T if the decomposition observed in Theorem 28.2.2 for semi-Fredholm operators holds, that is, $X = M \oplus N$, $T \mid M$ is semi-regular, $T \mid N$ is nilpotent, and N is finite-dimensional. In the literature the operators which admit a Kato decomposition are called *essentially semi-regular*; see Müller (1994). A semi-regular operator is obviously another example of an operator which admits a Kato decomposition. Note that not every semi-Fredholm operator is semi-regular; a semi-Fredholm T is semi-regular precisely when its *jump* $j(T) = 0$; see Part IV, Chapter 24 for the definition and also West (1987).

It is not difficult to show that, if (M, N) is a Kato decomposition for T, then the pair $(N^{\perp}, M^{\perp})$ is a Kato decomposition for T'. The proof of this fact requires methods of standard duality theory (Exercise).

Lemma 28.2.3 *Suppose that (M, N) is a Kato decomposition for $T \in \mathcal{B}(X)$. Then:*

(i) $K(T) = K(T \mid M)$ *and* $K(T)$ *is closed;*
(ii) $\ker T \mid M = K(T) \cap \ker T$;
(iii) $K(T) = T^{\infty}(X)$.

Proof (i) To prove the equality $K(T) = K(T \mid M)$, we need only to show the inclusion $K(T) \subseteq M$. Let $x \in K(T)$ and, according to the definition of $K(T)$, let (u_n) be a sequence of X and $\delta > 0$ be such that $x = u_0$, $Tu_{n+1} = u_n$, and

$$\|u_n\| \leq \delta^n \|x\| \quad (n \in \mathbb{Z}^+).$$

Clearly, $T^n u_n = x$ $(n \in \mathbb{N})$. From the decomposition $X = M \oplus N$, we know that $x = y + z$, $u_n = y_n + z_n$, with $y, y_n \in M$ and $z, z_n \in N$ for every $n \in \mathbb{N}$. Then $x = T^n u_n = T^n y_n + T^n z_n$, hence, by the uniqueness of the decomposition, $y = T^n y_n$ and $z = T^n z_n$. Let P denote the projection of X onto N along M. From the estimate

$$\|((T|N)P)^n\|^{1/n} \le \|(T \mid N)^n\|^{1/n}\|P^n\|^{1/n} = \|(T \mid N)^n\|^{1/n}\|P\|,$$

we infer that $(T\,|N)P$ is quasi-nilpotent, since, by assumption, $T \mid N$ is nilpotent, and hence quasi-nilpotent. Therefore, if $\varepsilon > 0$, there is a positive integer n_0 such that $\|(T \mid N\ P)^n\|^{1/n} < \varepsilon$ $(n > n_0)$. Now we have

$$\|z\| = \|T^n z_n\| = \|T^n P v_n\| = \|T^n P^n v_n\| \|(TP)^n v_n\| \le \varepsilon^n \delta^n \|x\| \quad (n > n_0).$$

Since ε is arbitrary, the last term of the previous inequality converges to 0, so $z = 0$ and hence $x = y \in M$.

The final assertion is a consequence of Proposition 26.1.12 since $T \mid M$ is semi-regular.

(ii) This equality is an immediate consequence of (i). Indeed, $K(T) \subseteq M$ and, since $T \mid M$ is semi-regular, from Proposition 26.1.12 and part (i) we obtain that

$$\ker T \mid M \subseteq (T \mid M)^\infty(M) = K(T \mid M) = K(T).$$

Hence

$$K(T) \cap \ker T = K(T) \cap M \cap \ker T = K(T) \cap \ker T \mid M = \ker T \mid M.$$

(iii) If $(T \mid N)^d = 0$, then we have

$$T^n(X) = T^n(M) \oplus T^n(N) = T^n(M) \oplus \{0\} \quad (n \ge d),$$

and this implies that $T^\infty(X) = (T \mid M)^\infty(M)$. The semi-regularity of $T \mid M$ then yields that $K(T \mid M) = (T \mid M)^\infty(M)$, and hence, by (i), we have $T^\infty(X) = K(T \mid M) = K(T)$. □

Remark 28.2.4 It should be noted that, if (M, N) is a Kato decomposition for $\lambda_0 I - T$, then T has SVEP at λ_0 if and only if $T \mid M$ has SVEP at λ_0. In fact, if T has SVEP at λ_0, then $T \mid M$ has SVEP at λ_0 since the SVEP at a point is inherited by the restrictions to closed invariant subspaces. On the other hand, if $T \mid M$ has SVEP at λ_0, the semi-regularity of $(\lambda_0 I - T) \mid M$ entails that the restriction $(\lambda_0 I - T) \mid M$ is injective, and therefore

$$\ker(\lambda_0 I - T) \mid M = K(T) \cap \ker(\lambda_0 I - T) = \{0\}.$$

Thus, by Theorem 27.2.3, T has SVEP at λ_0. Note that this argument also shows T has SVEP at λ_0 if and only if $(\lambda_0 I - T) \mid M$ is injective. An obvious consequence of this fact is that a semi-regular operator T has SVEP at λ_0 if and only if $\lambda_0 I - T$ is injective, or, equivalently, $\lambda_0 I - T$ is bounded below.

Lemma 28.2.5 *Let $T \in \mathcal{B}(X)$, and suppose that X is the direct sum of two closed subspaces M, N each invariant under T, for which $T \mid M$ is semi-regular. Then $T \mid M$ is surjective if and only if $T' \mid N^\perp$ is injective.*

Proof Suppose first that $T(M) = M$ and $x' \in \ker T' \mid N^\perp = \ker T' \cap N^\perp$. For every $m \in M$, there exists $m_1 \in M$ such that $Tm_1 = m$. We have

$$\langle m, x' \rangle = \langle Tm_1, x' \rangle = \langle m_1, T'x' \rangle = 0\,,$$

and hence $x' \in M^\perp \cap N^\perp = \{0\}$.

Conversely, suppose that $T(M) \subset M$ and $T(M) \neq M$. By assumption, $T(M)$ is closed since $T \mid M$ is semi-regular, and hence, via the Hahn–Banach theorem, there exists $z' \in X'$ such that $z' \in T(M)^\perp$ and $z' \notin M^\perp$. Now, from the decomposition $X' = N^\perp \oplus M^\perp$, we have $z' = n' + m'$ with $n' \in N^\perp$ and $m' \in M^\perp$. For every $m \in M$, we obtain

$$\langle m, T'n' \rangle = \langle Tm, n' \rangle = \langle Tm, z' \rangle - \langle Tm, m' \rangle = 0\,.$$

Hence $T'n' \in T'(N^\perp) \cap M^\perp = N^\perp \cap M^\perp = \{0\}$, so that

$$0 \neq n' \in \ker T' \cap N^\perp.$$

□

Theorem 28.2.6 *Suppose that $\lambda_0 I - T \in \Phi_\pm(X)$. Then the implications* (i) *and* (ii) *of Proposition 28.1.5 are equivalences.*

Proof To show the equivalence in (i) of Proposition 28.1.5, we need only prove that, if T has SVEP at λ_0, then $p(\lambda_0 I - T) < \infty$. Let (M, N) be a Kato decomposition for $\lambda_0 I - T$, and let $k \in \mathbb{N}$ be such that $((\lambda_0 I - T) \mid N)^k = 0$. Suppose that $p := p(\lambda_0 I - T) = \infty$. Then $(\lambda_0 I - T)|M$ is not injective; otherwise, if $(\lambda_0 I - T)|M$ were injective, for $n \geq k$ we would have

$$\ker(\lambda_0 I - T)^n = \ker(\lambda_0 I - T) \mid M)^n = \ker(\lambda_0 I - T) \mid N)^n = \{0\}\,,$$

and hence $p(\lambda_0 I - T) < \infty$. Hence $(\lambda_0 I - T)|M$ is not injective, and therefore T does not have the SVEP at λ_0.

Analogously, to show the equivalence in (ii) of Proposition 28.1.5, we need only prove that, if T' has SVEP at λ_0, then $q(\lambda_0 I - T) < \infty$. Suppose that

$q(\lambda_0 I - T) = \infty$. Then $\lambda_0 I - T \mid M$ is not surjective; for otherwise, for every positive integer $n \geq k$, we would have

$$(\lambda_0 I - T)^n(X) = (\lambda_0 I - T)^n(M) \oplus \{0\} = M,$$

and hence $q(\lambda_0 I - T) < \infty$. Now, by Lemma 28.2.5, it follows that $\lambda_0 I' - T' \mid N^\perp$ is not injective, and hence T' does not have SVEP at λ_0. □

Recall that T is said to be a *Riesz–Schauder* operator if T is a Fredholm operator having both ascent and descent finite. Note that, by Lemma 28.1.2, a Riesz–Schauder operator has index 0.

The next result shows that the class of operators without SVEP is rather large, since every Fredholm operator having index greater than 0 belongs to the class.

Corollary 28.2.7 *If $\lambda_0 I - T \in \Phi_\pm(X)$, where X is a Banach space, and if T has SVEP at λ_0, then $\lambda_0 I - T$ is upper semi-Fredholm and*

$$\operatorname{ind}(\lambda_0 I - T) \leq 0.$$

Additionally, if also T' has SVEP at λ_0, then $\lambda_0 I - T$ is a Riesz–Schauder operator.

Proof The first assertion is clear since the condition $p(\lambda_0 I - T) < \infty$ entails that $\alpha(\lambda_0 I - T) \leq \beta(\lambda_0 I - T)$ by Proposition 28.1.4. The second assertion is also clear because, if both the operators T and T' have SVEP at λ_0, then $p(\lambda_0 I - T) < \infty$ and $q(\lambda_0 I - T) < \infty$. Consequently, $p(\lambda_0 I - T) = q(\lambda_0 I - T)$ by Theorem 28.1.3, and therefore, by Proposition 28.1.4,

$$\alpha(\lambda_0 I - T) = \beta(\lambda_0 I - T) < \infty.$$ □

It should be noted that the last assertion of Corollary 28.2.7 applies to every decomposable operator T on a Banach space, that is, if T is semi-Fredholm and decomposable, then T is a Riesz–Schauder operator.

Proposition 28.2.8 *Let $\lambda_0 I - T \in \Phi_\pm(X)$ and suppose that (M, N) is a Kato decomposition for $\lambda_0 I - T$. Then the following properties are equivalent:*

(i) *T has SVEP at λ_0;*
(ii) *$H_0(\lambda_0 I - T) = N$;*
(iii) *$H_0(\lambda_0 I - T)$ is finite-dimensional;*
(iv) *$H_0(\lambda_0 I - T)$ is closed;*
(v) *$H_0(\lambda_0 I - T) \cap K(\lambda_0 I - T) = \{0\}$;*

In this case, set $p = p(\lambda_0 I - T)$. Then

$$H_0(\lambda_0 I - T) = \mathcal{N}^\infty(\lambda_0 I - T) = \ker(\lambda_0 I - T)^p .$$

Proof There is no loss of generality if we suppose that $\lambda_0 = 0$.

Clearly, (ii) $\Rightarrow$ (iii) $\Rightarrow$ (iv), and, from Proposition 27.2.7, we know that the implications (iv) $\Rightarrow$ (v) $\Rightarrow$ (i) hold, so we need only prove that (i) $\Rightarrow$ (ii).

(i) $\Rightarrow$ (ii) First note that, if the operator T admits a Kato decomposition (M, N), then $H_0(T) = H_0(T \mid M) + H_0(T \mid N)$. The inclusion

$$H_0(T) \supseteq H_0(T \mid M) + H_0(T \mid N)$$

is obvious. In order to show the opposite inclusion, let us consider an arbitrary element $x \in H_0(T)$, and let $x = u + v$, with $u \in M$ and $v \in N$. Since $T \mid N$ is quasi-nilpotent, we have $N = H_0(T \mid N) \subseteq H_0(T)$. Consequently,

$$u = x - v \in H_0(T) \cap M = H_0(T \mid M),$$

and therefore $H_0(T) \subseteq H_0(T \mid M) + H_0(T \mid N)$. Hence

$$H_0(T) = H_0(T \mid M) + H_0(T \mid N) = H_0(T \mid M) + N .$$

Now, suppose that T has SVEP at 0. Then, as observed in Remark 28.2.4, $T \mid M$ is injective and therefore, by Proposition 26.2.9,

$$H_0(T \mid M) = \overline{\bigcup_{n=1}^{\infty} \ker(T \mid M)^n} = \{0\} .$$

This shows that $H_0(T) = N$.

To prove the last assertion, observe first that the inclusions

$$\ker T^n \subseteq \mathcal{N}^\infty(T) \subseteq H_0(T)$$

hold for every $T \in \mathcal{B}(X)$ and $n \in \mathbb{N}$. Let $k \in \mathbb{N}$ be such that $(T|N)^k = 0$. Then $H_0(T) = N \subseteq \ker T^k$, and hence $H_0(T) = \mathcal{N}^\infty(T) = \ker T^k$. From this, it follows that $p := p(T) \leq k$, and therefore $\ker T^k = \ker T^p$. □

Proposition 28.2.9 *Let $\lambda_0 I - T \in \Phi_\pm(X)$, and suppose that (M, N) is a Kato decomposition for T. Then the following assertions are equivalent:*

(i) *T' has SVEP at λ_0;*
(ii) *$K(T) = M$;*
(iii) *$K(T)$ is finite-codimensional;*

(iv) $X = H_0(\lambda_0 I - T) + K(\lambda_0 I - T)$;
(v) $H_0(\lambda_0 I - T) + K(\lambda_0 I - T)$ *is norm-dense in* X.

In this case, set $q = q(\lambda_0 I - T)$. *Then*

$$K(\lambda_0 I - T) = T^\infty(\lambda_0 I - T) = (\lambda_0 I - T)^q(X).$$

Proof Also here we suppose that $\lambda_0 = 0$.

(i) $\Rightarrow$ (ii) We know that the pair $(N^\perp, M^\perp)$ is a Kato decomposition for the semi-Fredholm operator T', thus $T' \mid N^\perp$ is semi-regular, $T' \mid M^\perp$ is nilpotent, and further $\dim M^\perp = \operatorname{codim} M < \infty$. From the assumption, T' has SVEP at 0, so the semi-regularity of $T' \mid N^\perp$ entails that $T' \mid N^\perp$ is injective, and hence, by Lemma 28.2.5, $T \mid M$ is surjective. This shows that $M = K(T \mid M) = K(T)$.

The implication (ii) $\Rightarrow$ (iii) is obvious.

(iii) $\Rightarrow$ (i) Since $K(T) = T^\infty(X)$, the space $T^\infty(X)$ has finite codimension. Because $T^\infty(X) \subseteq T^q(X)$ for every $q \in \mathbb{N}$, we then conclude that $q(T) < \infty$, so T' has SVEP at 0 by Proposition 28.2.6.

(i) $\Rightarrow$ (iv) Assume that T' has SVEP at 0, or, equivalently, that $q < \infty$, where $q = q(T)$. Then $K(T) = T^\infty(X) = T^q(X)$. Moreover, we see that $X = \ker T^q + T^n(X)$ $(n \in \mathbb{N})$ by Lemma 28.1.2(ii), and therefore it follows that $X = H_0(\lambda_0 I - T) + K(\lambda_0 I - T)$.

The implication (iv) $\Rightarrow$ (v) is obvious, while the implication (v) $\Rightarrow$ (i) has been proved in Corollary 27.2.9.

The last assertion is clear. □

Let $\sigma_{ap}(T)$ be the *approximate point spectrum* of T and let $\sigma_{su}(T)$ be the *surjectivity spectrum* of T; for the definitions, see Part IV, Chapter 22. Obviously, these spectra are non-empty sets since they both contain $\sigma_K(T)$, and hence, by Theorem 26.2.11, they contain also the boundary of $\sigma(T)$. From Remark 28.2.4 it easily follows that, if T has SVEP, then $\sigma_K(T) = \sigma_{ap}(T)$. We have already observed that T has SVEP at every point which is not a limit point of $\sigma(T)$. What happens if we consider instead of $\sigma(T)$ some distinguished parts of the spectrum? A simple argument shows the following implications:

$$\sigma_{ap}(T) \text{ does not cluster at } \lambda_0 \Rightarrow T \text{ has SVEP at } \lambda_0\,, \qquad (28.2.1)$$

and

$$\sigma_{su}(T) \text{ does not cluster at } \lambda_0 \Rightarrow T' \text{ has SVEP at } \lambda_0\,. \qquad (28.2.2)$$

Indeed, if $\sigma_{ap}(T)$ does not cluster at λ_0, then there is an open disc $\mathbb{D}(\lambda_0, \delta)$ centred at λ_0 such that $\lambda I - T$ is injective for every $\lambda \in \mathbb{D}(\lambda_0, \delta)$ with $\lambda \neq \lambda_0$. Let $f : \mathbb{D}(\lambda_0, \varepsilon) \to X$ be an analytic function defined on another open disc $\mathbb{D}(\lambda_0, \varepsilon)$ centred at λ_0 for which the equation $(\lambda I - T)f(\lambda) = 0$ holds for every $\lambda \in \mathbb{D}(\lambda_0, \varepsilon)$. Obviously, we may suppose that $\varepsilon \leq \delta$. It is clear that $f(\lambda) \in \ker(\lambda I - T) = \{0\}$ for every $\lambda \in \mathbb{D}(\lambda_0, \varepsilon)$ with $\lambda \neq \lambda_0$. Thus $f(\lambda) = 0$ for every $\lambda \in \mathbb{D}(\lambda_0, \varepsilon)$ with $\lambda \neq \lambda_0$. From the continuity of f at λ_0, we conclude that $f(\lambda_0) = 0$. Hence $f \equiv 0$ in $\mathbb{D}(\lambda_0, \varepsilon)$, and therefore T has SVEP at λ_0.

The second implication is an immediate consequence of the equality $\sigma_{su}(T) = \sigma_{ap}(T')$.

Note that neither of the implications (28.2.1) and (28.2.2) may be reversed in general. Indeed, if λ_0 is a non-isolated boundary point of $\sigma(T)$, then $\sigma_{ap}(T)$ and $\sigma_{su}(T)$ cluster at λ_0, but, as observed before, T and T' have SVEP at λ_0.

An example of an operator T having the SVEP and such that every spectral point is the limit of points of $\sigma_{ap}(T)$ may be found among the unilateral weighted right shift operators. Indeed, there exist unilateral weighted right shift operators T on $\ell^p(\mathbb{N})$ which have SVEP and for which $\sigma(T) = \sigma_{ap}(T) = \sigma_{su}(T)$ and $\sigma(T)$ is a closed ball centred at 0 with radius $r > 0$; see Laursen and Neumann (2000).

The next result shows that the implications (28.2.1) and (28.2.2) are actually equivalences when $\lambda_0 I - T$ is semi-Fredholm.

Theorem 28.2.10 *Suppose that $\lambda_0 I - T \in \Phi_\pm(X)$. Then the following equivalences hold:*

(i) *T has SVEP at $\lambda_0 \Leftrightarrow \sigma_{ap}(T)$ does not cluster at λ_0;*
(ii) *T' has SVEP at $\lambda_0 \Leftrightarrow \sigma_{su}(T)$ does not cluster at λ_0.*

Proof (i) We need only to prove that, if T has SVEP at λ_0, then $\sigma_{ap}(T)$ does not cluster at λ_0. Evidently, we may suppose that $\lambda_0 = 0$. Assume that T has SVEP at 0 and $T \in \Phi_\pm(X)$. Because $\Phi_\pm(X)$ is open there exists $\varepsilon > 0$ such that $\lambda I - T$ is semi-Fredholm, and hence has closed range, for every $|\lambda| < \varepsilon$. Then $\lambda \in (\mathbb{D}(0, \varepsilon) \setminus \{0\}) \cap \sigma_{ap}(T)$ if and only if λ is an eigenvalue for T. Now it is easy to see that $\ker(\lambda I - T) \subseteq T^\infty(X)$ for every $\lambda \neq 0$, so that every non-zero eigenvalue of T belongs to the spectrum of the restriction $T \mid T^\infty(X)$.

Finally, suppose that 0 is a cluster point of $\sigma_{ap}(T)$. Let (λ_n) be a sequence of non-zero eigenvalues which converges to 0. Then $\lambda_n \in \sigma(T \mid T^\infty(X))$ for every $n \in \mathbb{N}$, and hence $0 \in \sigma(T \mid T^\infty(X))$ since the spectrum of an operator is closed . But T has SVEP at 0, and thus $T \mid M$ is injective. From part (ii) of Lemma 28.2.3, we have

$$\{0\} = \ker T \mid M = \ker T \cap T^\infty(X),$$

so the restriction $T \mid T^\infty(X)$ is injective.

On the other hand, from the equality $T(T^\infty(X)) = T^\infty(X)$, we know that $T \mid T^\infty(X)$ is surjective, so that $0 \notin \sigma(T \mid T^\infty(X))$, a contradiction.

(ii) The equivalence follows from the first part once we have observed that always $\sigma_{ap}(T') = \sigma_{su}(T)$. □

The result of Theorem 28.2.10 is quite useful for establishing the membership of cluster points of some distinguished parts of the spectrum to the *semi-Fredholm spectrum* $\sigma_{sf}(T)$ and to the *essential spectrum* $\sigma_e(T)$, which are defined as follows:

$$\sigma_{sf}(T) := \{\lambda \in \mathbb{C} : \lambda I - T \notin \Phi_{\pm}(X)\}$$

and

$$\sigma_e(T) := \{\lambda \in \mathbb{C} : \lambda I - T \notin \Phi(X)\}.$$

Clearly $\sigma_{sf}(T) \subseteq \sigma_e(T)$. A first application is given in the following result, which improves a classical Putnam theorem about the non-isolated boundary points of the spectrum as a subset of the essential spectrum.

Corollary 28.2.11 *For every operator $T \in \mathcal{B}(X)$ on a Banach space X, every non-isolated boundary point of $\sigma(T)$ belongs to $\sigma_{sf}(T)$, and in particular to $\sigma_e(T)$.*

Proof If $\lambda_0 \in \partial\sigma(T)$ is non-isolated in $\sigma(T)$, then $\sigma_{ap}(T)$ clusters at λ_0 because $\partial\sigma(T) \subseteq \sigma_{ap}(T)$. But T has SVEP at every point of $\partial\sigma(T)$, and so, by Theorem 28.2.10s, $\lambda_0 I - T \notin \Phi_{\pm}(X)$. □

Corollary 28.2.12 *Suppose that $T \in \mathcal{B}(X)$, where X is a Banach space, has SVEP. Then all cluster points of $\sigma_{ap}(T)$ belong to $\sigma_{sf}(T)$. Analogously, if T' has SVEP, then all cluster points of $\sigma_{su}(T)$ belong to $\sigma_{sf}(T)$.*

Proof Suppose that $\lambda_0 \notin \sigma_{sf}(T)$. Since T has SVEP, and in particular has SVEP at λ_0, then $\sigma_{ap}(T)$ does not cluster at λ_0, by Theorem 28.2.10.

Analogously, suppose that $\lambda_0 \notin \sigma_{sf}(T)$. Since T' has SVEP at λ_0, then $\sigma_{su}(T)$ does not cluster at λ_0, again by Theorem 28.2.10. □

All the results established above have interesting applications. Recall that the *upper semi-Fredholm spectrum* and *lower semi-Fredholm spectrum* are defined by

$$\sigma_{uf}(T) := \{\lambda \in \mathbb{C} : \lambda I - T \notin \Phi_{+}(X)\},$$

$$\sigma_{lf}(T) := \{\lambda \in \mathbb{C} : \lambda I - T \notin \Phi_{-}(X)\},$$

respectively. Clearly,

$$\sigma_{sf}(T) \subseteq \sigma_{uf}(T) \subseteq \sigma_{ap}(T). \tag{28.2.3}$$

In the next result we consider a situation which occurs in some concrete cases.

Proposition 28.2.13 *Let $T \in \mathcal{B}(X)$ be an operator for which $\sigma_{ap}(T) = \partial\sigma(T)$ and every $\lambda \in \partial\sigma(T)$ is not isolated in $\sigma(T)$. Then*

$$\sigma_{sf}(T) = \sigma_{uf}(T) = \sigma_{ap}(T) = \sigma_K(T).$$

Proof The operator T has SVEP at every point of the boundary as well as at every point λ which belongs to the remaining part of the spectrum since $\sigma_{ap}(T)$ does not cluster at this λ. Hence T has SVEP. By Corollary 28.2.12, we have $\sigma_{ap}(T) \subseteq \sigma_{sf}(T)$, and hence, from the inclusions (28.2.3), we conclude that $\sigma_{sf}(T) = \sigma_{uf}(T) = \sigma_{ap}(T)$.

Finally, $\sigma_K(T) = \sigma_{ap}(T)$ because T has SVEP, so the proof is complete. □

Corollary 28.2.14 *Suppose that $\sigma_{su}(T) = \partial\sigma(T)$ and that every $\lambda \in \partial\sigma(T)$ is not isolated in $\sigma(T)$. Then*

$$\sigma_{sf}(T) = \sigma_{lf}(T) = \sigma_{su}(T) = \sigma_K(T).$$

Proof From the assumption, we obtain $\sigma_{su}(T) = \sigma_{ap}(T') = \partial\sigma(T) = \partial\sigma(T')$, and so we can apply Proposition 28.2.13. The statement then follows once we have observed that $\sigma_{sf}(T) = \sigma_{sf}(T')$, that $\sigma_{lf}(T) = \sigma_{uf}(T')$, and finally that $\sigma_K(T) = \sigma_K(T')$. □

Proposition 28.2.13 applies to arbitrary non-invertible isometries. In fact for these operators we have $\sigma(T) = \overline{\mathbb{D}(0, r(T))}$ and $\sigma_{ap}(T) = \partial\mathbb{D}(0, \nu(T))$, where $\nu(T)$ is the spectral radius of T; see Proposition 1.3.2 of Laursen and Neumann (2000).

Proposition 28.2.13 also applies to the *Cesáro operator* C_p defined in Part IV, Example 24.1.3. As noted in T. L. Miller V. G. Miller, and Smith (1998), the spectrum of the operator C_p is the closed disc $\overline{\mathbb{D}(p/2, p/2)}$ and

$$\sigma_e(C_p) = \sigma_{ap}(C_p) = \partial\mathbb{D}(p/2, p/2).$$

From Proposition 28.2.13, we also have

$$\sigma_{sf}(C_p) = \sigma_{uf}(C_p) = \sigma_K(C_p) = \partial\mathbb{D}(p/2, p/2).$$

Proposition 28.2.15 *Let X be a Banach space, and suppose that $\lambda_0 I - T$ in $\Phi(X)$ has index 0. Then the following statements are equivalent:*

(i) *T has SVEP at λ_0;*
(ii) *T' has SVEP at λ_0;*

(iii) $p(\lambda_0 I - T) < \infty$;
(iv) $q(\lambda_0 I - T) < \infty$;
(v) $\mathcal{N}^\infty(\lambda_0 I - T) \cap (\lambda_0 I - T)^\infty(X) = \{0\}$;
(vi) $X = \mathcal{N}^\infty(\lambda_0 I - T) + (\lambda_0 I - T)^\infty(X)$;
(vii) $X = \mathcal{N}^\infty(\lambda_0 I - T) \oplus (\lambda_0 I - T)^\infty(X)$;
(viii) $\sigma(T)$ *does not cluster at* λ_0;
(ix) $\lambda_0 I - T$ *is a Riesz–Schauder operator.*

Proof If $\alpha(\lambda_0 I - T) = \beta(\lambda_0 I - T) < \infty$, then

$$p(\lambda_0 I - T) < \infty \Leftrightarrow q(\lambda_0 I - T) < \infty.$$

The equivalence of the statements (i)–(ix) then easily follows by combining the preceding results.

Two other important spectra originating from the theory of Fredholm operators are the Weyl spectrum and the Browder spectrum. The *Weyl spectrum* $\sigma_w(T)$ of $T \in \mathcal{B}(X)$ is defined to be the complement of those complex $\lambda \in \mathbb{C}$ such that $\lambda I - T \in \Phi(X)$ and $\operatorname{ind}(\lambda I - T) = 0$. The *Browder spectrum* $\sigma_b(T)$ of T is defined to be the complement of those complex $\lambda \in \mathbb{C}$ such that $\lambda I - T$ is Riesz–Schauder. Obviously,

$$\sigma_{sf}(T) \subseteq \sigma_e(T) \subseteq \sigma_w(T) \subseteq \sigma_b(T).$$

The next result extends to decomposable operators on Banach spaces some classical results valid for normal operators on Hilbert spaces.

Theorem 28.2.16 *Let X be a Banach space, and let $T \in \mathcal{B}(X)$. Suppose that T or T' has SVEP. Then $\sigma_b(T) = \sigma_w(T)$. If both T and T' have SVEP, then*

$$\sigma_{sf}(T) = \sigma_e(T) = \sigma_w(T) = \sigma_b(T).$$

In particular, these equalities hold for every decomposable operator.

Proof Suppose first that T has SVEP. By Proposition 28.2.15, T is a Fredholm operator having index 0 if and only if T is a Riesz–Schauder operator, and hence $\sigma_b(T) = \sigma_w(T)$. The same argument works also in the case that T' has SVEP since T and T' are simultaneously Weyl or Browder, so that $\sigma_w(T) = \sigma_w(T')$ and $\sigma_b(T) = \sigma_b(T')$. If both T and T' have SVEP then, from Corollary 28.2.7, we easily obtain that $\sigma_{sf}(T) = \sigma_e(T) = \sigma_b(T)$. □

28.3 Exercises

1. Let T be a bounded linear operator on a Banach space. Each of the two conditions '$p(\lambda_0 I - T) < \infty$' and 'the space $H_0(\lambda_0 I - T)$ is closed' implies that T has SVEP at 0. Prove that neither of these two conditions implies the other. *Hint*: consider the operator T defined in Exercise 27.3.4, and the operator $S : \ell^2(\mathbb{N}) \to \ell^2(\mathbb{N})$ defined by $Sx := (x_2/2, \dots, x_n/n, \dots)$, where $x = (x_n) \in \ell^2(\mathbb{N})$.

2. Show that every multiplier T on a semisimple Banach algebra satisfies the condition that $p(\lambda I - T) \le 1 \ (\lambda \in \mathbb{C})$.

3. A bounded operator T on a Banach space X is *paranormal* if

$$\|Tx\|^2 \le \|T^2 x\| \|x\| \quad (x \in X).$$

An operator $T \in \mathcal{B}(X)$ on a Banach space X is *totally paranormal* if $\lambda I - T$ is paranormal for every $\lambda \in \mathbb{C}$. Show that every totally paranormal operator has SVEP. *Hint*: show that $p(\lambda I - T) \le 1 \ (\lambda \in \mathbb{C})$. Alternatively, show that $H_0(\lambda I - T) = \ker(\lambda I - T) \ (\lambda \in \mathbb{C})$.

4. An operator T satisfies a *polynomial growth condition* if there exists $K > 0$ and $\delta > 0$ for which

$$\|\exp(\mathrm{i}\lambda T)\| \le K(1 + |\lambda|^\delta) \quad (\lambda \in \mathbb{R}).$$

Let $\mathcal{P}(X)$ denote the class of all operators which satisfy this condition. The polynomial growth condition may be reformulated as follows: $T \in \mathcal{P}(X)$ if and only if $\sigma(T) \subseteq \mathbb{R}$ and there are constants $K > 0$ and $\delta > 0$ such that

$$\|(\lambda I - T)^{-1}\| \le K(1 + |\mathrm{Im}\,\lambda|^{-\delta})$$

for all $\lambda \in \mathbb{C}$ with $\mathrm{Im}\,\lambda \ne 0$; see Theorem 1.5.19 of Laursen and Neumann (2000). Show that $p(\lambda I - T) < \infty \ (\lambda \in \mathbb{C})$. *Hint*: show that, if $T \in \mathcal{P}(X)$ and if $c := \mathrm{Im}\,\lambda > 0$, then

$$(\lambda I - T)^{-1} = -\mathrm{i} \int_0^\infty \mathrm{e}^{\mathrm{i}\lambda t}\, \mathrm{e}^{-\mathrm{i}tT}\, \mathrm{d}t\,.$$

Assume then that $\lambda = 0$ and put $m := [\delta] + 1$, where δ is as above. Show that $p(T) \le m$ for every $\lambda \in \mathbb{C}$. See Barnes (1989) for details. Note that $\mathcal{P}(X)$ coincides with the class of all generalized scalar operators having real spectra; see Theorem 1.5.19 of Laursen and Neumann (2000).

5. Let λ_0 be an isolated point of $\sigma(T)$, and denote by P_0 the spectral projection associated with λ_0, as in Part I, Chapter 4. Show that $\ker P_0 = K(\lambda_0 I - T)$

and that $P_0(X) = H_0(\lambda_0 I - T)$. The proof of the second equality is not obvious; see Proposition 49.1 of Heuser (1982).

6. Let R and L denote the right and the left shift operator, respectively, on $\ell^2(\mathbb{N})$. Show that the operator $T := L \oplus R \in L(\ell^2(\mathbb{N}) \times \ell^2(\mathbb{N}))$ is a Fredholm operator with ind $T < 0$ and that T does not have SVEP at 0.

7. Show that, if $T \in \mathcal{B}(X)$ is essentially semi-regular, then there exists a disc $\mathbb{D}(0, \varepsilon)$ for which $\lambda I - T$ is semi-regular for all $\lambda \in \mathbb{D}(0, \varepsilon) \setminus \{0\}$. *Hint:* if (M, N) is a Kato decomposition for T and $T_0 := T \mid T^\infty(X)$, take

$$\varepsilon := \min\{\gamma(T \mid M), \gamma(T_0)\}.$$

8. Show that, if $T \in \mathcal{B}(X)$ is essentially semi-regular, then there exists a finite-dimensional subspace F of X such that $\mathcal{N}^\infty(T) \subseteq T^\infty(X) + F$. The converse is true if we assume that $T(X)$ is closed; see Müller (1994).

9. Suppose that $T, S \in \mathcal{B}(X)$ are commuting operators for which TS is essentially semi-regular. Show that both T and S are essentially semi-regular. *Hint:* use the result of Exercise 8, and see the remark after Proposition 26.1.4.

10. Let $T \in \mathcal{B}(X)$, where X is a Banach space, be essentially semi-regular. Then the operator $\widetilde{T} : X/T^\infty(X) \to X/T^\infty(X)$ induced by T is upper semi-Fredholm and also $p(\widetilde{T}) < \infty$.

11. Let f be an analytic function defined on an open set containing $\sigma(T)$ such that f is non-constant on the connected components of $\rho(T)$. Show that $f(\sigma_{ap}(T)) = \sigma_{ap}(f(T))$ and $f(\sigma_{su}(T)) = \sigma_{su}(f(T))$. *Hint*: show first that $\sigma_{su}(T) = \bigcup_{x \in X} \sigma_T(x)$, and then use remark (ii) before Theorem 27.2.10.

12. Let X be a Banach space, and take $T \in \mathcal{B}(X)$. Show that

$$\sigma_b(p(T)) = p(\sigma_b(T))$$

for every polynomial p. *Hint*: use Theorem 27.2.10. It should be noted that this equality holds if we replace p by an analytic function defined on an open neighbourhood U of $\sigma(T)$; see Gramsch and Lay (1971).

13. Let X be a Banach space, and take $T \in \mathcal{B}(X)$. Suppose that T is essentially semi-regular and quasi-nilpotent. Show that X is finite-dimensional and T is nilpotent.

28.4 Additional notes

Most of the material of this chapter is a sample of results of Aiena and Monsalve (2000, 2001), Aiena, Colasante and Gonzalez (2001), and Aiena and Rosas (2003). In these papers the results have been established in the more general situation of *Kato type operators*, that is, operators on a Banach space for which there exists a decomposition $X = M \oplus N$, where M and N are closed T-invariant subspaces, such that $T \mid M$ is semi-regular and $T \mid N$ is nilpotent. Some of these results, in the special case of semi-Fredholm operators, may be found in Finch (1975) and Schmoeger (1995). Note that some of the equivalences of Proposition 28.2.6, Proposition 28.2.8, and Proposition 28.2.9 have been established in West (1987) by using methods of Fredholm theory, without involving local spectral theory.

The condition that $p(\lambda I - T) < \infty$ for all $\lambda \in \mathbb{C}$ has been investigated in Laursen (1992) by different methods. In this paper the reader may find some interesting open problems. Finally, results very close to those established in the corollaries of Theorem 28.2.10 may be found in Section 3.7 of Laursen and Neumann (2000).

References

Aiena, P. and Monsalve, O. (2000). Operators which do not have the single valued extension property, *J. Math. Anal. Appl.*, **250**, 435–48.

Aiena, P. and Monsalve, O. (2001). The single valued extension property and the generalized Kato decomposition property, *Acta Sci. Math. (Szeged)*, **67**, 461–77.

Aiena, P. and Rosas, E. (2003). The single valued extension property at the points of the approximate point spectrum, *J. Math. Anal. Appl.*, to appear.

Aiena, P., Colasante, M. L., and Gonzalez, M. (2002). Operators which have a closed quasi-nilpotent part, *Proc. American Math. Soc.*, **130**, 2701–10.

Aiena, P., Miller, T. L., and Neumann, M. M. (2001). On a localized single valued extension property. Preprint, Mississippi State University.

Apostol, C. (1984). The reduced minimum modulus, *Michigan Math. J.*, **32**, 279–94.

Barnes, B. A. (1989). Operators which satisfy polynomial growth conditions, *Pacific J. Math.*, **138**, 209–19.

Caradus, S. R., Pfaffenberger, W. E., and Yood, B. (1974). *Calkin algebras and algebras of operators in Banach spaces*, New York, Dekker.

Colojoară, I. and Foiaş, C. (1968). *Theory of generalized spectral operators*, New York, Gordon and Breach.

Dunford, N. (1952). Spectral theory II. Resolution of the identity, *Pacific J. Math.*, **2**, 559–614.

Dunford, N. (1954). Spectral operators, *Pacific J. Math.*, **4** (1954), 321–54.

Dunford, N. and Schartz, J. T. (1958, 1963, 1971). *Linear operators, Vols. I, II, III.* New York, Wiley-Interscience.

Finch, J. K. (1975). The single valued extension property on a Banach space, *Pacific J. Math.*, **58**, 61–9.

Förster, K. H. (1966). Über die Invarianz eineger Räume, die zum Operator $T - \lambda A$ gehören, *Arch. Math. (Basel)*, **17**, 56–64.

Goldberg, S. (1966). *Unbounded linear operators: theory and applications*, New York, McGraw-Hill.

Gramsch, B. and Lay, D. (1971). Spectral mapping theorems for essential spectra, *Math. Ann.*, **192**, 17–32.

Heuser, H. (1982). *Functional analysis*, Chichester, Wiley-Interscience.

Kato, T. (1958). Perturbation theory for nullity, deficiency and other quantities of linear operators, *J. Analysis Math.*, **6**, 261–322.

Kato, T. (1966). *Perturbation theory for linear operators*, Berlin–Heidelberg–New York, Springer–Verlag.

Laursen, K. B. (1992). Operators with finite ascent, *Pacific J. Math.*, **152**, 323–36.

Laursen, K. B. and Neumann, M. M. (2000). *An introduction to local spectral theory*, London Mathematical Society Monograph, **20**, Oxford, Clarendon Press.

Laursen, K. B. and Vrbová, P. (1989). Some remarks on the surjectivity spectrum of linear operators, *Czechoslovak Math. J.*, **39** (114), 730–9.

Mbekhta, M. (1987). Généralisation de la décomposition de Kato aux opérateurs paranormaux et spectraux, *Glasgow Math. J.*, **29**, 159–75.

Mbekhta, M. (1990). Sur la théorie spectrale locale et limite des nilpotents, *Proc. American Math. Soc.*, **110**, 621–31.

Mbekhta, M. and Ouahab, A. (1994). Opérateur s-régulier dans un espace de Banach et théorie spectrale, *Acta Sci. Math. (Szeged)*, **59**, 525–43.

Miller, T. L., Miller, V. G., and Smith, R. C. (1998). Bishop's property (β) and Cesáro operator, *J. London Math. Soc.* (2), **58**, 197–207.

Müller, V. (1994). On the regular spectrum, *J. Operator Theory*, **31**, 363–80.

Searcóid, M. Ó and West, T. T. (1989). Continuity of the generalized kernel and range of semi-Fredholm operators. *Math. Proc. Cambridge Philos. Soc.*, **105**, 513–22.

Saphar, P. (1964). Contribution á l'étude des applications linéaires dans un espace de Banach, *Bull. Soc. Math. France*, **92**, 363–84.

Schmoeger, C. (1990). Ein Spektralabbildungssatz. *Arch. Math. (Basel)*, **55**, 484–9.

Schmoeger, C. (1995). Semi-Fredholm operators and local spectral theory, *Demonstratio Math.*, **4**, 997–1004.

Vrbová, P. (1973). On local spectral properties of operators in Banach spaces, *Czechoslovak Math. J.*, **23** (98), 483–92.

West, T. T. (1987). A Riesz–Schauder theorem for semi-Fredholm operators, *Proc. Royal Irish Acad.*, **87A**, N.2, 137–46.

Index of symbols

Subject index

Printed in the United States
by Baker & Taylor Publisher Services

Printed in the United States
by Baker & Taylor Publisher Services